PRACTICAL PERMACULTURE FOR HOME LANDSCAPES, YOUR COMMUNITY, AND THE WHOLE EARTH

PRACTICAL PERMA- CULTURE

FOR HOME LANDSCAPES, YOUR COMMUNITY, AND THE WHOLE EARTH

Jessi Bloom &
Dave Boehnlein

With illustrations by Paul Kearsley

TIMBER PRESS
Portland, Oregon

(Frontispiece) Ecological landscapes can produce many environmental functions as well as beautiful human habitat.

(Opposite) Sweet potato vines can grow as ground cover, yielding delicious edible tubers.

Published in 2015 by Timber Press, Inc.

Illustrations copyright © 2015 by Paul Kearsley: see page 335.

Photo and illustration credits appear on page 335.

The Haseltine Building
133 S.W. Second Ave., Ste. 450
Portland, Oregon 97204-3527
timberpress.com

Printed in China
Sixth printing 2022

Book design by Laura Shaw Design, lshawdesign.com
Cover design by Debbie Berne

Library of Congress Cataloging-in-Publication Data
Bloom, Jessi.
Practical permaculture for home landscapes, your community, and the whole earth/Jessi Bloom and
Dave Boehnlein; with illustrations by Paul Kearsley.—1st edition.
pages cm
Includes index.
ISBN 978-1-60469-443-7
1. Permaculture. 2. Permaculture plants. I. Boehnlein, Dave.
II. Title.
S494.5.P47B575 2015
631.5'8—dc23
2014020736

A catalog record for this book is also available from the British Library.

We dedicate this book to our mothers, who nurtured and supported our growth, and to the Bullock family, whose work inspired change in our lives.

Contents

Preface

Each reader may come to this book on a different path, but we all have something in common: we're human, in this together, and we belong to the intricate web of life that Chief Seattle describes. We are all born into that connection and are hardwired to feel it, but depending on our experience in life, we may be pulled away from it. Humans are inherently always looking for a better quality of life—for health, happiness, comfort, and financial stability, among other things. However, it is easy to look ahead and see that many of the choices we as a species have made in that quest are not going to sustain us.

We are at a crossroads and need to start making changes. Many of us rely on systems that are beyond our control—the industrial food system, municipal waste and water systems, the energy grid, to name a few. But we don't have to rely on those systems—we can be more self-sufficient and take back the reins. We can remember that human life has an ecological purpose and function, which we, the authors, hope this book will raise awareness of.

We authors, Jessi and Dave, have similar stories. In the 1980s we were growing up on opposite ends of the country, spending our time in nature's playground. Tromping through forests, building forts, and damming small streams was a part of our daily lives. We both explored the outdoors through hiking, fishing, camping, and swimming in natural bodies of water, and we both developed deep emotional connections to the land we grew up on. Little did we know in our youth we were "second-growth" permaculturists in the making. At similar times in our lives, we both witnessed our childhood stomping grounds getting bulldozed in the name of development and were completely distraught. We went on to pursue different studies and careers related to the environment. Jessi's passion was dedicated to ecological design, while Dave devoted his career to outdoor education. However,

both of us ended up making permaculture not only our profession but also the basis of our lifestyles.

We first crossed paths in 2010 at the Northwest Flower and Garden Show in Seattle. Dave was on the hunt for professionals with whom he could collaborate and looking to create a network of permaculturists who were in the field doing good work. That year at the show, Jessi had designed and built an urban permaculture demonstration garden that won numerous awards for sustainability and aesthetics. The irony is that the word *permaculture* was so misunderstood that it was taken out of the name of that garden because show managers didn't want to confuse the public. Since that initial meeting, we have collaborated on many projects including designs, workshops, trampoline sessions, and now this book.

Humankind has not woven the web of life. We are but one thread within it. Whatever we do to the web, we do to ourselves. All things are bound together. All things connect.

—CHIEF SEATTLE

Our intention is to make permaculture approachable and inspirational. We want to empower you to start making positive changes, now. We promise that the way you experience the world around you will be richer because of your increased understanding of and participation in nature. At some point, as your connection to the natural world strengthens, you will feel a tap on the shoulder and Mother Nature will say, "You're working for me now." Have you gotten the call yet? If not, get ready for a wild ride.

We have written this book for a wide range of readers. We cover a huge range of climate types as well as scopes and sizes of land. We want people in rural areas as well as urban folks to be able to use this book. The first section, Permaculture Basics, equips you with basic information about permaculture, from the definition to the guiding ethics and principles behind it. This part also outlines basic earth science you'll need to know before jumping into the design process. The second section of the book walks you through the entire design process, from observation to analysis and assessment to developing a master plan and planning how to implement the design and maintain what you have created. The third section of the book covers the systems that should be included in any permaculture design: systems for managing soil fertility, water, waste, energy, shelter, food production, and animals and wildlife. The fourth section describes fifty rock-star plants for permaculture landscapes. The fifth section focuses on invisible structures, the economic and social underpinnings of any permaculture design. At the back of the book are lists of resources and suggestions for further reading that we highly recommend you take advantage of, since this book is just a starting point.

PERMACULTURE BASICS

Permaculture is a term first coined by Bill Mollison and David Holmgren in the mid-1970s. It has been defined in a huge variety of ways, with the most common being a contraction of the words *permanent* and *culture*. It's also been defined as extreme organic gardening and as a lifestyle that includes being barefoot and spiritual, doing circle dances and full-time yoga. Permaculture can certainly include these elements, but it's not defined by them. So what, exactly, is permaculture?

In a sense, permaculture is a way of life that's ahead of its time while also taking us back to how our ancestors lived, sustainably and within their ecological means. That doesn't mean we can't live in comfort and utilize current technology in a permaculture lifestyle. Permaculture gives us the tools and techniques to live sustainably while still having our needs met in a lifestyle rich with healthy food, comfortable housing, and renewable energy and resources.

Permaculture is similar to both architecture and engineering in that it is a design approach first and foremost. Whether we are designing a house, a chicken coop, a garden, a bowling alley, or a schoolyard, permaculture is a process that starts with a problem and finds solutions. In permaculture, design decisions are first based on ethics and then incorporate the logic of natural systems. Mimicking nature's patterns makes our lives more sustainable and less reliant on resources outside of our control.

In contrast to living in a wasteful, consumerist manner that depletes our resources and doesn't leave future generations much to work with, permaculture is about building resilience and using only what we need and what we have access to—in other words, living within our ecological means. It's about building fertility and abundance. It's about designing and building systems in our lives that work together to provide food and water and energy, that reuse waste, and that make life easier. It's about making these systems beautiful and inspiring, and by doing that, making our lives healthier and better overall.

In a nutshell, then, *permaculture* can be defined as meeting human needs through ecological and regenerative design.

This section lays the foundation for your understanding of permaculture by first discussing permaculture ethics and principles, and then considering how nature itself works. Beginning to think in these terms is a prerequisite for making permaculture design choices.

Permaculture Ethics and Principles

(left) In this suburban garden, the space is used to maximize yields and ecological function.

(right) Placing rainwater tanks between the garden and the chicken run means minimal piping for water to reach both places at CSC Youth House Garden, Corvallis, Oregon.

With all the apocalyptic scenarios available to worry about today, from environmental disasters to peak oil to medical pandemics, it's clear that we should be concerned about our future. We need to start asking ourselves some questions: Where are we headed as a society? How are our choices today going to impact future generations? Are there alternatives to our high-speed lifestyles? Is there a way to live responsibly that doesn't feel like we're sacrificing what we have become accustomed to? Permaculture encourages us to really explore these questions. It offers a set of ethical guidelines and basic principles to keep foremost in mind as we make choices about how to design our lives. This chapter outlines these guidelines and principles after considering why we need permaculture.

WHY PERMACULTURE?

We all have the right to live happily in a healthy world free of pollutants and corrupt systems. Unfortunately, that's not the world we find ourselves in. Instead, we face actual or potential disruptions to natural, economic, and social systems because of choices that have been made and are being made that don't take into account the next seven generations or our fellow species on the planet.

Here are a few questions we would like everyone to think about:

- Economics: Would you be able to meet your basic needs if suddenly your life savings, income, and property values plummeted to a tenth of what they are today?
- Resilience: If all movement or transportation of resources (such as food or water) to your area were cut off for two weeks, could you manage?
- Health: How would you feel if someone in your family became terminally ill from toxins introduced into the environment by humans?
- Climate: Would people in your area be able to thrive if the climate where you live became extremely erratic (higher high temperatures, lower low temperatures, periods of both drought and flooding)?
- Environment: How would you feel if the natural place to which you feel most connected were destroyed to make way for a parking lot or a strip mine?

The only ethical decision is to take responsibility for our own existence and that of our children.

—BILL MOLLISON

Answering those questions evokes some powerful emotions and points to why we think permaculture design is the way to go.

- Permaculture empowers us to take control of our lives and teaches us the skills we need to take care of ourselves.
- It makes us think about the big picture.
- It forces us to think outside the box and puts us in an ecological frame of mind that asks, "What would nature do?"
- It's holistic and combines the best of all design fields.
- It makes us feel connected to the place where we live.
- Anyone can do it—we don't need a fancy degree to get started.
- We can easily start to put it into action.
- It's not prescriptive or formula based, so it allows different people in different places to arrive at different solutions.

PERMACULTURE ETHICS

Permaculture design choices are based on a few simple ethical guidelines. Permaculture's basis in ethics is a big part of what makes it different from other design fields. Permaculture literature generally agrees on the first two ethical guidelines, but the third (and sometimes fourth and fifth) can vary a bit. What we're offering you here is the approach we find most useful for designers. Note that whenever one ethic seems to be in conflict with another, the first ethic trumps the rest.

Urban farmer Will Allen has dedicated his life to developing Growing Power, a nonprofit that educates people and grows food in underserved communities.

Care for the earth

It may seem obvious, but being good stewards of the planet comes first when we are designing permaculture systems. The intrinsic value of functional ecosystems, as well as other living beings, is a big part of this. We have a vested interest in maintaining functional ecosystems for our own health and prosperity. The more we damage ecosystems, the poorer the air and water quality become for us and our children. Thus, to help, or at the very least to do no harm, should be a core tenet of the way we impact the earth with our landscape designs, just as "first, do no harm" is a guiding principle of medical practice.

We want to avoid damaging intact, functional ecologies in the name of fulfilling our dreams of self-reliance, productive food systems, and homestead living. A part of being good stewards of the earth means leaving alone and protecting healthy, functioning ecosystems. From a landscape perspective, we quickly realize that permaculture design is best used to regenerate degraded landscapes. Such landscapes are often lacking in biodiversity, structural complexity, and resilience in the face of disturbance. These landscapes may include conventional agricultural fields, lawns, eroded sites, logged lands, polluted sites, and plantations.

Care of people

The permaculture systems we design should also address the needs of people. In the developing world and the inner city, we can see plenty of examples of the basic needs of people not being met. For these folks, environmental stewardship may be a lower priority than food, water, and income. We can't expect everyone to go the extra mile for environmental stewardship when their basic needs are unmet. That's why our systems need to address the needs of people. By making sure people's needs are met through our design solutions, we offer them the capacity to be better stewards of the earth.

For instance, what if a design for a park in a low-income community incorporated lots of edible landscaping? The local residents might find that they have a vested interest in being good stewards of that park since it now helps provide for their needs in a direct way. If that same park were filled only with low-maintenance ornamentals or grass, it might be more likely to collect litter and degrade over time because people would be less likely to take an active stewardship role in something that didn't provide more tangible benefits for them. That's when the idea of environmentalism becomes something for everyone, not just the fortunate few who don't have to worry about food, clean water, and sanitation.

When designing systems that take care of people, we don't just want to consider ourselves, our families, and our neighbors; we want to extend this concept to future generations. If we extend this ethical mandate to include our children, grandchildren, and so forth, our designs start to look very different. All of a sudden it's worth planting trees that won't produce a harvest in our lifetime. The slow task of building healthy, rich topsoil through organic practices instead of looking for the chemical "quick fix" becomes more sensible. We start to be able to wrap our minds around ecological changes as they relate to geological time scales. We can think bigger and we can set into motion actions with impacts that will be felt well into the future. Many design decisions that seem difficult become much clearer when we ask ourselves which decision would best serve our grandchildren.

From a broader perspective, humans and nature aren't really as separate as we often think. We're actually as much a part of nature as ants, mushrooms, or apple trees. That's why, in reality, caring for the earth actually means caring for people, too, and that's why the first ethic wins if they ever seem to be in conflict.

Careful process

It's not just about the end goal, but also about how we get there. The approaches we choose and the context we set with our design decisions are just as important as the end product. This ethic essentially serves as a bit of a catchall for the other ethical concerns we want to address with our designs. It breaks down into two separate concerns: redistribution of the surplus (to the ends of caring for the earth and people) and self-regulation of consumption and growth.

With regard to the first concern, one goal of the systems we design should be to create abundant yields. After all, we'd all much rather live in a world of abundance than a world of scarcity. This means that permaculture systems often generate surpluses—of food, biomass, electricity, time, knowledge, and such. If we don't handle these surpluses properly, many of them can actually become pollutants. Therefore, we have an ethical mandate to redistribute those surpluses to turn potential problems into elegant solutions.

If you grow more food than you can eat, you can turn it into money by selling it at a farmers' market. You can use the income generated to make your land more sustainable.

What this ethic is really about is taking the surpluses a system produces and reinvesting them where they will do some good. That can mean rolling them back into the system (for example, composting food waste and putting that compost into the garden) or sharing them with someone else to help increase their capacity to care for the earth (for example, giving a neighbor seedling nut trees to plant). By using our surpluses to make sure the needs of those around us are met, we benefit in many ways. First, if the needs of those around us are met, they are in a better position to join us in our efforts to care for the earth. Second, if the needs of those around us are met, we find ourselves living in communities characterized by hope and pride rather than desperation and apathy. However you work it out, all surpluses of the system should be reinvested in caring for the earth and people.

The second concern, self-regulation of consumption and growth, requires us to look at our behaviors as much as our design choices. Sometimes the best designs require fundamental behavior changes to decrease consumption. Therefore, we need to pay close attention to the assumptions regarding consumption that go into our designed systems. For instance, before figuring out how to increase water supply on an arid site, it makes sense to look at methods for decreasing consumption of water first.

Similarly, many of us now realize that one of the biggest problems with capitalism is that it's based upon assumptions of never-ending growth. In nature, nothing grows forever. Body mass, population size, and temperature all increase, decrease, and level off at various points in time. We need to design systems with the flexibility to grow, shrink, or achieve a steady state (at least for a while). We certainly don't want to continue designing our world in such a way that it ceases to function when continual growth is in question.

In essence, we can choose to apply self-regulation to consumption and growth, or else nature will ultimately do it for us—in the form of epidemics, disasters, or famine. Luckily, we have a capacity for independent thought that goes way beyond instinct. We alone in the animal kingdom can actively choose to regulate our own growth and consumption, which will increase the stability of our systems and avoid boom-and-bust cycles.

This is where the concept of carrying capacity becomes important. *Carrying capacity* is the maximum number of individuals of a species that can be supported by a given environment. When the population is below carrying capacity, it tends to grow; when it passes carrying capacity, it starts to decline. We must understand a site's baseline carrying capacity and then either work within that carrying capacity or design ways to increase it. Either way, understanding and modifying our expectations around growth and consumption are first steps toward sustainable living.

Transitional ethic

The transitional ethic is your lucky, bonus ethic. It appears in only a few pieces of the existing permaculture literature, but we find it to be extremely important. Essentially, the transitional ethic says that no one is going from zero to sustainable overnight. Making that transition takes time.

What does that mean for using nonsustainable technologies like excavators and lawn mowers? If we consider the embodied energy in those types of things and we consider that they already exist, we must ask whether the worse sin is to use them or to let them go to waste. *Embodied energy* refers to the total energy involved in the production of an object. For instance, the embodied energy in a bulldozer includes the energy spent on mining the metals, smelting the steel, manufacturing the parts, shipping the parts to a factory, assembling the machinery, and shipping the bulldozer to the sales floor.

Given that a bulldozer has a huge amount of embodied energy, especially as compared to a shovel, the issue becomes *how* we will use it. Using a bulldozer as part of a strip-mining operation to fill someone's pockets is ethically questionable. Using a bulldozer to install water-collecting earthworks in an arid landscape to aid in revegetation and aquifer recharge makes more sense. Using nonsustainable technologies may be appropriate if they are being used to set up systems for sustainability that will last long into the future.

Permaculturists avoid designing systems that are reliant upon nonsustainable technologies in perpetuity. From a permaculture perspective, this is where the concept of appropriate technology fits in. *Appropriate technology* refers to any application of knowledge and skills that is considered to be appropriate for a given situation. What is appropriate in one circumstance may not be so in another.

Typically, appropriate technologies are small-scale, labor-intensive, energy-efficient, environmentally sound, and locally controlled.

Not only must we examine whether the use of unsustainable or polluting technology is appropriate in terms of our end goal; we must also look at the technologies that we will be relying on for the long term. Permaculturist Douglas Bullock often says that to determine whether a technology is appropriate, we must look at whether the end user can appropriate it. In other words, the people who will be using it must able to understand, rebuild, repair, and recycle the technology. If this is not the case, we need to think long and hard about whether a technology has a long-term future in our systems even if it is being used as a short-term solution.

To this way of thinking, many of the high-tech climate-control systems found in cutting-edge green homes may not be appropriate because the people living in

The Bullock family has been developing their permaculture site on Orcas Island, Washington, for more than thirty years. Their focus has been on getting their own systems established first, then teaching others what they learn.

those buildings and their neighbors likely do not possess the skills to repair them. While those complicated systems may be used in the interim, a long-term shift toward passive solar design and low-tech heating and cooling devices such as efficient woodstoves and swamp coolers may be better options.

The transitional ethic is also the one that keeps competent permaculturists humble. There is no more effective way of turning people off to permaculture than to act judgmental about where one has landed on the road to sustainability. Looking down one's nose at someone else for not composting their poop is not helpful; welcoming permaculture newcomers aboard and asking how one can help them take the next step is.

Start by making sure your own needs are met within a regenerative context and only then focus on your family, then your friends, then your community, and so on. Putting your energy toward solving global issues when your own needs and impact at home aren't addressed just doesn't make sense. We must take care of ourselves and make sure the impact of our existence is a positive one before we can take care of others and help them have a positive impact.

Beware of those who walk through a site and are vocal about what features are and aren't permaculture. Without understanding the fullness of a given situation, no one can make competent judgments about another's design or site, especially if they're only looking at the project at one stage of its development. This is why we like to say that permaculture design offers us a perspective, not a prescription.

SYSTEMS THINKING AND PERMACULTURE

We often refer to systems when we talk about permaculture (water systems, energy systems, food systems, and so on). In its simplest form, a system is a bunch of parts (elements) arranged such that their relationship to one another (their function) allows some sort of job to get done or some goal to be accomplished (purpose). For instance, a bicycle is a simple system composed of a bunch of elements (handlebars, chain, wheels, and so forth) put together in such a way (handlebars connected to frame, frame connected to wheels) that they function to accomplish the purpose of transportation. We can see the same concept when looking at the parts of the human body. A pile of organs sitting on a table does not make a person. However, when those organs relate to each other in just the right way and each performs its functions, *we* are the result.

When all the elements of a system come together in the right way, the whole becomes more than the sum of its parts and emergent properties appear. For example, the emergent properties of a human may include the ability to cook Thai food, tell a joke, or write haiku. Essentially, systems thinking is all about exploring those relationships between elements that allow unique system properties to emerge.

Permaculture is a process of holistic design that forces us to think in systems. Not only must we think about the relationships among all the parts in, let's say, our landscape, but we must also cross conventional boundaries between disciplines and look at how our landscape, buildings, social systems, and so on all relate to

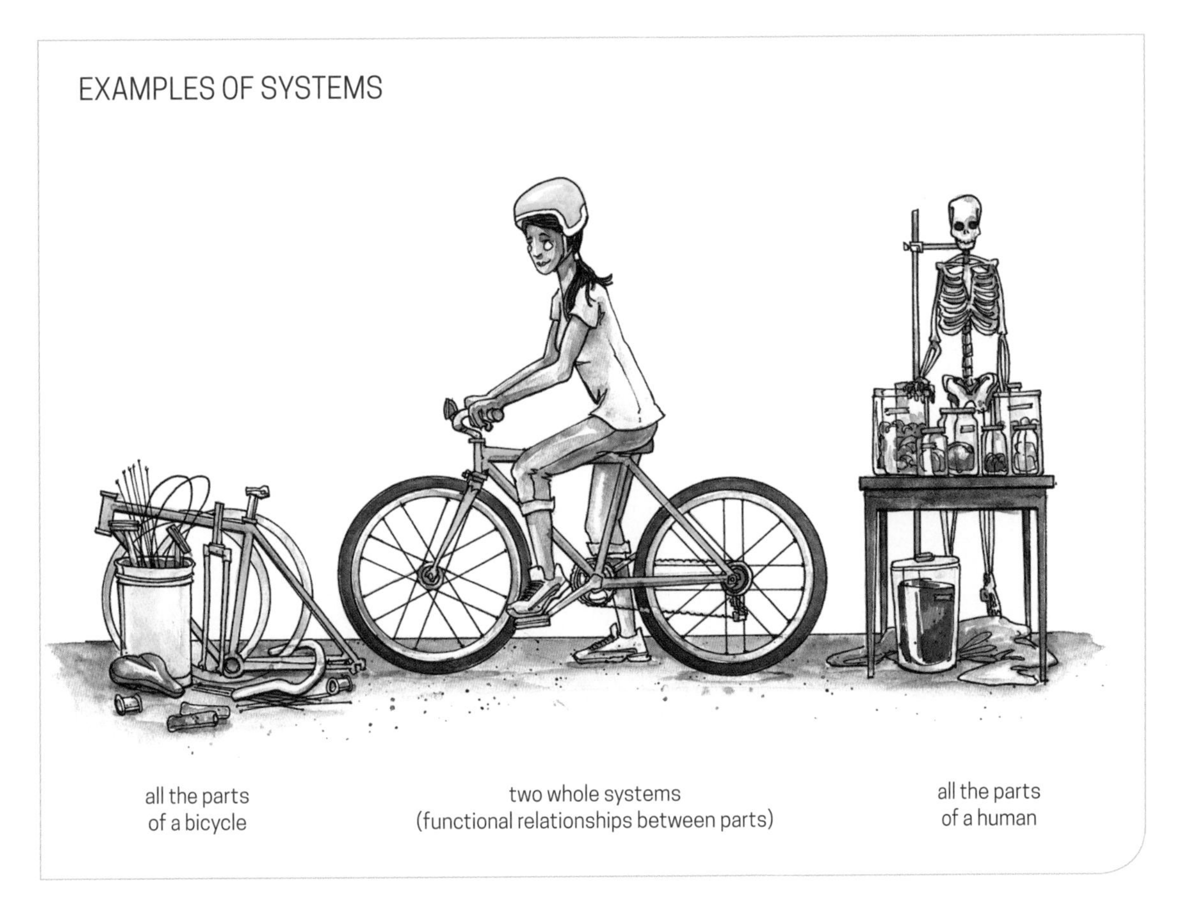

one another. That means the landscape is no longer just the pretty accessory that surrounds the house but is just as important as the house. In fact, in some cases we may decide to change the design of our house, energy system, or social structure to accommodate some other part of the design.

A number of permaculture design strategies follow from systems thinking. Let's look at some of these strategies.

Anticipate limiting factors.
Looking at limiting factors allows us to see where the leverage points are to make changes in a system. For instance, if the limiting factor for plant growth in your garden is nitrogen, adding all the potassium in the world won't help it to grow better. However, we must also understand that the nature of limiting factors is that they are dynamic. If we add nitrogen to our garden, at some point it will no longer be the limiting factor, but something else will take its place (such as potassium, manganese, or calcium).

We could look at any complex system and think of it as having layers of limiting factors. Therefore, our designs must both address the limiting factors in play at present and try to predict what limiting factors will come. Returning to our example, if we apply an amendment to our garden with an array of nutrients that closely matches our soil's deficiencies, we can avoid experiencing many limiting factors altogether. Thinking about limiting factors will allow us to address upcoming limiting factors before they actually become an issue. We must always remember, though, that there are environmentally bounded limits to growth for all of our systems (think carrying capacity), and if we disregard these, our systems will have unintended negative consequences.

Develop a holistic context.

Traditionally, Western culture has used reductionism as a way to simplify decision making. Sometimes this is helpful, as trying to take everything in the universe into account could mean you never actually get around to making a decision, but sometimes reductionist thinking can go too far. For example, a reductionist approach might look at body weight alone as an indicator of human health. A holistic approach, by contrast, would look at weight, fitness, diet, genetic factors, lifestyle choices, and emotional well-being, among other factors.

When we think holistically, instead of looking at how many parts we can throw away when making a decision, we think about how many parts we can keep in the equation without becoming overwhelmed. In design, this may mean that answering a question like Should I graze cattle here or plant a forest? may start with answering the question How do I want my life to be and what do I want for my children and grandchildren? What are the overarching goals? Allan Savory, founder of holistic management, refers to this process as developing a holistic context from which all of your decisions flow. Once you have a holistic context, you can check each decision in your design to see whether it moves you toward what you really want. If it doesn't, try a different option.

Design systems with closed loops.

If you think back to your days in high school earth science or physics, you might vaguely recall learning about the laws of thermodynamics. The first two have specific importance for permaculture designers. The first law of thermodynamics basically says that neither matter nor energy can be created nor destroyed. They can only change form. This is important for permaculture designers to understand because it holds within it the idea that there is no "away." Nothing ever disappears. Sometimes it leaves our system (for example, the trash we put out by the curb), but it just goes into someone else's system. In permaculture, we design systems with closed loops as best we can. For example, instead of throwing away food waste, we can compost it (that is, change its form) and reapply it to our garden. That way we don't export waste that needs to be dealt with elsewhere and we don't have to import fertilizer to grow our vegetables.

Make use of by-products.

The second law of thermodynamics says that whenever something is used to do work or whenever it changes form, some of the energy is lost as heat or light. For example, if you put your hand near an incandescent light bulb, you feel heat. The goal of that bulb is to create light so we can see, but turning electricity into

At Growing Power in Milwaukee, Wisconsin, this greenhouse is heated by the biological activity of the compost piles in the corners. Dave took this picture on a day in January with a high of 17°F outside, and the greens looked just fine.

light actually involves losing a lot of the energy as heat. If we keep this in mind as designers, we can do our best to make use of that by-product.

For instance, in composting, some of the energy in decomposing food waste gets lost as heat, which you know if you've ever turned a biologically active compost pile on a cold day and watched the steam rise. As designers, our job is figure out how to either minimize this loss or, better yet, take advantage of it. What if you put your compost pile inside of a greenhouse in the winter to keep it warmer than it would otherwise be? Will Allen of Growing Power in Milwaukee, Wisconsin, did just that. In a greenhouse with no active heating strategy other than compost piles in the corners, Will Allen produces hardy crops of salad greens all winter.

Avoid shifting the burden to the intervener.

Basically, shifting the burden means that if we design systems that don't regulate themselves through their functional interconnections, we must supply some of those regulating functions. If our systems don't function without our intervention, we set ourselves up for more work. Taking on the burden for system function can lead to a feedback loop where more and more intervention is required over time to keep systems functioning. Therefore, we should strive to design systems that are largely self-maintaining. Even systems that require lots of work from us during establishment should be designed so that the system will provide for most of its own needs as it matures.

Ask yourself how much maintenance your systems require and whether those systems will require more or less maintenance in the future. Then rethink any systems that will require more maintenance in the future to help them become more self-maintaining. And make sure the systems you've designed address root problems and not just symptoms.

Maximize positive emergent properties.

Emergent properties of a system show up when each element in the system is performing its function. Emergent properties can be expected or unexpected, good or bad. The key for permaculture designers is to look at each aspect of our designs through a holistic lens so we can try to anticipate as many emergent properties of the systems we design as possible. When designing it is also important to periodically check back with our overarching mission, vision, and objectives so we can make sure the systems we are designing are creating the sort of emergent properties that will move the project in the right direction. Keeping these things in mind enables holistic designers to maximize the positive emergent properties and lessen the negative ones.

PERMACULTURE PRINCIPLES

We like to think of the principles of permaculture design as filters for our design decisions. We can use them to test different solutions to design problems to see which ones will lead us closer to our vision and to sustainability. The lists of permaculture principles found in various publications range in length from four to forty. What follows is a distillation of the principles that we have found to be most broadly applicable to permaculture design in our experience. We also offer a few questions for each principle that you can use to work through your design.

Locate elements for functional interconnection.

Putting elements in the right place in the landscape and in relation to one another can create beneficial relationships and allow systems to function. These functional interconnections between elements result in systems with closed loops, which are

In this garden, greens are being grown between bean poles and a greenhouse to provide shade in the summer, which prevents them from bolting.

more efficient and lower maintenance than those without. For instance, a greenhouse attached to the sunniest side of the house may help heat the house through solar gain and create a productive growing space. These benefits would be downplayed or lost if the greenhouse were on the shady side of the house.

Questions to ask:

- Does this element relate to those around it in a way that makes sense?
- If this element were moved somewhere else, what would be the potential benefits and/or limitations?

Choose elements that have multiple functions.

The elements we include in our designs should have multiple functions, and they should be used to the fullest. When we have a choice of different elements that could do a particular job, we should aim to use the ones that do more for us and the surrounding ecosystem. For instance, if we wanted to create a privacy screen between our yard and our neighbor's, we could plant a mixed-species hedge. This

could provide food, flowers, medicine, habitat for wildlife, and a nectar source for pest predators as well as screening. This is preferable to a wooden fence that doesn't do much other than block the view.

Questions to ask:

- Does each element in my design provide multiple functions?
- Are there any elements I could swap for others to provide more functions?
- Am I taking full advantage of all the functions offered by the elements in my design?

This sauna stove located in a greenhouse not only heats the sauna for people but also for dehydrating produce; it also has a flat spot on the upper surface to hold a teapot.

Design for resilience.

Each essential function in our designs (for example, potable water, income, food production) should be supported by multiple elements. We want to make sure that our systems don't stop functioning in the face of disruption. As a rule of thumb, we like to have backups for our backups. The more essential a function, the more backups we want to have. For instance, if the water main breaks during a hot, dry spell, we need to have other ways to irrigate our crops such as cisterns, rain barrels, and, if necessary, bucket brigades from the nearest water source. All of these approaches require some forethought in the initial stages of design. Applying this kind of "resilience thinking" to each aspect of a design leads to a more robust system overall.

Questions to ask:

- What kinds of events might disrupt each system in my design (for example, disasters, service disruptions, loss of employment)?
- Will my systems keep functioning in the face of these disruptions?
- If not, how can I add at least two layers of job redundancy to improve system resilience?

Obtain a yield.

We want to design systems that provide yields for us. In fact, we want to make sure that our systems don't just yield in the distant future, but throughout their lifespan. Our systems should start yielding early in their development and continue to increase yields as they grow until they reach maturity. For example, in the landscapes we design, first-year yields are primarily annual crops. In the second year, berry crops come online while we still harvest lots of annuals. The third year brings light yields from tree fruits. In the fourth year, we might add in yields from mushroom logs. Eventually, perhaps five to ten years along, we can start to receive yields from nut trees and small-diameter timber crops. The key is planning ahead to make sure we won't be working hard with nothing to show.

We must also make sure that we recognize all the yields that our systems provide and take advantage of them when we can. This means acknowledging yields that may not have a direct economic value such as carbon sequestration, aesthetics, and wildlife habitat.

Questions to ask:

- Will my systems provide yields throughout their development?
- Can I make any changes to my design to insure both short- and long-term yields?
- Do I have any gaps in the development of my system or annually where I receive no yield?

Look for small-scale, intensive solutions.

Whenever possible, we should aim to use solutions that are small-scale so they are more easily manageable and easier to adjust and control. Even for designs that may eventually grow to have large positive impacts, it is important to start small. That way, initial failures will also be small. We can use these failures as learning experiences to perfect our designs before we scale up. After all, we can often learn the same lesson from a $100 mistake that we can from a $1,000 mistake. Perhaps just as important, when people fail big they tend to quit. Failing small means you are more likely to try again.

Intensive land use means using the least land possible to supply our needs. Ultimately, if we design intensive systems and meet our needs on less land, we can afford to leave more land to return to functional wilderness ecosystems.

Questions to ask:

- Regardless of how grand my ultimate vision may be, am I starting small?
- Could I start smaller?
- What will be the consequences if this idea fails?
- Am I making the most of the land I'm using or could it provide more functions?
- Could I achieve my goals in less space?

Immediately after planting a new perennial landscape, Jessi likes to plant winter squash or pumpkins to make a quick ground cover that provides a yield the first year.

Mimic nature and use biological resources.

The way things are done in nature should provide many clues as to how we might get the same jobs done within a context of sustainability. Mimicking nature doesn't just mean copying the forms we see in nature, but more important, copying the functions that result from those forms. In other words, emulating relationships found in nature is where we want to focus our energy. Often this means using biological resources. The wonderful thing about biological resources, ranging from yeast to lettuce to cattle, is that they are inherently regenerative. They can use energy from the sun (either directly or indirectly) to make more of themselves. The same cannot be said for nonbiological resources such as tractors, concrete, and even solar panels. Therefore, when looking for the solution to a problem, we should make sure we explore biological solutions first and use them when practicality allows.

For instance, to mulch a garden we can use a biological material like wood chips or we can use black plastic. Both will maintain soil moisture, prevent the sun from baking the soil, and minimize weed problems. However, the black plastic is made from petroleum, which is a nonrenewable resource. It will eventually degrade and need to be thrown away. Conversely, the wood chips will slowly break down. When they're gone, more wood chips can be produced locally. What's more, the breakdown of the wood chips adds organic material to the soil, whereas the breakdown of the black plastic makes a mess.

Questions to ask:

- Is this how this problem or situation would be dealt with in a natural system?
- Does my design rely on nonbiological resources?
- If so, as the design matures are the nonbiological resources replaced by biological ones?

Strive for diversity.

Whether we're talking about biodiversity or cultural diversity, increasing it is almost always an important part of our designs. This principle is intimately tied to several of the others. High diversity can help to maintain a dynamic sort of stability in a system. Economically speaking, people often talk about diversifying their portfolios. Diversity helps to build resilience in a system. More diversity in a landscape means more potential for functional interconnection, more potential yields, and improved ability to withstand losing any one element. More diversity in the cultural makeup of your community means more perspectives, more ideas, and more opportunities for learning.

A good example of using diversity in our landscapes is to plant more than one species together (an arrangement known as polyculture). Let's say we plant an orchard with many different species of fruit trees. Within each species we plant a range of different varieties with different characteristics. If a disease strikes the apples, other fruits will still produce for us. Perhaps even some of the apple varieties will prove to be more resistant than others. If we are selling fruit, we won't have all our eggs in one basket; our bad apple year will be buffered by our crops of cherries, plums, pears, and apricots.

(left) This garden at Bullock's Permaculture Homestead uses multiple vertical layers to increase the number and diversity of plants in the area, mimicking what we might find in a natural ecosystem.

(right) This border at Jessi's house includes a variety of plants, each of which supports different types of insects and has different functions, ranging from food and medicine to beauty and weed control.

Questions to ask:

- Does my landscape design increase biodiversity (at the species and genetic levels)?
- Does my social system design value and honor diversity?
- Does it encourage greater diversity?
- Are the economic opportunities in the design diverse or are they all contingent on one element?

Solve problems creatively.

Permaculturists have embraced the idea that every problem contains the seed of its own solution, or simply, the problem is the solution. There is some truth to this, but it can be difficult to see when we're in the middle of dealing with the problem. Adjusting our perspective can help. When a problem seems unsolvable, it is important to look at our assumptions and make sure they're valid. Think outside the box. Think at both larger and smaller scales. Perhaps most important, make sure you're asking the right questions.

We could look at a tree falling down as a problem, or we could see it as an opportunity to use the wood for firewood or mushroom production.

Espaliered fruit trees can be grown in tight spaces but also require a lot more maintenance. This may be fine, but we need to be aware that it will take more work.

Questions to ask:

- Does my solution address the core problem or just a symptom?
- Am I trying to work around something that can't be corrected? Is it worth it?
- Are there other problems that could help solve this one?
- Is there a solution that would involve less work?

Make sure an element's use is in harmony with its nature.

Stress is the result when elements in a system don't have their needs met, when they compete with one another, or when they are being forced to perform unnatural functions. When our designed systems are in harmony, this is not the case. Therefore, when designing our systems we must make sure that we are using elements in a way that is in line with their nature. Whenever we force an element to perform an unnatural function, it results in more work for us.

For example, if we were to plant an apple on standard rootstock in a container on the back porch, the tree would soon become root-bound and would need to be dug up and root pruned frequently. We would also have to prune the top severely several times a year to prevent it from blocking our view. All of this disturbance might cause the apple to become more susceptible to pests and disease, which we would have to manage. After all this extra work, the calories we spent on tree care would probably outweigh the calories in the apples harvested.

If, on the other hand, we chose a dwarfing rootstock for the container-grown apple on the back porch, the apple's characteristics would be a natural match for the conditions available. We would likely have less work (also known as more hammock time) and greater success.

Questions to ask:

- Are the elements in my design stressed?
- If I used different elements in place of stressed elements (different species, varieties, breeds, makes, models), would I have a more harmonious result?
- If I put the element somewhere else, would it be less stressed?
- Will any aspects of my design result in more work for me in the long term?
- How could I alleviate that?

Manage edges.

In this context we're using the term *edge* as a synonym for surface area or perimeter. When creating our designs we must always pay attention to edges or the places where two distinct things come together (land and water; forest and field; one cultural group and another). These edges are often very dynamic—that is, conditions found at edges support individuals unlike those found in either of the things coming together. Fusion cuisine results at the edge of two different cultures' culinary traditions; cattails grow at the edge between deep water and land. Some of these edges are beneficial, but others create more work for us. Therefore, we strive to increase beneficial edges and minimize problem edges.

For instance, when fencing we often try to minimize the perimeter of the fenced area while maximizing the area enclosed. This minimizes the materials we use and the time we spend walking the fence line to check for damage. On the

(left) At permaculturist Jude Hobbs's house in Cottage Grove, Oregon, this lawn area has curved edges to allow for more points of access to the productive garden.

(right) At her house in Seattle, garden mentor Robin Haglund catches rainwater from her greenhouse roof and uses it in her landscape to increase productivity.

other hand, if we were putting in a pond and we really wanted to grow lots of cattails, we would want to make the pond edge extremely sinuous to maximize the amount of edge for our chosen crop.

Questions to ask:

- What kinds of edges am I dealing with?
- Would it be beneficial to increase or decrease them?
- Would increasing or decreasing the edge in question create more or less work?
- Does my management of edges align with the project's overarching vision and goals?

Cycle and recycle energy.

Electricity, money, time, steel, potatoes, and potentially love are all just different forms of energy from which we can benefit. This principle asks us to look at how we can keep beneficial forms of energy in our systems as long as possible. How can we take full advantage of an energy source and then use it again and again until it is no longer useful? Part of the goal here is to create as many closed loops as possible. In other words, how can we take the waste products of one system and turn them into the resources of another one? Also, as parts of our system degrade, we should look at how to use them for the next highest use. Finally, we must figure out how each resource in our design responds to being used. Does use cause it to increase or decrease, or is it unaffected? Is it lost when unused? Does it create pollution or system degradation when used?

The idea of next highest use is a way to take maximum advantage of the resources we consume before the end of their useful lifespan. For instance, if our old farmhouse is getting old, we can choose to rebuild, reusing any timbers that are still good. Any timbers with rot can be cut into good and bad pieces, and the good pieces can be used to make chairs for the kitchen table. Some years down the road, if one of the rungs on the chair breaks, we can carve spoons out of the wood. When the spoons break or get ratty, we can add them to the wood chips on our garden paths or use them as kindling for a cooking fire. The ash remaining from the fire can then be used to make soap. Notice how this is very different from tearing down the original house and turning the timbers directly into firewood. This same thinking can be applied to everything from water to heat to electricity. Ask yourself, "How can I get another use out of this before I throw it away?"

A great example of energy cycling and recycling can be found at Bullock's Permaculture Homestead. Solar pumps lift irrigation water from a pond at the lowest point on their property to the highest point. From there the water feeds by gravity down to irrigation systems in the gardens. Water escaping the root zone of plants reenters the water table and eventually finds its way to the pond again. From there it cycles again and helps to supply much needed water for a homestead with a three-to-four-month dry season. If that water weren't cycled back into the system, it would overflow out of the pond and enter Puget Sound, where it would mix with salt water and become less useful.

Questions to ask:

- Am I using each resource in my design as much as possible?
- Am I closing loops with the systems I design?
- Does my design take advantage of resources' next highest uses?
- Am I making use of resources that degrade when unused?
- Am I avoiding using polluting resources as much as possible?

Learning from Nature

A fundamental understanding of ecology is critical to permaculture design. After all, how can we create designs that mimic nature if we don't understand how nature functions? Ecology is the science of looking at living organisms and how they interact with and relate to each other in the habitat in which they live. The spectrum of what ecology covers is vast—everything from the microorganisms found in streams to carnivorous mammals living in the Amazon rainforest. It's important to consider how significant the relationships and connections are among animals (including humans), plants, soil, air, and water. Thinking in this context gives us a much broader and more holistic perspective when creating our own designs and solving problems.

Breadfruit is an important staple crop in many tropical locations. A large, fast-growing evergreen that casts dense shade, breadfruit prefers to grow in deep, fertile, well-drained soils. These are all aspects of its species niche.

CORE ECOLOGICAL CONCEPTS

Ecology encompasses many specialties, including but not limited to soil science, wildlife biology, ocean physics, climatology, and botany. Here we highlight some core concepts of ecology that are useful to think about during the design process. Some important questions to keep in mind when you are stuck with a challenge are: What would nature do? How would this play out if nature took its course? What can I do to mimic natural processes?

Niche

The term *niche* describes the lifestyle of a species or its unique job description within the community or ecosystem in which it lives. Niche can be expressed in two ways. One is the species niche, which is what makes that species unique. Think of it as that species' baseball card stats, where the stats would include food sources, types of habitat, survival and adaptation strategies, and what it can provide to other species in the ecosystem. Niche can also be expressed as the community niche, which defines the specific role in which the species contributes to the community—like a job description. Nitrogen fixer is one job in a community that could be filled by many species; a more specific role is a canopy-dwelling insect that eats leaves.

Some species, such as humans, can be defined as generalists, which are widely adaptable and less likely to suffer when their habitat changes. Specialists, on the other hand, have very specific jobs and depend on conditions that are just right in order to survive. Through a pattern of interspecies relationships and cooperation, the resources of a given habitat are divided up among species. However, depending on the job of the species and on the environment, competition for resources will exist if other species in the area have the same niche. In our designs, we want to minimize competition and maximize cooperation by mimicking beneficial relationships found in nature.

Ecological succession

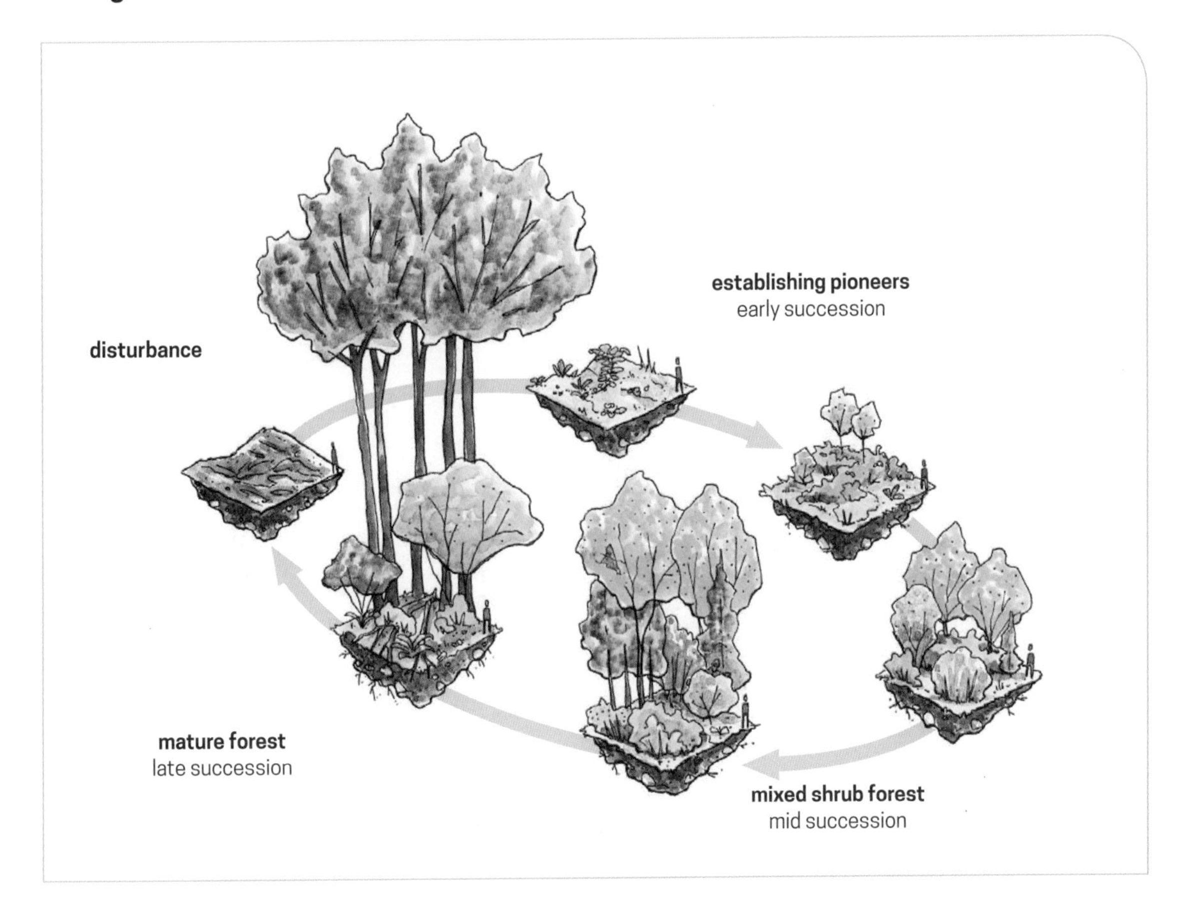

Ecological succession is a core concept in ecology that relates to the dynamic process or sequence of development in an ecosystem. These changes often follow a predictable pattern in the classic linear view of succession. However, variable site conditions and further expected or unexpected disturbance events can change the direction in which an ecosystem develops.

Disturbance can include human-caused types such as land clearing and pollution, as well as natural occurrences such as fire, lava flow, or landslide.

* Primary succession is when the ecosystem starts developing after a severe disturbance has left no ecological community in place. This is usually the result of volcanic eruption, desertification, glaciation, or severe erosion from either natural causes or intentional site scraping for land development.

- Secondary succession occurs after a previous biological community has been disturbed but leaves behind legacies such as organic matter in the soil, seed banks, and organisms from that previous ecosystem. This often follows forest fires, wind blowdowns, or clear-cutting.

After disturbance, a group of plants known as pioneer species that are highly tolerant of extreme or poor growing conditions usually moves into the newly disturbed space. These are often nitrogen fixers or plants that produce a lot of biomass; they create better living conditions to facilitate the success of future species. The niche strategy of pioneer species is that they are fast growing and sometimes spread aggressively. These plants can often be labeled weeds, but they are simply acting as a Band-Aid to disturbed sites.

During the intermediate stages of succession, the plant communities grow more complex with time. This transition is most often depicted as going from grassland to a mixed shrub layer that turns into a deciduous forest and then into a coniferous forest. However, vegetation will vary with differing site conditions. The niche strategy for survival for these mid-succession species is that they are competitors.

The large-leafed plant is burdock (*Arctium lappa*), a biennial pioneer species that produces many seeds and spreads them by attaching its burrs to passing animals. In this system the burdock is grown in a barrel for easier harvest of its long roots, which are both edible and medicinal.

The late successional community is traditionally seen as a stable ecosystem, but it is certainly not static. The plants are taller, grow more slowly, and in many cases are able to support a more diverse population of organisms in comparison to earlier stages. The matrix of late-successional species varies widely from region to region. Mountain ranges offer a different mix of conditions and species adapted to those conditions than a low valley nearby. The niche strategy associated with late successional species is that they are stress tolerators.

Often succession is illustrated in a linear fashion, from disturbance through several phases to a climax. However, that image lacks the complexity of the real world. At any point during the development of an ecological community, patches within it can be set back by disturbance. Then the process starts over again but not necessarily always in the same direction. Most landscapes hold a mosaic of patches at different stages of succession. The same disturbance can impact an ecosystem differently in different areas. That means two important features of succession must be acknowledged: (1) there is no single climax ecosystem that a landscape moves toward; (2) succession can actually move forward, backward, and laterally within a landscape. However, it tends to move in the same direction within individual patches, or ecological units within which conditions are largely the same.

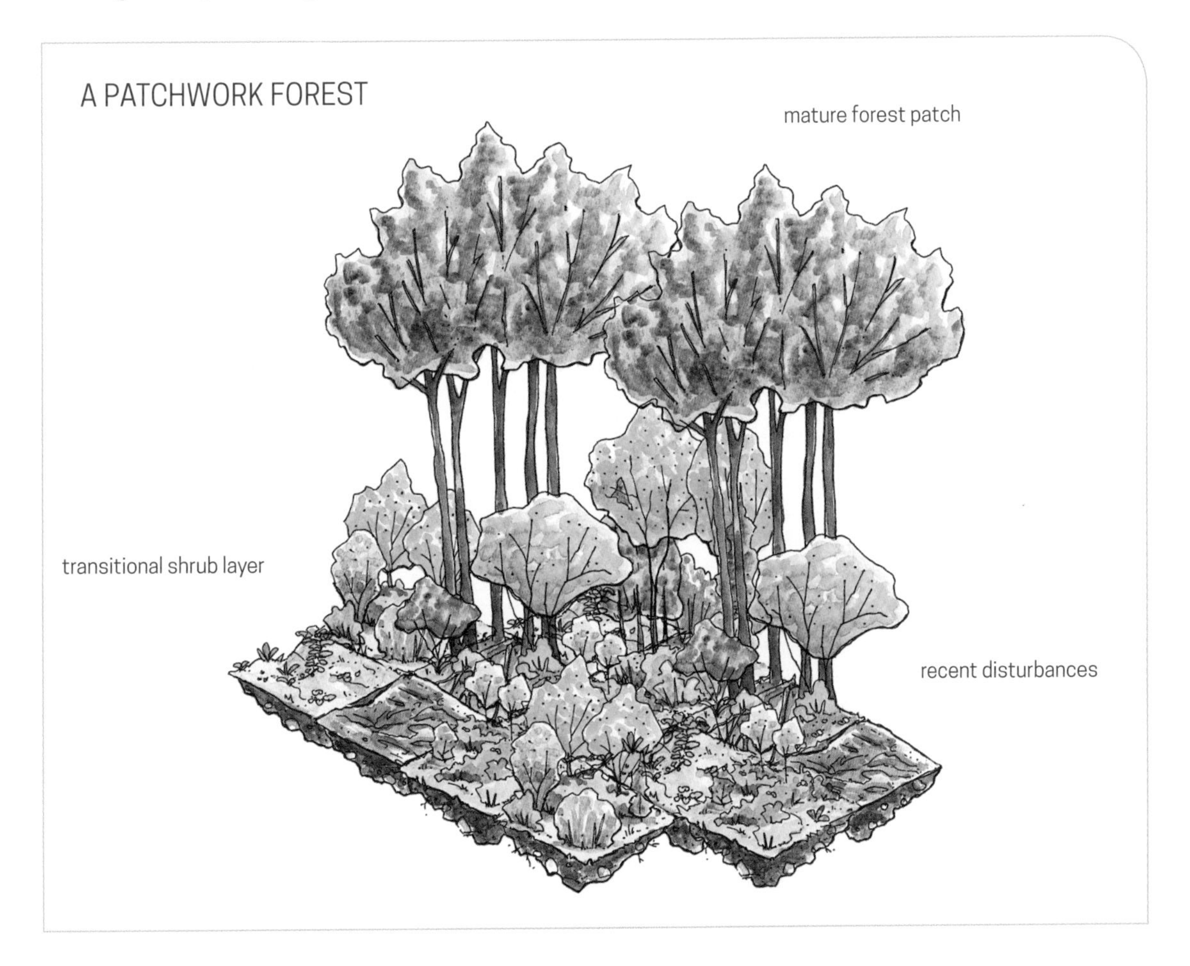

As permaculture designers, we need to understand how our gardens can be looked at in different stages of succession. In many cases, we can accelerate these stages through careful selection and management of species. However, we must have a clear idea of which successional stage we would like to achieve. We are not always rushing toward late succession. In fact, most abundant food plants tend to be early to mid-successional species.

Bioregion

A bioregion is an area of land defined by geological boundaries or ecological similarities rather than political boundaries. Bioregions often follow watershed boundaries and encompass a distinct area that shares unique characteristics of flora, fauna, geology, and climate in comparison to its neighboring regions. From the human standpoint, looking at the local bioregional resources and how we are using them is important in living sustainably. If we look at human history within a bioregion, we find embedded local traditions that are adapted to the local resources available. This includes food production, architecture, and health systems. We can often replicate those cultural patterns in our designs to make them appropriate to our bioregion.

Biodiversity

The diversity of species is usually an indication of ecosystem health. Natural ecosystems contain a wide variety of organisms—humans, plants, animals, fungi, bacteria, and so forth—that interact in the web of life. The more species within an ecosystem, the more potential for beneficial interaction we have. The same goes for genetic diversity within a species. In our designs, it should be a goal to increase or maintain biodiversity to achieve greater stability. Note that doesn't mean static stability (the ecosystem never changes), but dynamic stability (the system changes, ebbs, and flows but never fails). Species may come and species may go. The same goes for individuals, but if there are many species and genetically different individuals present, the whole system will not collapse with those changes. This means the ecosystem, on the whole, is more adaptable.

Interaction between species allows resources to be exchanged—whether those are food and nutrients, oxygen, shelter, or predation. One organism may protect or be symbiotic with another. The more species we have, the more opportunity there is. Genetic diversity within a species is also an important component of biodiversity. If we are going to plant ten apple trees and we choose to plant all 'Fuji', they all will share the same weaknesses. That means if a disease strikes and 'Fuji' is susceptible, we may lose all our apples. Instead, we can plant ten different cultivars; then, if one or more proves resistant to the disease, we still get a harvest. What's more, we can select cultivars that produce apples with different qualities and that produce at different times of the year.

Monocultures and polycultures

Monoculture is the agricultural practice of growing one species in large quantity to minimize the need for labor inputs and create easier and more cost-effective main-

tenance. For instance, an industrial farm might grow 2 acres of carrots. However, monocultures preclude biodiversity. The carrots in our example will be more susceptible to diseases and pests, such as carrot rust fly, because the pests have a near-unlimited supply of food. That means their populations will grow much larger than they would in nature. This is a huge part of the reason industrial agriculture is so reliant upon chemicals. How else can you do battle with massive swarms of ravenous insects?

Permaculture designers favor polyculture, the practice of growing multiple species in an agricultural setting in the same space and at the same time. This emulates nature's example of biodiversity and makes it more difficult for pests and diseases to take hold. For example, if our carrots are being grown in smaller plots mixed with other vegetables and possibly even perennials, it will be much more difficult for the carrot rust fly to find a carrot. Even better, all those other plants may support predators that eat carrot rust flies! Growing plants in polycultures also means a greater diversity of products served up throughout the year. Instead of just one big harvest of carrots, we can have an entire smorgasbord of fruits and vegetables.

(left) A variety of flowers and vegetables are grown together at Bullock's Permaculture Homestead, creating a profusion of colorful production.

(right) At Bullock's Permaculture Homestead, more than eighty varieties of apples are grown. This picture shows a fall harvest of storage apples that will be eaten in ripening order all the way through early spring.

These beehives at Kailash Ecovillage in Portland, Oregon, are a home for beneficial insects, in this case honeybees, which provide valuable pollination services.

Ecosystem services

The term *ecosystem services* refers to the benefits that functional, healthy ecosystems provide. Examples of ecosystem services include pollination, water filtration, carbon sequestration, and erosion prevention. As designers, it is in our best interest (and the interest of other life forms and future generations) to strive to make sure that the landscapes we design provide plenty of ecosystem services for the benefit of all. It is also important that our designs do not impede ecosystem services from happening.

YOUR ECOLOGICAL CONTEXT

An important part of learning from nature is understanding your ecological context—that is, the environment you live in. Your ecological context encompasses your climate, your soil, your watershed, and your surrounding human settlement pattern. Learning some earth science will help you better understand your context and make appropriate design choices.

In Houston, Texas, the climate is hot in the summer and mild in the winter, with an average of 50 inches of rainfall spread throughout the year. At this residence in Houston, a huge variety of plants, such as Mediterranean herbs and pomegranates, can be grown.

Your climate

Understanding your climate is essential to making good decisions as a permaculture designer. In fact, many of the least functional and least efficient designs we see in the world are a result of ignoring climate. While there are many aspects to climate, three stand out as important for permaculture designers:

- Temperature. What are your average highs and lows? What are your extreme highs and lows?
- Precipitation. How much rain do you get? Does it generally come as sprinkles or torrential downpours? How does it relate to your evaporation rate? How much of it comes as snow or ice?
- Seasonality. When does your precipitation occur? When are cold snaps most likely to occur? What other climatic factors change with the seasons?

You can easily see these aspects of your climate by looking up (or creating) a climatogram for your area. A climatogram shows precipitation and temperature extremes throughout the year to give you an idea of your climatic patterns. It overlays a line graph for temperature highs and lows with a bar graph for precipitation, with a temperature scale on the left, a precipitation scale on the right, and the months labeled along the bottom.

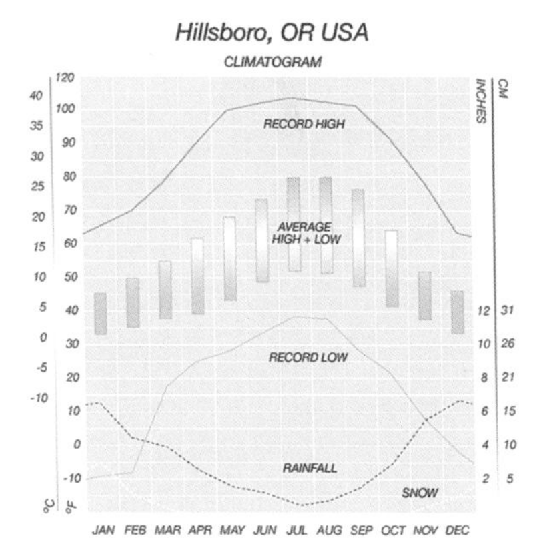

A climatogram for Hillsboro, Oregon

Some other resources that may help you to understand your climate better include the USDA Plant Hardiness Zone Map and the Köppen Climate Classification System.

The USDA map (planthardiness.ars.usda.gov/phzmweb/) tells you your average annual minimum temperature. For example, Seattle is in zone 8b, which means our average annual minimum temperature falls between 15 and 20°F. Note, however, that temperatures can and do dip below that. This information is useful as a tool to determine which plants will survive and thrive in your climate.

The Köppen system divides the entire earth into different classifications based on temperatures, precipitation, and seasonality. For instance, Seattle is classified as Csb: warm-summer Mediterranean climate. The Köppen system applies fairly broad brushstrokes to climate. For instance, Tampa Bay, Florida, falls in the same climate zone as parts of New York City. However, it is still useful for identifying the broad climatic conditions in a location.

While much further climatic subdivision is possible, design decisions often ride on which of these broad classifications a project falls into:

Temperate. This includes most places at midlatitudes with four distinct seasons. Ocean and air currents, proximity to large bodies of water, and topography can cause some variation here. The temperatures here tend to be moderated as compared to those in the polar regions or the tropics. Day length in summer and winter is noticeably different, creating seasonal temperature variation. Within the temperate zones are areas far from the moderating influences of large bodies of water or topographic extremes referred to as continental climates (for example, Kansas City, Missouri, United States; Kiev, Ukraine; Ulaanbaatar, Mongolia).

Tropical. These are the places that fall between the Tropic of Cancer and the Tropic of Capricorn (approximately 23° north and south latitude, respectively). They tend to be hot, with the coolest temperatures found at higher elevation. Day length remains relatively constant throughout the year. The term *subtropical* loosely denotes places near the tropic lines that have tropical conditions for part of the year (for example, Cape Town, South Africa; Los Angeles, California; and Barcelona, Spain).

Arid. This modifier could be applied to any of the previously described climates. However, it refers specifically to precipitation. In arid climates, evaporation exceeds precipitation. This is important because there is a whole set of techniques that may be necessary for a permaculture system to thrive in arid conditions that are not necessary where precipitation is more plentiful. Examples of arid climates can be found in Tucson, Arizona, United States; northern Africa; northwest China; and all of interior Australia.

Polar. These are places within the polar circles, where daylight is scarce in winter and the growing season may be insignificant.

Putting all this information together for your site will result in a climate profile that you can use in a variety of ways during the design process.

Your microclimates

Microclimates are areas within a climate zone that are a little different from the area that surrounds them. These areas, which can be very small (next to the dryer vent on the side of your house) or quite large (the north-facing side of a hill that runs for several miles), can differ by being warmer, colder, wetter, drier, windier, calmer, and so on. In design, the patterns of the overall climate are the rules we must work within; however, microclimates allow us to use our creativity to bend or break those rules. Microclimates are affected by an area's aspect, solar orientation, airflow, and vegetation.

Aspect. The aspect of a slope is the direction it faces. In the Northern Hemisphere, a slope with a south aspect faces the equator and receives more sun than a slope with a north aspect, which faces the North Pole; it's just the opposite in the Southern Hemisphere. Depending on your climate and latitude, each of these aspects lends itself to a different use. For instance, in temperate climates in the middle to high

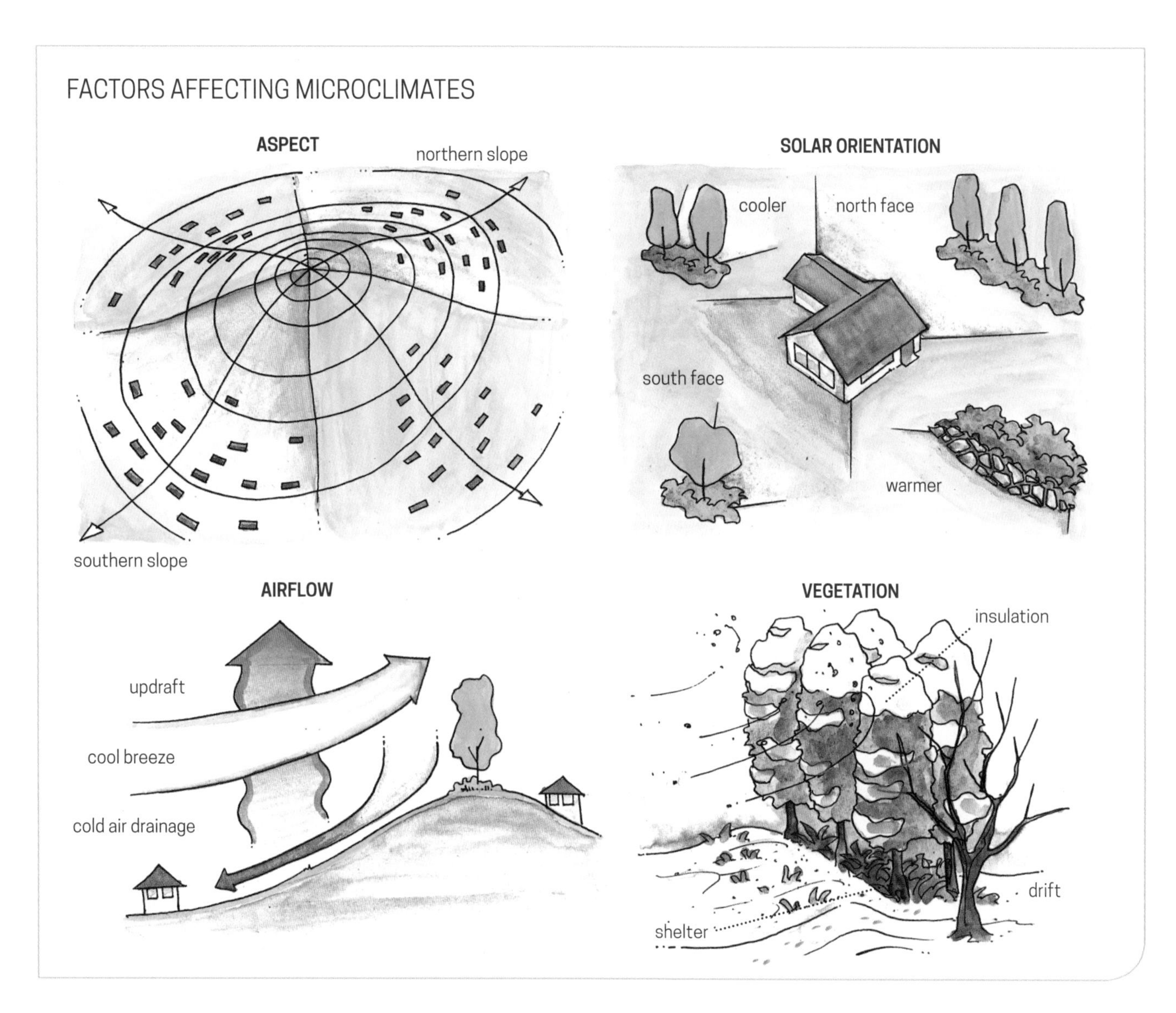

latitudes, aspects oriented toward the equator offer some of the best opportunities for winter sun, passive solar design, and gardens. Closer to the equator, north and south aspects matter less. However, east aspects are cooler in the late afternoon when the sun is at its most intense, and west aspects tend to get very hot in the afternoons. Essentially, if your land has areas with different aspects, you should take note and plan uses for each aspect that match the conditions found there.

Solar orientation. Solar orientation refers to where something is in relation to the path traveled by the sun. As designers, we need to look at existing and planned landscape elements with an understanding of how the path traveled by the sun will interact with them. At the equator, the sun travels a relatively constant path across the sky throughout the year; as we move toward the poles, the variation in the sun's path throughout the year becomes more extreme. That's why the South

Pole experiences twenty-four hours of darkness on June 21 and twenty-four hours of daylight December 21, and why the opposite is true at the North Pole.

Understanding the path of the sun where you live can help immensely in planning to take advantage of the free solar energy that hits the earth. For instance, if you had a rectangular house near Chicago, Stockholm, or Tokyo, it would be beneficial to orient the broad side toward the south and cover it with windows. That way, more of the house would get hit by low-angle winter sun that could do part of the job of heating the house in the winter. In this case, it would make sense to put a root cellar on the north side of the house so it would receive minimal heat from the sun. This would work exactly the opposite in the mid-high latitudes of the Southern Hemisphere. Clearly, this is closely linked to aspect, but even on flat land, solar orientation can make a huge difference in microclimate effects.

Airflow. The way air moves across the land can also have major impacts on the climatic conditions there. While extremely windy conditions can be detrimental anywhere, in cold-winter climates they can be especially damaging and cause loss of much-needed heat. In hot, humid climates, still areas with no airflow can be stifling and oppressive, and can create conditions for fungal outbreaks. In many climates, cold night air drains into low spots. Depending on the location, this can mean that cherry trees situated in these areas produce because their chill requirements are met, or it can mean the garden gets hit with frost a full month earlier than it would if it were located just a few feet higher on the slope.

The midslope is often referred to as the thermal band because it stays warmer than both the frost pocket below and the exposed hilltop above. Updrafts occur when the land heats up under the sun's rays. As the hot air rises, cooler air moves in to take its place. On slopes, this usually means breezes blow uphill in the late afternoon, an effect known as convection. As you can see, understanding how air moves on your site and looking for the spots that are different can provide opportunities for good design.

Vegetation. The vegetation on-site can have profound impacts on microclimate. In temperate climates, evergreen forests generally stay warmer in the winter than surrounding areas due to the insulating and wind-stopping effects of the trees. Deciduous forests have a similar effect but to a lesser degree. Vegetation standing between an element and the prevailing wind can block it, which may be good or bad depending on what you are trying to do. In snowy regions, this can result in snowdrifts, which can block your door or provide free spring irrigation. Some trees, such as western red cedar (*Thuja plicata*), can actually create dry spots under their dense evergreen foliage. This is important to know before you think about what to plant there. The key is to look at the existing vegetation on a site and determine how it is modifying climatic conditions.

For starters, identifying existing microclimates on a site is essential for good design. From there you can decide what you want to do about them. As we get into the design section, you will see that each microclimate can be enhanced, buffered, or neutralized. Once you've worked with the existing microclimates on the site, you can start to look at creating some of your own through design.

Your soil

Many of your design decisions will be based on knowing your soil and its capabilities. Knowing your soil will also be a factor in managing your site, something we'll discuss later in the book.

(left) This olive has been planted next to a south-facing rock for additional heat in a marginal climate. It is growing successfully in this warm microclimate.

(right) This row of trees at New Forest Farm in Viola, Wisconsin, stops the wind and causes snow to drift on the leeward side. When that snow melts, the water infiltrates into the land.

Healthy soils have a few different components: they consist of 40 percent pore space, 30 percent minerals, 15 percent water, 12 percent biota, and 3 to 12 percent organic material. Much of the soil is pore space, which means healthy soils are well aerated. Water, which is necessary for soil life and plants to flourish, also accounts for a large percentage of a healthy soil.

To gain a better understanding of what's happening in your soil, let's take a closer look at its physical properties, such as texture, consistence, and color. These properties determine many things, from drainage characteristics to nutrient holding capacity. These aspects of soil can be hard to change or may require a lot of management to get what you want from your soil. Later in the book we'll look at nutrients and soil life when we discuss managing your soil's fertility.

Soil texture. To assess soil texture, look at the relative amount of sand (coarse particles), silt (fine particles), and clay (super-fine particles) in your soil. In the field,

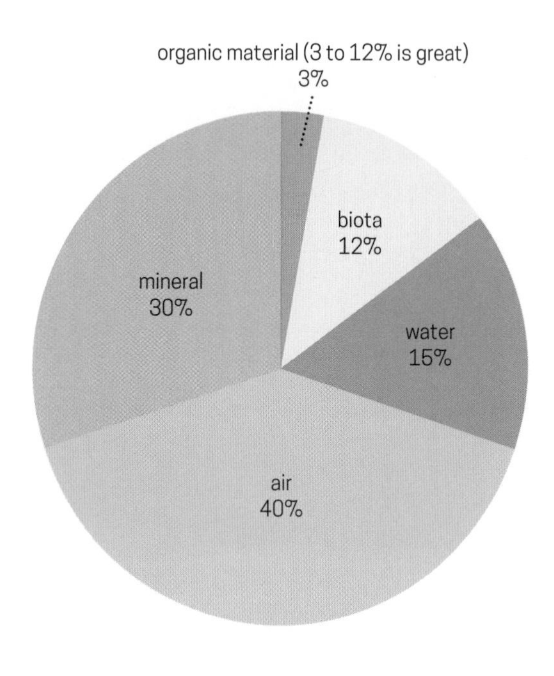

Components of healthy soil

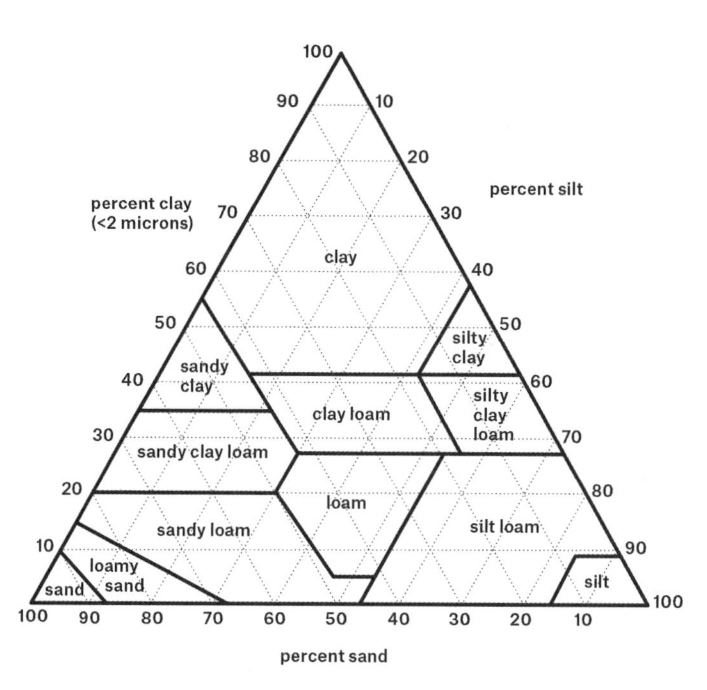

Soil texture classification chart

you can assess this by feel. Get the soil moist and rub it between your fingers. You may feel more than one texture at the same time.

- Gritty indicates the presence of sand.
- Smooth indicates the presence of silt.
- Slippery indicates the presence of clay.

Another way to assess your soil's texture is to use the shaker jar test:

1. Dig up a sample of your soil from the top 12 inches; try to get a slice through all the layers.
2. Remove any rocks and organic particles, place the soil in a jar, and fill it with water.
3. Shake the contents vigorously until everything goes into solution
4. Allow the contents to settle for a few days.

Once the soil has settled in the jar, you can see the layers: sand drops to the bottom, silt stays in the middle, and clay is at the top. Make a rough estimate of the percentage of each and look up your soil type on the soil texture classification chart here. You can then look up the more detailed properties associated with that type of soil.

Consistence. A term from soil science, consistence refers to how loose, light, and airy (or conversely, compacted) your soil is. If your soil doesn't even hold together (like a handful of playbox sand), it could be described as loose. If it gives when you press on it, like a sponge, we call it friable. This is the sweet spot. Moving toward the other end of the spectrum, if a small clod of moist soil is barely crushable between your fingers, it can be described as firm or very firm, generally indicating compaction and/or a lack of organic material and microbial activity.

Color. Different colors indicate the presence of different minerals in your soil. For instance, red soils are typically rich in iron, while yellow and olive soils are rich in aluminum or manganese. Dark coloration generally indicates organic matter. Gray indicates a lack of oxygen, usually due to a high water table or frequent inundation. For climates where moisture is not spread evenly throughout the year, it is important to look for mottled soils. For instance, gray soil with patches of red generally indicates inundation for at least part of the year. This could be due to compaction or a high water table. Either way, it is important to know so you can plant the right species there or modify the soil conditions.

Horizons. If you dig a hole in your landscape, you will likely notice that the soil actually appears in layers. These are called soil horizons. In undisturbed soils they tend to occur in a predictable pattern:

O—the organic layer at the top of the soil profile; the least broken-down organic material is usually at the very top (in a forest this would be the recently fallen leaf litter), and below that the organic material is typically broken down to varying degrees.

A—where the organic material gets mixed into the existing soil matrix; usually dark in color, containing most plant roots.

E—not present in all soils; appears as a layer that is light in color when water percolating through the soil leaches the nutrients out of this layer, taking them deeper into the soil.

B—where nutrients end up after they leach out of the A and E horizons; the last chance for plants to take up those nutrients before they leave the system.

C—mineral soils with no organic material; at this layer, all weathering is chemical as no other influences are present.

D—bedrock or parent material, often (but not always) the original nutrient source on a landscape; in some landscapes this is buried too deep to find, while in others it is right at the surface.

Dark coloration in soils typically indicates the presence of organic material and potential fertility. The 2-liter bottle is being used as an inexpensive hot cap in this image.

If land has been severely disturbed, you won't see these horizons. For instance, the top layers are not intact in an agricultural field that is ploughed annually. On

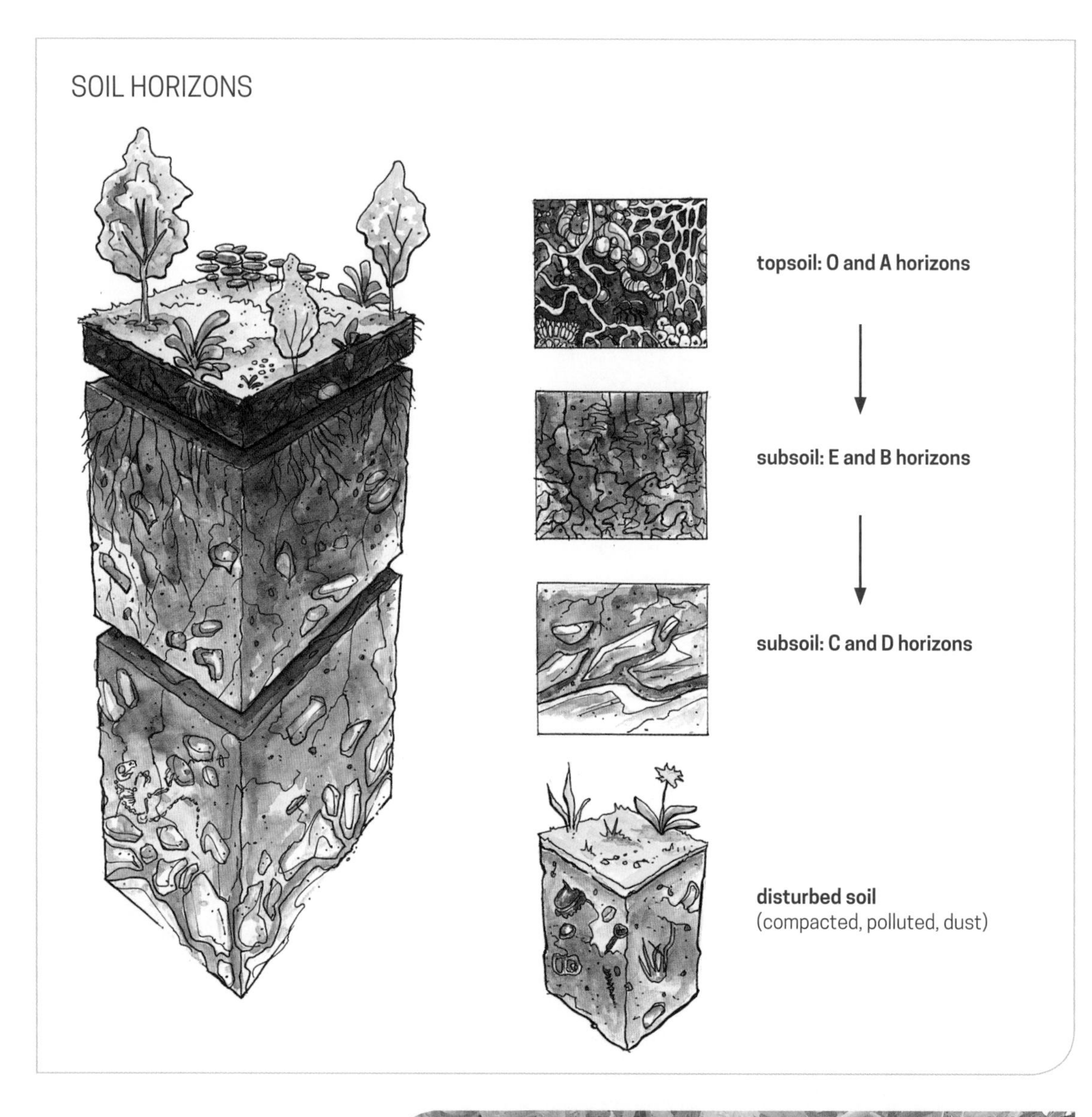

Daikon radishes produce very large roots that are excellent at breaking up soils with mild compaction.

urban sites that have had fill dumped, the entire site often looks like a uniform mishmash. It takes a long time for horizons to develop.

Compaction. As you dig into your soil, you may also find some hard and/or impermeable layers. These can occur at any level and indicate compaction. This can lead to drainage issues and can also be a barrier to roots.

Infiltration rate. Infiltration is the process of surface water being absorbed and entering into the soil layer of the earth. Water will pass into soil at different rates depending on the soil type, soil saturation levels, and the topography of the area. In urban areas, we have problems where water is unable to penetrate into the soil layer because there are so many impervious surfaces such as asphalt and concrete. You can do your own infiltration test (also known as a perc test) to understand the percolation or drainage of certain soils or areas within your site and to determine if they are appropriate for the uses you have in mind for them.

To do this test, you will need a digging tool (preferably a post-hole digger, although a shovel will work just fine), a measuring device (a tape measure or yardstick), water to fill your perc hole, a timer or watch, and patience—this test could take a few hours.

PERC TEST

1. In each of the areas where you are assessing infiltration, dig a hole approximately 2 feet deep and 10 to 14 inches wide.
2. Fill the hole with water and let it drain to simulate saturated conditions. Depending on your soil type, you may need to do this a few times. It is important to assess your soil's drainage abilities during the wettest time of the year.
3. Once the soil is saturated, fill the hole with water a final time and then time the drainage and record the depth it has dropped every hour.
4. Calculate your minutes per inch (MPI) by dividing the amount of time in minutes by the distance in inches the water level drops. For example, a rate of 3 inches per hour would correspond to a rate of 20 MPI (60 minutes ÷ 3 inches = 20 MPI)

15 or less MPI (4 inches or more per hour) = free draining

20–30 MPI (2 to 3 inches per hour) = fast drainage

30–60 MPI (1 to 2 inches per hour) = average drainage

60 or more MPI (1 inch or less per hour) = poor drainage

From above, it is often easy to see broad watershed patterns in the landscape. It's important to our permaculture designs to understand watersheds from the ground level as well.

Your watershed

Every piece of land belongs to a watershed, defined as an area of land that sheds or drains runoff from precipitation (rain or snow) downhill from the highest geographical barriers such as hills, ridges, and mountains to a specific low point, generally a tributary outlet to a larger river or a lake. The stream order in a watershed is a classic example of a hierarchical pattern found in nature. Starting with sheet flow, water accumulates into rills and runnels, then into creeks and streams, finally becoming rivers ending in estuaries. Watersheds exist at different scales, both large and small, and in different states of health.

As rain falls from the sky and moves downhill, increasing in volume, that water can become degraded and have detrimental effects on the land and our quality of life as a result. Some examples of degradation are landslides from deforestation, sedimentation in streams from erosion, lack of infiltration from excessive impervious surfaces, and water pollution from various industrial and agricultural practices. In a healthy watershed, the steep slopes are vegetated, the soil is absorbing and infiltrating rainfall, and waterways provide habitat and are pollution-free. Realistically, most watersheds have a mix of healthy and damaged areas.

It is important to consider the entire watershed in any permaculture design project because it gives us a sense of where we are in the big picture. We can begin to understand how we are affected by activity upstream and how we are affecting activity downstream. By designing with a holistic viewpoint in mind, we are able to protect and better utilize our water supply and other shared resources. We can also improve degraded watersheds in a relatively short time frame through restoration and conservation.

A few important watershed features to understand and assess are aquifers (groundwater), springs and artesian wells, stormwater, and erosion.

Aquifers (groundwater). An aquifer is an underground body of water moving through a layer of permeable rock. This groundwater can be withdrawn for human use but should be recharged through infiltration to keep the aquifer alive. An aquifer that is sealed and is no longer being recharged is considered a nonrenewable source of water; water from one of these aquifers is known as fossil water.

Springs. Springs are a natural occurrence where groundwater from an aquifer is pushed to the earth's surface. The water can surface as seepage through the soil, or in channelized fissures or fractures in rock, or even from a cave. Humans have long used this resource of fresh water, sometimes even regarding it as sacred. Under the right conditions of pressure and elevation, springs can become artesian wells, a valuable source of drinking water even today.

Stormwater. Stormwater is rain that falls to the earth and lands on manmade surfaces (such as rooftops, parking lots, and freeways). It then typically leaves the site where it fell. It can carry with it pollutants such as oil, fertilizers, and pesticides. Along with pollution, stormwater can carry with it other problems: the volume increases during peak storm events and oftentimes overloads waterways, leading

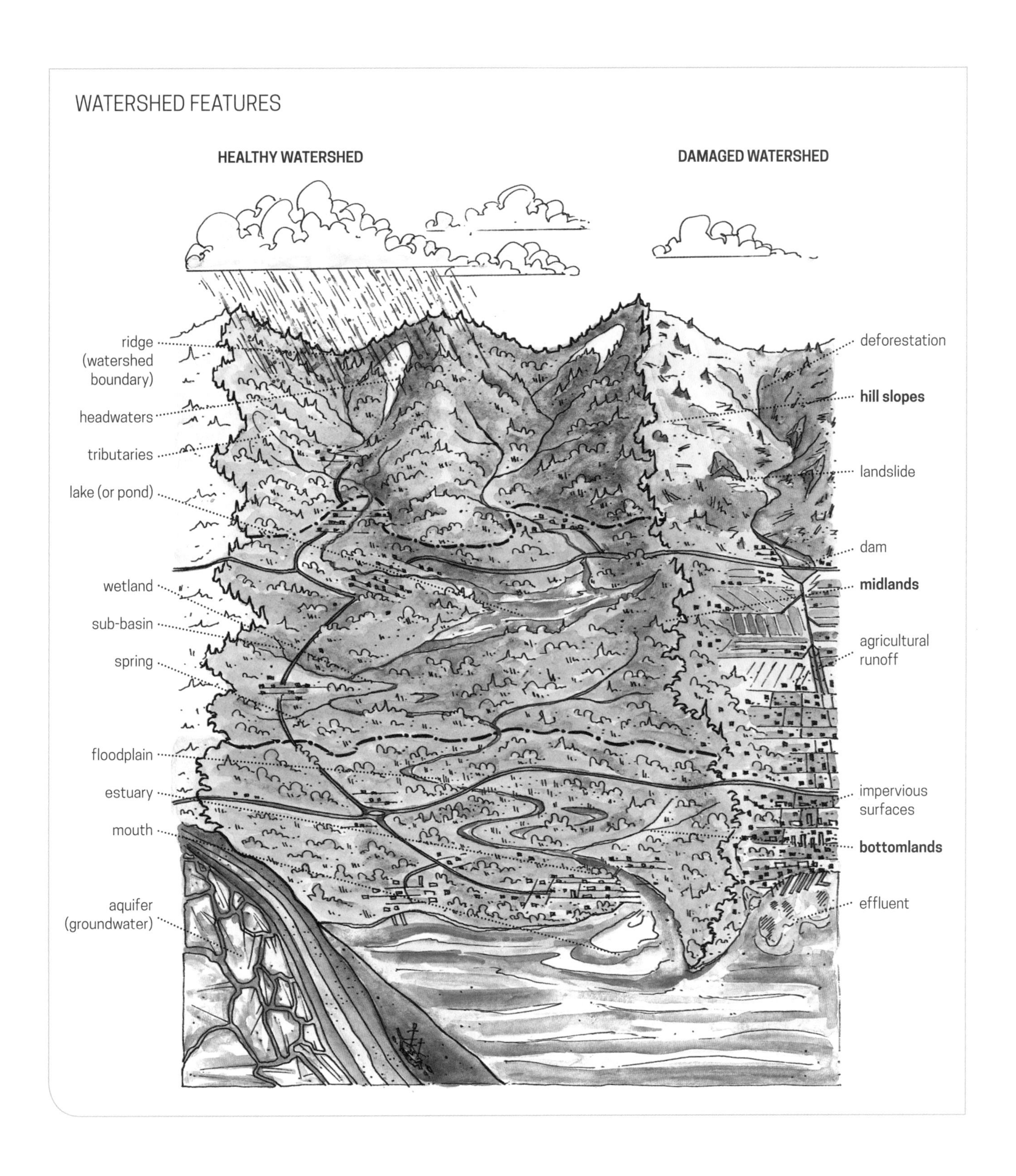

WATERSHED FEATURES
HEALTHY WATERSHED
DAMAGED WATERSHED
ridge (watershed boundary)
headwaters
tributaries
lake (or pond)
wetland
sub-basin
spring
floodplain
estuary
mouth
aquifer (groundwater)
deforestation
hill slopes
landslide
dam
midlands
agricultural runoff
impervious surfaces
bottomlands
effluent

to loss of habitat and sedimentation. The heat from the impervious surfaces is carried into waterways, increasing their temperatures.

Erosion. Erosion is the natural process of soil and rock being moved from one place and deposited to another on earth's surface. Erosion can be caused by rain, ice, waves, and wind. Excessive erosion is one of the leading causes of environmental degradation; it can cause hillsides and river banks to collapse, and soils to lose nutrients. Erosion can be minimized in your own landscape with careful planning and implementation of a good design.

Surrounding human settlement pattern

Just as the design decisions you make will be influenced by your climate type, soil type, and relation to watershed, factors relating to human settlement will influence your design. While permaculture design is certainly applicable to just about any situation, the density and character of your location will define many of your available resources, opportunities, and constraints. Whether you live in a natural setting, a rural community, a small town or village, a suburban neighborhood, an urban neighborhood, or a setting that doesn't fall squarely into any of these categories, the idea is to focus on taking advantage of the resources those areas have to offer and doing your best to eliminate or work around the obstacles.

Natural setting

Resources and benefits

- lots of ecosystem services
- high biodiversity
- wild harvesting opportunities (mushrooms, plants, game, and so forth)
- excellent models upon which to base design decisions

Potential constraints

- degradation and destruction from human disturbance
- limited understanding of how natural systems function
- vehicle dependence or isolation from community

Rural community

Resources and benefits

- space for large-scale techniques
- affordability and availability of land
- neighbors with land-based knowledge
- interface with wildlife and healthy ecosystems
- less-restrictive land use laws

Potential constraints

- vehicle dependence or isolation from community
- lack of economic opportunities to generate financial resources for land management
- less cultural diversity
- difficulty of managing large acreage

Small town or village

Resources and benefits

* easier sharing of resources due to conglomeration of people
* greater walkability, security, and sense of community
* little opportunity to get away with negative behaviors in anonymity
* enough exposure to nature to receive ecosystem services and direct experience

Potential constraints

* lack of the land resources of a rural setting
* lack of the cultural and transportation resources of an urban setting

Suburban neighborhood

Resources and benefits

* human-scale amount of land to manage (less than 2 acres)
* economic resources to help projects get off the ground
* accessibility of resources from both urban and rural landscapes

Potential constraints

* lack of community (many people don't know their neighbors)
* landscape designed for automobiles
* restriction of design opportunities by local politics

Urban neighborhood

Resources and benefits

* huge waste stream that can be tapped for resources
* abundance of heat created by concrete and asphalt
* broader mix of culture and art due to larger and more diverse population
* walkability and public transportation options

Potential constraints

* higher levels of air, water, and soil pollution
* more crime and negative human behavior due to greater anonymity
* lack of enough land to meet the needs of all the inhabitants
* limitation of design options by existing infrastructure

NATURE'S PATTERNS

As we think about permaculture design, we can also learn from nature's patterns. A pattern is defined as an arrangement of repeating or corresponding parts. Many different patterns show up in nature—large and small, visible and invisible, exemplified in space and in time. Nature uses patterns to move, collect, and disperse matter and energy. Ultimately, patterns are nature's elegant way of solving problems efficiently and effectively. If we apply nature's patterns appropriately in our designs, we should be able to tap into that efficiency and effectiveness.The key word here is *appropriately*. Applying patterns from nature will yield beneficial results only if the appropriate pattern is applied for any given situation. Applying the wrong pattern can actually create more problems than solutions. Large spiral annual gardens are often a pain because getting to the center with a hose or wheelbarrow involves walking in circles for a long time (and pulling a hose around a spiral is no fun at all). A network pattern or a network overlaid on a spiral may be a better design for easy access in an annual garden. Therefore, as we design it is crucial that we pay attention to the patterns that address the situation in nature most analogous to our own situation.

What follows is a basic, nonexhaustive library of patterns from nature. For each pattern, we give examples of that pattern showing up in nature as well as some design applications to help drive home the point that our designs get better when we use the right pattern at the right time.

Branching

branches in a tree streams in a watershed circulatory system

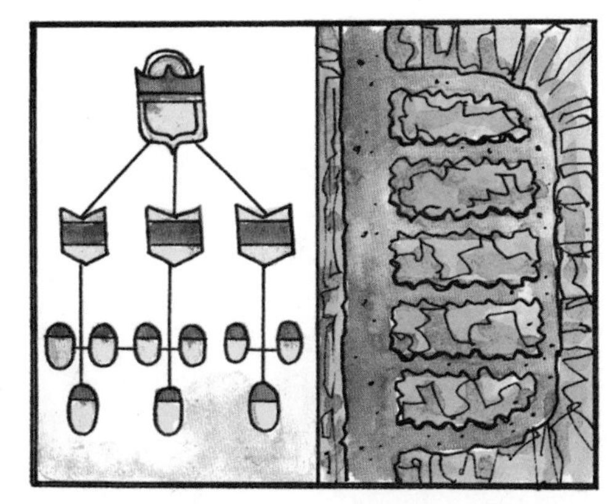

fire department organization pathway hierarchy

Branching patterns are great examples of ways nature collects and disperses both energy and materials. The branches of a tree allow energy photosynthesized in the leaves to be collected and redistributed via the trunk. Creeks come together to form streams, which lead to rivers and so on as water is collected from an entire watershed. The circulatory system in the human body is a way to collect de-oxygenated blood, recharge it, and send it back out to bring oxygen to the farthest reaches.

In design we use branching patterns all the time in the form of hierarchies. An access hierarchy allows us to have paths of increasing widths and durability for more intensive uses. Imagine a paved road that can accommodate a dump truck with offshoots that are somewhat smaller to accommodate pickup trucks. These may have offshoots that accommodate garden carts or wheelbarrows, which may in turn have offshoots that accommodate a single person. This can be incredibly useful when trying to make the most efficient use of space while allowing good access to all locations.

A good example of an invisible application of this pattern in design is the organization of a fire department. For every one district chief, there are typically a few battalion chiefs. Under the battalion chiefs are several lieutenants and under them a host of firefighters. This clear chain of command makes a lot of sense in this case. After all, no one wants the firefighters making decisions by consensus when their house is on fire. In this case the branching pattern is well applied to social design.

Waves and meanders

meandering river mushroom growth ocean waves

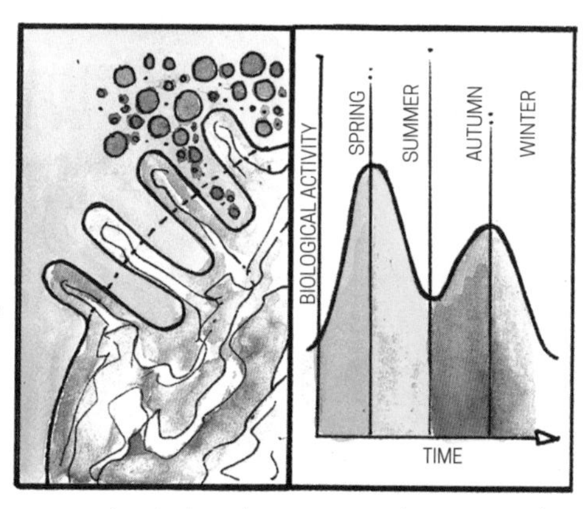

constructed pond seasonal energy needs

Waveforms appear frequently in nature and provide an example of how manipulating an edge can achieve a goal. When rivers come out of foothills and hit a flat floodplain, they often begin to meander a lot more. This happens because much of the energy in that water was played out when it was moving through steeper areas. These meanders are where rivers drop out most of their sediments, leading to some of the most fertile places on earth, floodplains. Many mushrooms burst from the ground and unfurl in a modified wave pattern called an Overbeck jet. This is where ominous mushroom clouds get their name. This pattern allows the mushroom to push up from the soil with minimal edge, thus minimal resistance. When it breaks the surface, it can unfurl to expose its gills and send out spores. When two fluids move past one another, waves are the result because they minimize turbulence. This means each fluid can move past more efficiently. Waves in bodies of water occur when wind passes over the surface (in which case air is acting as a fluid).

In design, one way we can use waves and their accompanying manipulation of edge is by changing the shapes of elements. A pond that is a perfect circle has minimal perimeter for maximum area. In some cases this is exactly what we want. However, if we want to grow more pond edge species, we can increase the pond edge using a wave pattern while still allowing it to hold the same amount of water.

Similarly, we can find an invisible pattern in the pulse of the seasons. Pulses are actually waves that occur over time, which we can see when we map them out on a graph. In this graph we see that relatively little biological activity happens in winter. When spring comes, plants start growing, animals wake up and begin seeking food and finding mates, and soil microbes quicken the pace of their decomposition work. During summer, things aren't as frantic as spring. Plants are either continuing to grow at a modest pace or they are in maintenance mode. Animals are raising young. When fall hits, everything in nature must get ready for winter. Extra energy is spent to harden off cell walls, store food, and shut down systems that will be damaged by cold. Then we return to the calm of winter. This waveform repeats year after year. The double pulse of energy expenditure in nature that happens in spring and fall matches up perfectly with time and energy required of people working with natural systems. Knowing this, we can plan the activity cycles of our lives appropriately or try to minimize extra work when we know our biological systems will need our attention.

Grids, networks, and tessellations

spiderweb honeycomb sunflower seeds

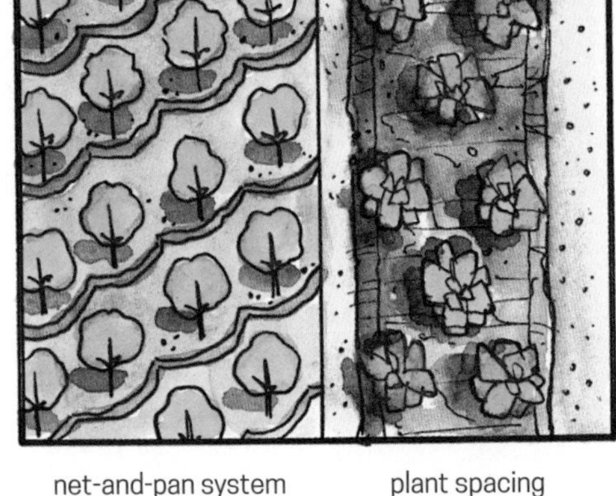

net-and-pan system plant spacing

Grids, networks, and tessellations all tend to maximize connections. They are also patterns of expansion and contraction (think of how your clothes, which are a network of threads, have the ability to stretch). In nature, the network pattern of a spiderweb is quite resilient. If part of the web is damaged, the bulk of it can still catch flies. Tessellations are repetitions of the same shape that, when put together, can fill up all space, leaving no voids. Hexagons in a honeycomb efficiently use space and maximize the volume of honey the bees can store while using the minimum amount of material to make the structure itself. Another example is the pattern of sunflower seeds in a flower head, which fit tightly together with no wasted space in between.

In landscapes, we often see orchards planted in grids. When dealing with arid landscapes, we can use a network pattern to help concentrate the small amount of rainfall where the trees are. By installing basins around the trees and a series of shallow drains to connect them, we can catch any rainfall hitting the landscape and route it to where it's needed most. This is called a net-and-pan system.

In the garden, we can also think about tessellations when we consider plant spacing. The crown shape of many plants is roughly round, so when we plant on a rectilinear grid, there is always some wasted space. If you triangulate the plants instead, you can fit more in the same amount of space. What if you thought about other ways to cause the different shapes of your plants to tessellate so you could fit more per acre?

Spirals

pinecone tendrils weather systems

drip irrigation outdoor shower

Spirals are patterns of strength and repetition. They lend themselves to situations where energy is going into something and then back out again, but not through. Snails go into their shells and then come back out the same way; they never pass through to a different exit. Most pinecones have their scales organized into double, overlapping spirals that run both directions. The tendrils of a climbing plant use a spiral shape to wrap around objects, enabling them to climb with firm support. Weather events such as hurricanes and cyclones are much larger examples of spirals in nature.

In landscapes, applying a spiral at the right time can get work done efficiently. For example, when laying out drip irrigation we often have spirals of pipe around trees. This allows for the tree to get water, but also other plants that may be planted together with the tree.

Because they lend themselves to going in and out again, spirals are also great for bathrooms and showers. We've designed many outdoor showers with a spiral piece of fencing that affords privacy while taking up a minimum of space and leaving no dead corner space that is of little use. Two such showers placed side by side would resemble the Overbeck jet of the mushroom and provide two space-efficient shower stalls with plenty of privacy.

Fractals

leaf veins twigs on a branch branches on a tree

A fractal is a pattern that repeats at smaller and/or larger scales. In essence, fractals are meta-patterns that can be applied to almost any other type of pattern. Once you learn to recognize them, you can see fractals everywhere. A head of romanesco is roughly cone shaped. Each segment of the cone is composed of cones, which are composed of cones. To use the branching pattern as an example, you can see how the same branching pattern appears in the veins of a leaf, the twigs on a branch, and the branches on a tree. In fact, the ratio of smaller branches to larger branches is often the same at each of these scales. Plant ecologists from the University of Arizona are even finding that in some tropical forests these same relationships apply to the entire forest, enabling them to make calculations of biomass production and carbon sequestration for entire landscapes by measuring a few trees.

For designers, fractals provide us with clues as to how a solution applied at one scale may be applicable at other scales. For instance, companion planting is a common vegetable gardening practice where plants that are beneficial to one another are placed together. This could be extrapolated to tree crops by planting companions with each tree in an orchard. Further, one could look at the beneficial relationships between one orchard block and another (or woodlot, or field) nearby. Continuing, one could look at the beneficial relationships between an entire permaculture landscape and other landscapes in the region. This kind of analysis is fractal application at its finest.

A pattern language

The concept of a pattern language was first conceived by renowned architect Christopher Alexander. In the context of permaculture design, a pattern language is a way to organize patterns that appear frequently in our designs and create a palette of interrelated options upon which we can draw repeatedly, as appropriate. The idea of using a pattern language to create a palette of time-tested solutions has been successfully applied to architecture, urban and regional planning, computer programming, and other design-based disciplines.

At this point no one has developed an all-encompassing pattern language specifically for permaculture, so as you gain experience using permaculture design you may want to pull together some of the most repeatedly useful patterns you've applied and write up when they are (and are not) applicable. This will give you a palette of options upon which you can draw. In volume 2 of *Edible Forest Gardens*, Dave Jacke and Eric Toensmeier have done this for forest gardening, and in his book *The Permaculture Handbook*, Peter Bane has done this for home-scale gardening. Christopher Alexander's work in *A Pattern Language* is also a useful starting point and source of inspiration for many of the problems faced by permaculture designers.

This herb spiral has several microclimates that are appropriate for plants that like different conditions (drier, wetter, hotter, colder).

THE PERMACULTURE DESIGN PROCESS

> *Design is a continuous process, guided in its evolution by information and skills derived from earlier observations of that process.*
>
> —BILL MOLLISON

Now that you're familiar with some basic permaculture concepts and have given some thought to how the natural world functions, it's time to get into the process of creating a permaculture design. The permaculture design process is about assembling components—concepts, materials, techniques, and strategies—into mutually beneficial relationships. Elements can be placed in a number of different arrangements, but the connections made between them is what builds systems that work effectively.

We take a master planning approach to the design process, which means breaking it down into steps that result finally in a master plan. In part 1 of the design process, you will analyze and assess the needs of your site and yourself. By bringing these sets of needs together, in part 2 you will use permaculture design methods to generate ideas that you can evaluate using the principles we discussed earlier. After that, in part 3, you will take the big picture master plan you've created and get down to the details of implementing it.

The design process can seem overwhelming. Depending on the site and the people involved, some parts of the design will be easier to complete than others. But don't shortchange yourself by limiting yourself to a simple design. Tackle the process one step at a time and it will become more manageable. You will find that the process gets easier as you go, and soon you will be looking through permaculture design lenses at every decision you make.

DESIGN PROCESS, PART 1: GATHERING INFORMATION

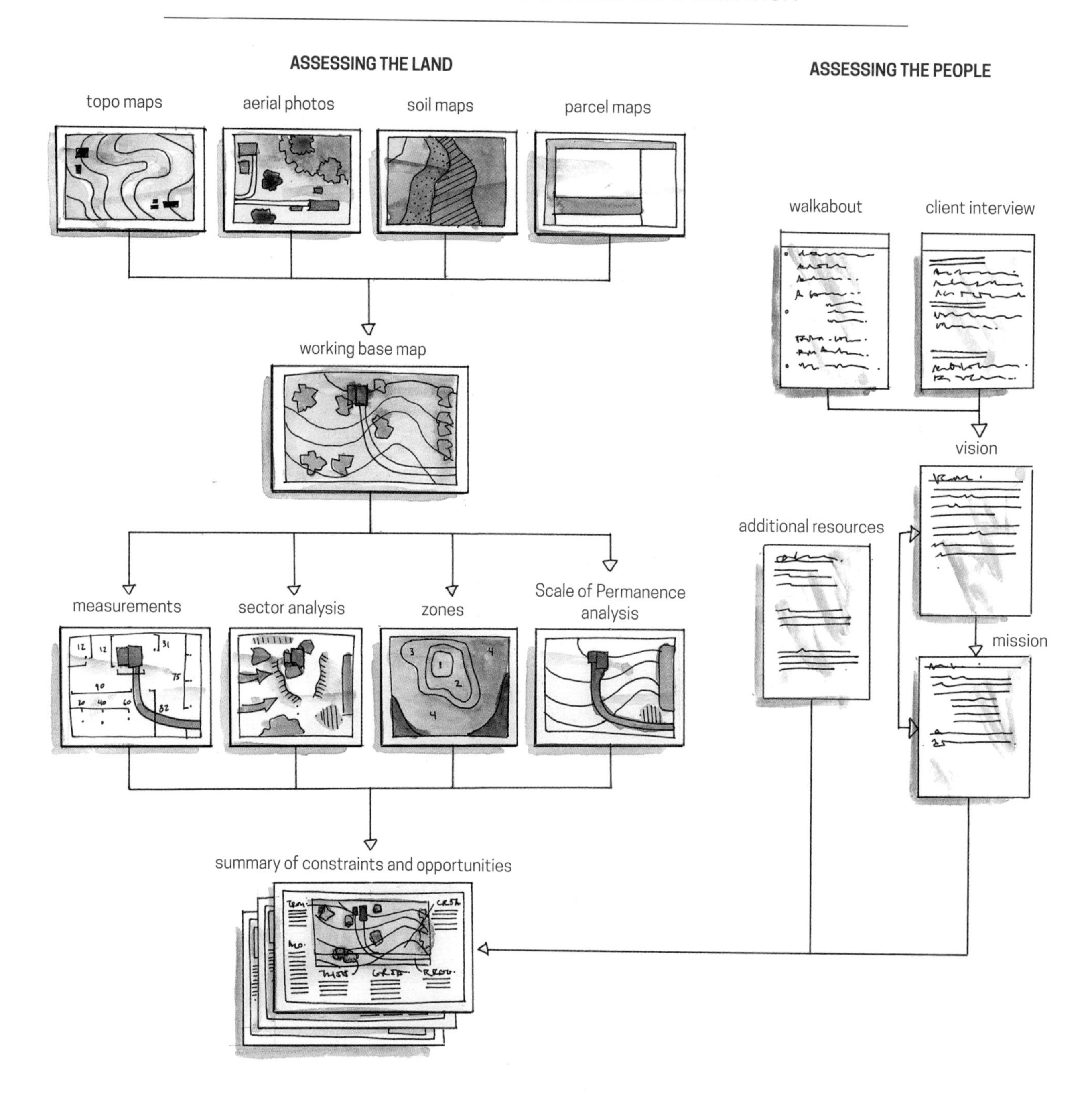

DESIGN PROCESS, PART 2: PUTTING THE DESIGN TOGETHER

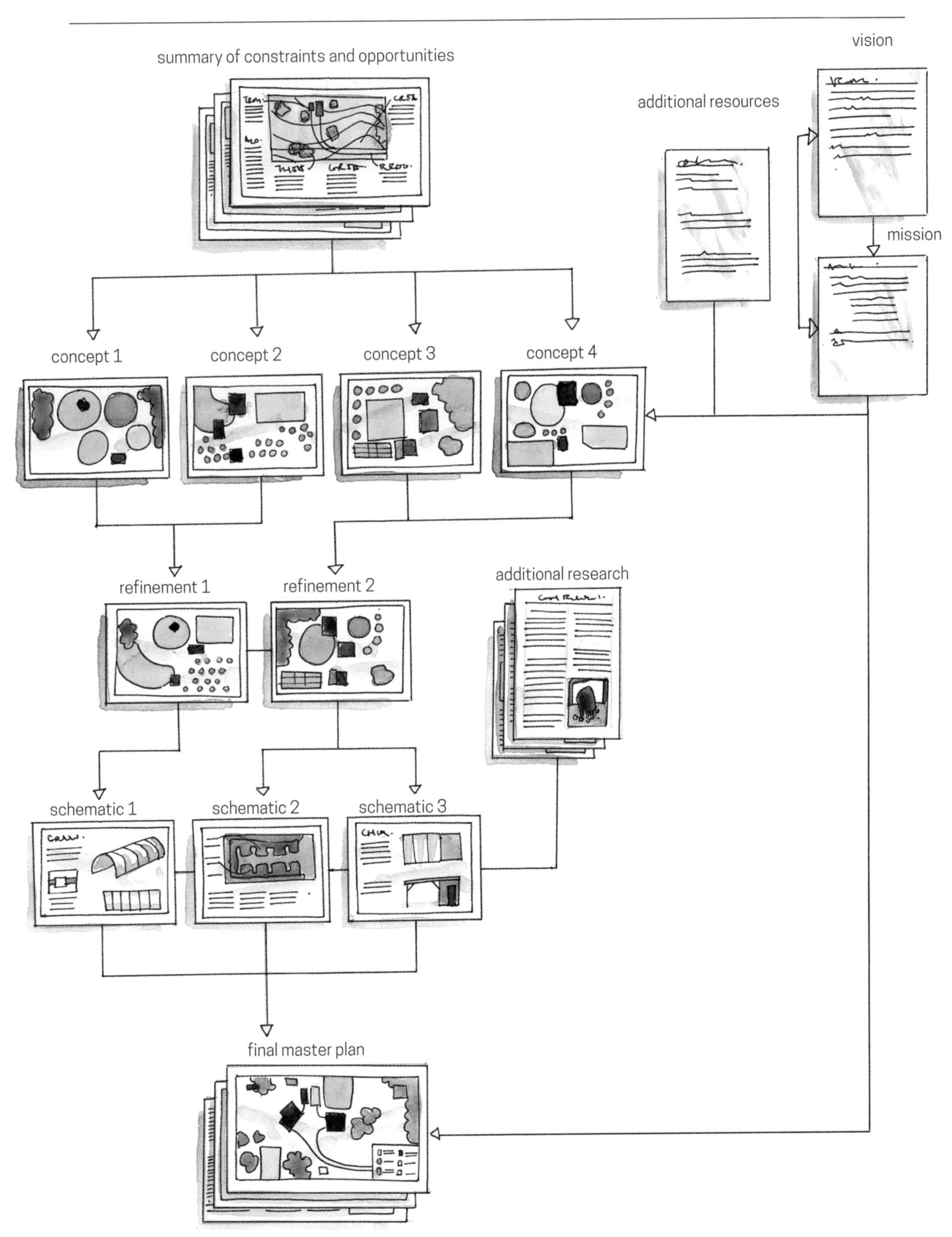

DESIGN PROCESS, PART 3: FIGURING OUT THE DETAILS

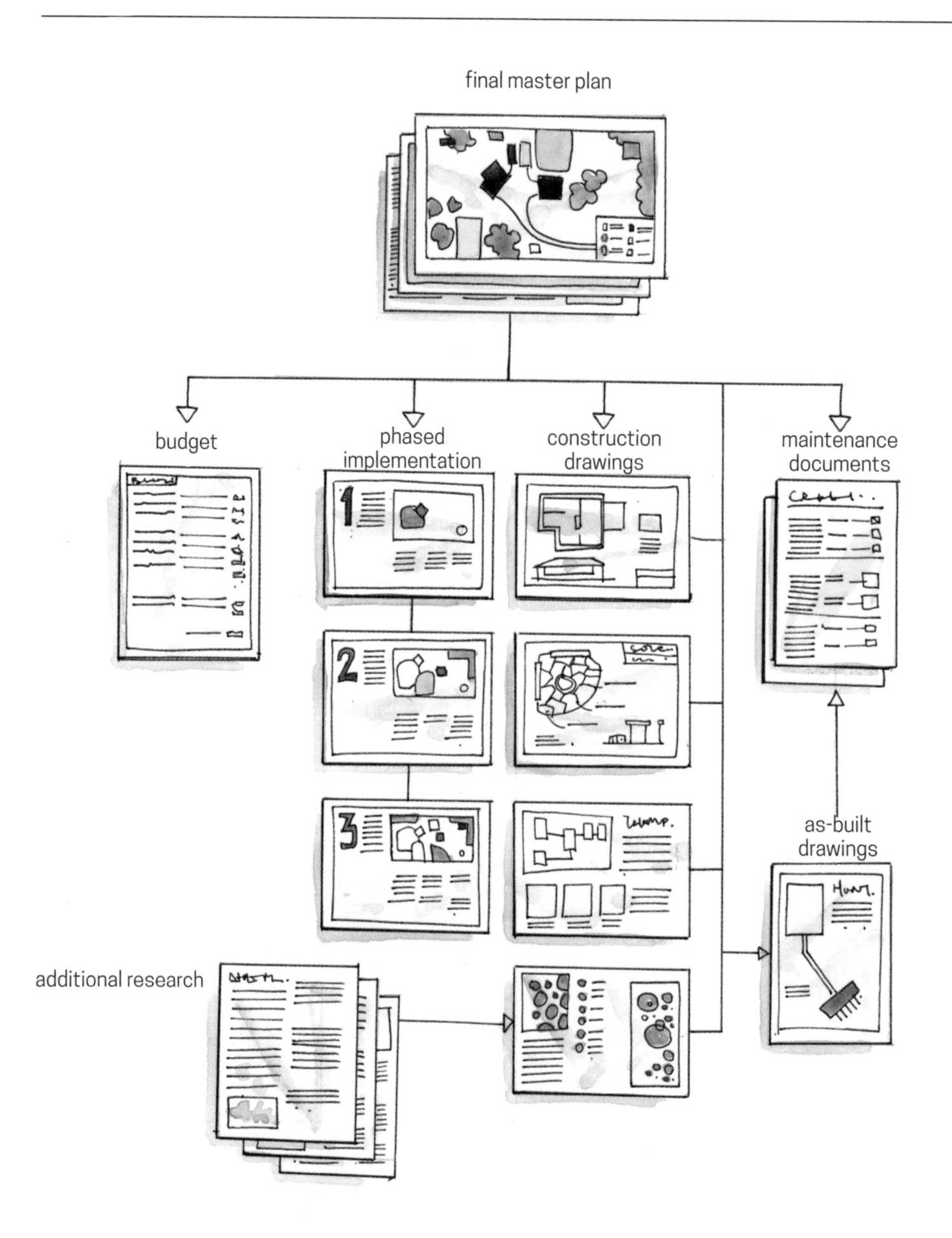

BLOOM HOMESTEAD

Permaculture design case study

Throughout the design section, we use a real-life case study to illustrate the products created at each step in the process. In the late 1990s, author Jessi Bloom was on the hunt for a piece of land that met specific criteria. It needed to be affordable, close to the Seattle metro area, large enough to keep animals and have a large garden. The property had to serve as a lifelong homestead to raise a family. It also needed to serve as a base for Jessi's ecological landscape design-build business. Space was needed for a nursery, construction materials, and parking for the company's vehicles. After many months of looking at properties throughout the Puget Sound region, Jessi found one that fit the bill.

Size: 2.33 acres, fenced

Access: via easement through neighboring property (no road frontage)

Location: unincorporated Snohomish County, Washington

Climate type: cool, Mediterranean

USDA hardiness zone: 7b

Rainfall: 48.5 inches a year

Snowfall: 8.4 inches per year

Precipitation days: 184

Sunny days: 160

Growing season: 165 days

Average July high: 77°F

Average January low: 32.8°F

Elevation: 401 feet

Existing site conditions: Mostly level pasture with many large conifers. No established garden. The northern portion of the property slopes toward the eastern corner. Buildings include an older double-wide mobile home with a small shop used by the former owner for a home-based business, a carport, and a couple of outbuildings used for storage and animals. Most of the land has been used as pasture, very little for outdoor living.

Gathering Information

Before you begin designing anything, it is critical to gather information so you have something upon which to base your design decisions. People commonly want to get started quickly and complete a design soon after they've acquired land. However, it's a good idea to start slowly and gather information for at least a year before making or acting on big decisions. You'll be gathering information about the site as well as the stakeholders and their needs and wants. The final product of this part of the design process will be your summary of constraints and opportunities.

Here are the steps and the products you'll generate in part 1 of the design process:

Step 1. Initial site observation—field notes

Step 2. Development of vision and objectives—mission, vision, and objectives

Step 3. Site analysis and assessment—base map, sector analysis, overlay sketches, summary of constraints and opportunities

STEP 1: INITIAL SITE OBSERVATION AND GETTING TO KNOW THE LAND

Before going to the site, you can do some basic research and even print out a map or soil survey to bring with you. However, before you start doing anything in depth, we recommend spending a little time walking the property with a mind geared up for observation, free of bias. You get only one chance to have a first impression. That means when you make your initial visit to the site it is important to take the opportunity to view it impartially. Even if you've already lived on your land for a while, you still can start observing it through permaculture design lenses.

Taking field notes

The goal of initial observation is to discover natural processes on-site and interpret them over time. Start a journal or log noting natural patterns: temperature extremes, seasonal growth, water flows, and the like. Make use of all your senses. What are those smells on the wind? Do different parts of the property feel warmer or colder? What sounds can you hear and what direction do they come from? Once you really get into the process of analyzing and assessing your site, these initial notes and impressions will provide useful information.

It is also important at this stage to make a conscious distinction between your observations and your interpretations. An observation is a fact, a statement of an observed phenomenon. An interpretation is a guess at why the phenomenon is happening. For instance, if I see a puddle in the yard, I might leap to the conclusion that the yard has poor drainage. That's an interpretation, not an observation.

The observation would be, "I see a puddle in the yard." The possible interpretations would include poor drainage, a leaky irrigation system, or a slight depression caused by compaction. When taking field notes, separate your observations from your interpretations so that you get the opportunity to research all the possible reasons for a phenomenon instead of jumping to conclusions.

Reading the landscape

Reading the landscape is an essential skill for permaculture designers to help figure out what the land needs and what it has to offer. Every piece of land has a story to tell. The better your understanding of this story, the more responsive your design can be to the specific situation.

One of the surest ways to learn how to read the landscape is to practice. There is no substitute for "dirt time." Pick up a set of maps and field guides for your local area and get out in the field. Comparing maps to what you see on the ground will help you to understand local geography and how it relates to natural patterns. Looking at how your site fits into the greater watershed will help you understand movement of water through your site. Field guides and natural history books will help you gain a better understanding of the players in your local ecosystem and their roles. Use all your senses and look at the landscape up close and far away. As your skill develops over time, you will eventually have a much better understanding of what the land is telling you.

Here are some things to look for when reading the landscape:

- Indicator plants tell you something about the environment. However, never make broad assumptions based on just one plant; a single reed on the site doesn't necessarily mean the drainage is poor. Look at all the plants and see if a pattern emerges. Plants can tell you about the desiccating effects of the wind, the presence of permanent or seasonal wetlands, the successional stage of the landscape, the soil pH, and whether the site experiences drought.
- Animal tracks and sign, such as scat, can tell you about wildlife activity on the site.
- Erosion evidence, such as bare soil, rills and gullies, and sedimentation, can tell you about problems you may need to consider in your design.
- Flagging at the tops of the trees may indicate prevailing wind direction.
- Plants and the behavior of water can sometimes indicate a change in soil type.
- Visible changes in topography can indicate a change in watershed (small and large).
- Microtopography and vegetation can indicate water movement patterns.
- Evidence of site history (such as barbed wire, old foundations, giant stumps) can help you piece together information about what has happened on the site at various points in the past.

* Material resources available on the site, such as rocks or clay, should be noted for use in your design.

Site observation is a step that you will revisit throughout the entire design process. By continuing to observe, you can continue to learn from your site for a long time.

READING THE LANDSCAPE

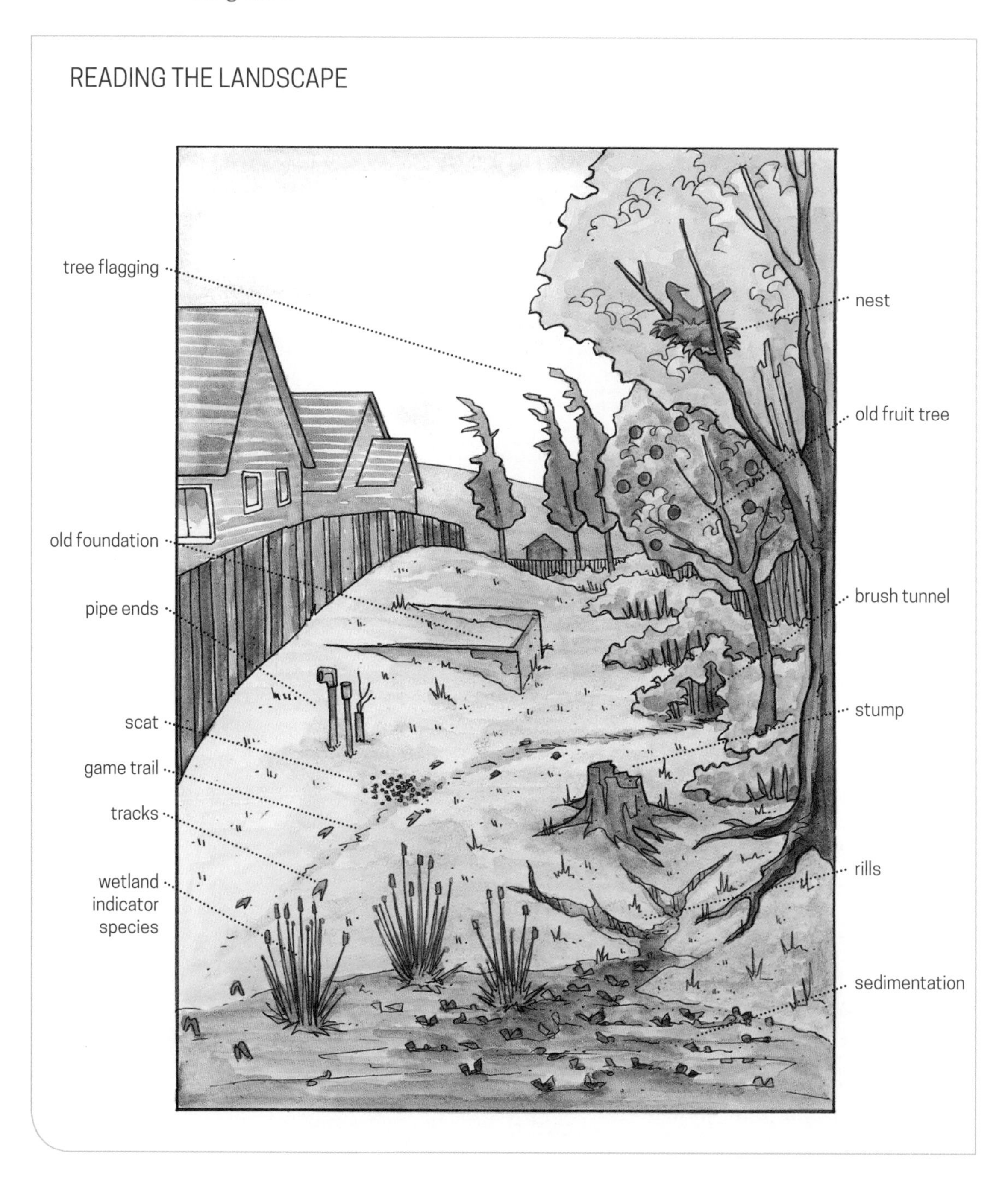

STEP 2: DEVELOPMENT OF VISION AND OBJECTIVES

To accomplish the ethical goal of making sure your design takes care of people, it is imperative that you gain a clear understanding of the needs and goals of the people for whom you are designing. Then you can move forward with everyone on the same page as you write up your mission, vision, and goals. Throughout this section we will assume you are designing your own site or a site with which you are directly involved.

Identifying stakeholders and their preferences

In most cases, multiple people will be affected by the design you create. These stakeholders may include spouses, partners, children, other residents, neighbors, community members, or employees, depending on the project. If you're designing your homestead in the country, your immediate family members are your stakeholders. If you're designing a community garden, all the people who have a plot there plus all the neighbors are stakeholders. When considering stakeholders, think about who will have the power to propel the project forward and who will have the power to slow or halt the project if they don't like where it's going. The design you create should minimally meet every stakeholder's needs and hopefully go way beyond that.

Begin by listing the current and planned residents along with their ages, skills and expertise, hobbies and interests, special considerations such as allergies or impaired mobility, and wish list items. Once you've identified the stakeholders in this way, get each person's responses to the "Stakeholder preferences" questions that follow. This will give you some frame of reference about the wants and needs of everyone involved. Then research the "Site specifics" questions, with stakeholder input where appropriate. The more detail you include, the more useful this will be. If you can't answer some of the questions yet, put them aside and research them. In fact, you can involve the other stakeholders in the process by having each person research different pieces and bring back what he or she has learned.

Stakeholder preferences

What vision and goals do you have for the property? (How long do you intend to be here? Does the property need to generate income for you?)

Do you have or want pets or other livestock?

Are there particular plants you do or do not want to include?

What kinds of vehicles, and how many, will you bring onto the property?

What resources (time, money, access to materials, friends, local community connections, and so forth) do you have available?

What is your installation plan (DIY, contractors, volunteers, a mix)?

Who has decision-making power and what is your process for making decisions?

What aesthetic improvements or changes do you desire?

Site specifics

Are there existing site plans or blueprints available?

What and where are the existing utilities (water source, electricity source, wastewater management, communications, trash, recycling, compost)?

What are the size, location, and condition of existing structures?

What are the location and features of known planned structures?

What are the known microclimates?

Which views would you like to preserve? Which do you want to block?

Which hardscaping elements (rocks, paths, patios, and such) would you like to keep or change?

What are the known challenges (nuisances such as noise, dust, car headlights, invasive species; design constraints such as setbacks, easements, legal restrictions; site characteristics such as damaging winds or drainage problems)?

Getting everyone on the same page

Getting answers to the list of questions will likely generate more questions. Try to get everyone on the same page before moving forward. We advocate a design process that keeps the stakeholders involved as much as possible. If there are multiple decision makers, input from all of them is necessary (even if they don't think so). Realistically, the best results will come from continually getting input throughout the process as opposed to disappearing with the information you collected initially and reappearing with a completed design a month later. If you do the latter, you run the risk of getting far off track and committing a crippling error during the design process.

Getting everyone on the same page is best accomplished with a guided discussion. The goal of this discussion is to help make sure you have an accurate understanding of everyone's goals and the information they have provided. Take good notes and ask clarifying questions. Ideally, everyone will participate and offer their perspective. If you notice anyone is not contributing or not being heard, you can check in with her or him individually or apart from the rest of the group. At the end of the discussion, it can be helpful to repeat back to the group what you've heard. That way they have an opportunity to clarify anything you may have misconstrued. Ultimately, the goal is to achieve a unified vision that meets the needs of everyone.

Jessi's design process involves many check-ins with stakeholders throughout the entire process. She doesn't move forward until she has buy-in from everyone at each phase. When there are many stakeholders, Dave occasionally uses a design charrette early in the process, which involves getting all stakeholders and information holders together in a meeting and collecting information and design ideas from everyone at the same time. For larger projects, it's a great way to work toward a result that is acceptable and exciting for everyone involved.

As you continue your design process, you will need to clarify lots of site and stakeholder specifics. Try to put some check-in points on the calendar in advance and have goals to reach by those dates. Don't hesitate to involve others in the research. When people have tasks and make contributions to the process, they take greater ownership of the end product. All of this will keep the design process moving and ensure progress is being made.

Writing up your mission, vision, and goals

Before you start focusing on the nuts and bolts of design, it is important to write up your mission, vision, and goals using the information you've collected so far. You may already have done this. Regardless of where you start, the goal is to have a clearly written vision for the project so everyone involved is aiming at the same target.

A mission statement answers the question, What are you trying to accomplish? This statement should be the ultimate distillation of the project, no longer than a couple sentences. A vision statement answers the question, What will this place look like if you achieve your mission? It should allow you to close your eyes and begin to see the place you are designing in your mind's eye.

Goals statements describe benchmarks that will tell you whether the mission is being accomplished and the vision brought to fruition. The literature on strategic development has much advice on goal development. However, simply put, goals should be SMART: specific, measurable, achievable, realistic, and time-bounded. An example of a well-written goal statement is: By the fifth year of establishment, our food forest will produce at least 200 pounds of food per year for the local food bank.

Doing a reality check

At any point during the process of gaining a better understanding of your stakeholders, you may find that their vision or wants are incompatible with the chosen property or with good earth stewardship. It's also possible that someone's expectations don't match the budget, skill sets, or time available. This is a difficult, and often uncomfortable, situation. Tough though it may be, in this case you need to give the stakeholders a reality check. If you don't, you may find yourself tasked with a project that is impossible to accomplish or causing harm to the earth.

Do your best to express your concerns to the group clearly and tactfully. Always approach these situations with the mindset that you may have misconstrued something that has been said. Feel free to propose modifications to the vision that would make the project more realistic. Ideally, you can reach an agreement and everyone will learn something in the process.

Mission, vision, and goals statements

Mission: To create a sustainable environment that is functional for a family, and the family's landscape business, which builds resilience and stability in the lives of those involved.

Vision: The property will become a lush homestead full of functional elements such as perennial food systems, animals grazing naturally, and colorful demonstrations of sustainable living from which people can learn. The landscape will contain a mixture of delicious food plants and stunning ornamentals that also support wildlife. The air will be abuzz with beneficial winged creatures such as dragonflies flitting about the water features, hummingbirds sipping nectar, and honeybees from the hives pollinating crops. A veritable Noah's Ark, this property will support a wide array of ecstatic pets and livestock, which will help maintain the landscape and support the inhabitants. The music of moving water will evoke a sense of peace and contemplation wherever one goes on the property. Verdant rain gardens will capture and clean excess stormwater that isn't used for irrigation. The tasks of everyday living will be fueled mainly by the sun, both directly and indirectly. A conservation-based approach to energy will showcase that living within our ecological means does not feel like a life of sacrifice. Exploring this landscape will awaken a sense of adventure with new outdoor rooms and sacred spaces to discover around each corner. Visitors will find a space to disconnect from the hustle and bustle of everyday life and connect with nature directly. The spaces essential to a working farm and landscape company will be colorful and fun, while allowing for right livelihood. These will be balanced by nooks dedicated to luxuriant relaxation. Any time of year, one will be able to use the sense of taste to experience what a productive landscape can really provide.

Goals:

- *Animals:* Lay out rotational grazing paddock system to accommodate livestock by 2001.
- *Vibe:* Construct at least one outdoor feel-good space for family and friends by 2001.
- *Structures:* Build house and all essential outbuildings, enabling us to live and work on-site by 2003.
- *Stormwater:* Eliminate all stormwater runoff through the use of infiltration strategies by 2005.
- *Fertility:* Produce all mulches and fertilizers necessary to support the landscape on the property by 2005.
- *Food production:* Grow all essential produce consumed by residents and guests by 2012.
- *Energy:* Achieve 80 percent energy independence for the house by 2015.

(top left) The barn at Jessi's is colorful and flanked by pots with edible plants. Photovoltaic panels and a rainwater catchment system meet the building's needs for electricity and water.

(top right) Adjacent to Jessi's kitchen, herbs and other edibles in containers are used daily. A color theme ties it all together and makes it a good place to relax. The flowers, mason bee house, and hummingbird feeder provide for wildlife close to home where the family can enjoy them.

(bottom left) This is one of the outdoor feel-good spaces where Jessi can spend time with family and friends.

(bottom right) The first annual garden Jessi created required protection from rabbits. She created a gabion-style fence with scrap materials that also absorbs extra heat to keep the area slightly warmer on cold nights.

STEP 3:
SITE ANALYSIS AND ASSESSMENT

Any permaculturist needs to be able to investigate a site and evaluate what she or he finds. Analysis and assessment involve collecting as much information as possible about the various aspects of the site through observation, research, and scientific measurement, and then evaluating that information to figure out what it means for your design. It's natural to want to plow ahead with a design after just a quick site analysis and assessment (site A + A from here on). After all, the design is the really fun part. However, make sure you don't shortcut this part of the process. The reasoning behind almost all your design decisions should point back to your site A + A. Your final design can only be as good as the A + A that informed it. An incomplete job of either analysis or assessment will lead to a design that can't live up to its potential.

How do you begin to record information for your site A + A? First, you'll need a base map that you can use as a template to create different types of site A + A maps as well as your final design. We present two different methods of analysis to generate information to note on your maps. The first, sector analysis, is a fairly quick and dirty way to get a lot of important contextual information all collected onto one sheet of paper. The second, analysis by the Scale of Permanence, is much more thorough. We recommend familiarizing yourself with both techniques and using the most appropriate one (or, more likely, your own unique hybrid) for your situation. For the most robust analysis, use both.

Before we describe these methods of analysis, let's take a look at the basics of how to represent the landscape on a piece of paper.

Measurement and mapping basics

Creating a map is actually fairly simple. The form most of our basic site maps will take is a plan view, which is essentially a bird's-eye view. Illustrating landscapes can also involve creating section views or elevations—essentially cross-sections through the design that show the vertical components to scale. This can be helpful to visualize how different heights of elements will impact one another. Finally, if you are artistically inclined, you can create perspective drawings, where you pick a location and draw the site as though you were actually looking at it from this location. This requires some pretty advanced artistic skill, but it can be worth it if you are trying to get someone to really understand the feel of the place.

First let's focus on plan views, as they will make up the bulk of the graphics you create for your property. The mapping process breaks down into a few simple steps that anyone can accomplish. Every map you create doesn't have to be accurate down to the inch, foot, or yard. Depending on the use of the map in question, "close enough" may be just fine. In fact, for everything up to the master planning phase, a less accurate map will work as you will be using it mostly to identify relationships more than exact placements.

Still, you should strive to make your map a scale drawing. This means you should define a scale (for example, 1 inch = 20 feet) and make your map accurate

(left) This perspective drawing of a food forest at an urban residence, by Sebastian Collett, helps to give people a better understanding of how the completed project will look and feel.

(below) This perspective drawing of a tropical home and yard takes the viewer high above the ground to get a better angle on what the project will look like.

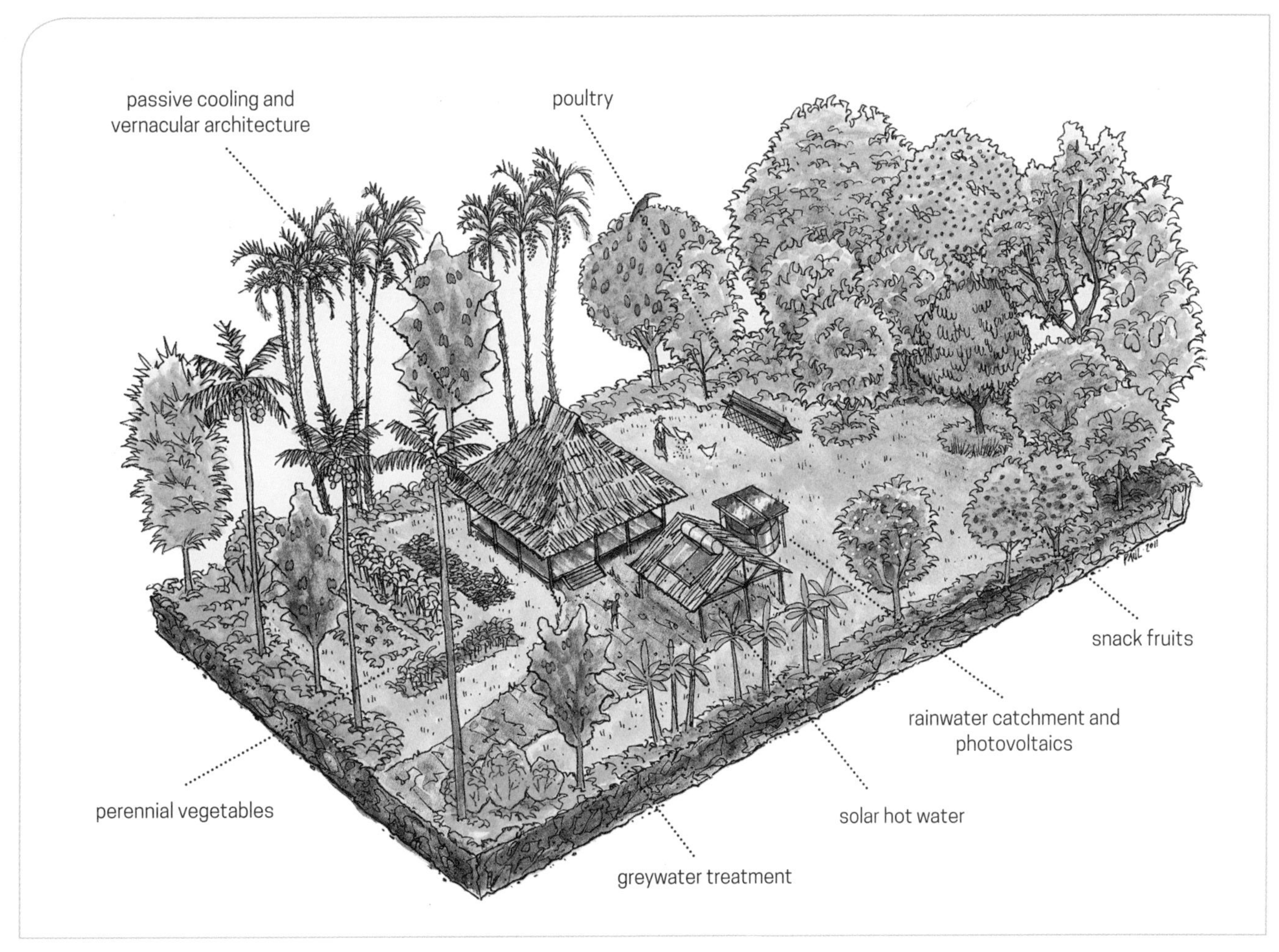

(left) Paul Kearsley uses a small, clipboard-sized map in the field while collecting data. Blueprint-size drawings can be difficult to handle in the field, so use the right size map for the situation.

(right) Tools to take various measurements on your land include a laser site level, a measuring tape (25-foot and 100-foot), a string line, a soil probe, and a surveyor's wheel. For rainy climates, a waterproof notebook and pencil can be invaluable as well.

enough so that when you measure something on the map you get a realistic approximation of the measurement in real life. The size of a house, the width of a tree's crown, and the length of a driveway should be proportional to each other in ways that resemble the real world.

When you get into figuring out details and creating working documents, you can shoot for more or less accuracy depending on the actual piece in question. For instance, a planting plan for a bed doesn't need to be accurate to the inch. However, it can be really important to get elevations down to the inch on a complex water map since an inch off can mean water flows from a storage tank onto your roof instead of the reverse!

In addition to measuring tapes and other devices, your body has a variety of built-in measuring tools. Most useful among these is your pace. Lay out a 100-foot measuring tape on flat, even ground and count how many paces it takes to go from one end to the other. Divide 100 feet by the number of paces to get the length of your pace. It can be helpful if your measuring pace is different from your usual walking pace so that it brings a certain consciousness about when you use it. It can also be useful to train yourself to use a pace of 3 feet (if you're taller) or 2 feet (if you're shorter) for easier math later on. However you do it, keep measuring your pace until you get it consistent, even when going up and down hills. Practice will help you build body memory, and you can actually be quite accurate with pacing.

For measuring smaller things, you can always use other parts of your body. Learn the approximate distance from your fingertips to your wrist, your fingertips to your armpit, your fingertips to the center of your chest, and so on. In the built environment you can also use your brain as a measuring tool by using objects of common size as measuring helpers. For instance, if you know the length of one cinder block in a building, you can just count the cinder blocks to get the total length of the building.

The whole point of mapping is to communicate ideas graphically, so you need to develop a clear graphic language for your maps. For instance, lines denoting different features like fences, property lines, and contour lines can be of different colors or weights; some can consist of crosshatches while others are dashed or dotted. Point features such as wells, sprinkler heads, and fence posts can be marked with a single dot, a star, or an X. Polygon features such as ponds, buildings, the crowns of trees, and pastures can be differentiated by way of shading, color, textures, patterns, and symbology. Especially if you're using maps to share information with others, be sure to include a title, a date, the name of the person who produced it, a scale bar showing how a unit of measure on the map relates to a unit of measure on the ground (for instance, 1 inch = 20 feet), a north arrow (or solar orientation), and a legend.

The base map

The first map you will need to make for your project is a base map. The job of the base map is to serve as the bare-bones layer onto which you can put all of the other information. You will likely end up making many copies of this to play with. We recommend putting together a base map in black and white with crisp, clear lines for ease of tracing and layering later. A base map should show buildings, permanent access roads or walkways, property lines, hardscape elements that you are sure will stay (such as arbors, fences, and outbuildings), large trees, and important or noteworthy off-site features (such as large trees that block sun, buildings near property lines). It can also include contour lines.

One excellent way to kick-start this process is to research existing map resources to use as a starting point. This could include satellite imagery, recent aerial photos, legal land description maps such as plat maps and parcel maps, topographical maps from the US Geological Survey and/or the Natural Resources Conservation Service, prior landscape plans, and surveys. When accuracy is not of paramount importance you can trace these existing resources to create a simple base map. Ground truthing is always a good idea when you do this. If you can't find much in the way of existing map resources or such maps are lacking some key elements, you will need to create a base map from scratch or at least add in some items.

For basic mapping it helps to identify what measurements you need to take before you start. Jessi takes note of which measurements she needs before going to the site so that she doesn't miss any important measurements. Begin with any rectilinear objects that may be on the property such as houses, outbuildings, and concrete patios. These can easily be plotted on graph paper, which lends itself to rectilinear shapes. These can also be traced from an aerial image on a computer screen for data collection purposes. Later these shapes can be traced onto other

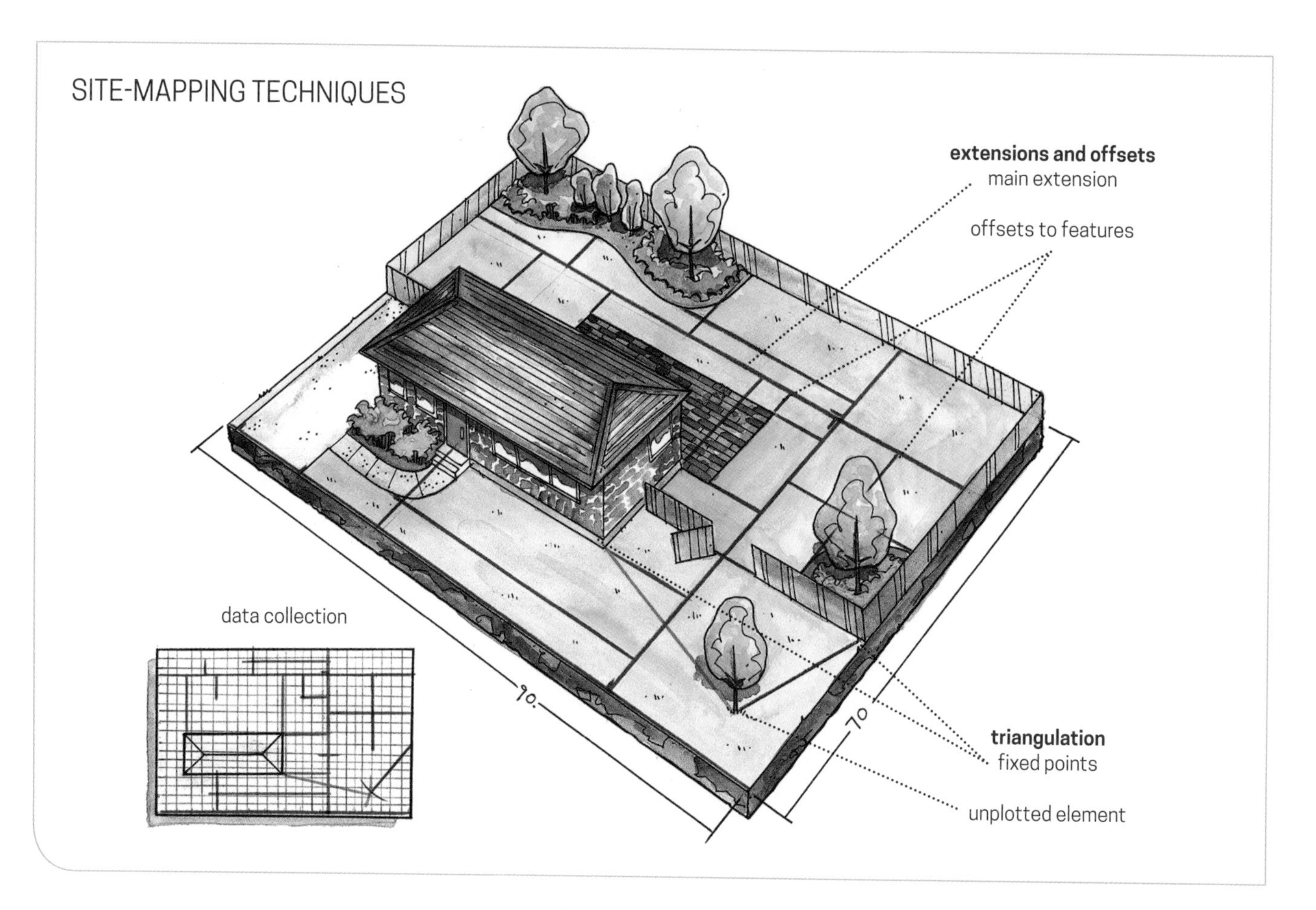

paper as needed. These rectilinear objects can be used as reference points to plot other landscape features through a variety of methods including extensions and offsets, triangulation, taking GPS readings, or taking measurements from satellite or aerial imagery.

In the technique known as extensions and offsets, you use straight lines that extend directly off of a building or rectilinear object (extension lines) as a way to quantify where other things are so you can plot them on your map. The illustration of site-mapping techniques shows red and blue transects, which run at a measured distance away from and parallel to the house. Use these to avoid having to measure right along the building. Place extensions or transects such that large portions of the property can be viewed while walking their length. From these lines, use perpendicular offsets that reach to key features of the landscape, such as building corners, large trees, and bed edges. You can use a framing square to make sure your offset lines are perpendicular.

For nonrectangular landscapes and features, triangulation often works better. Triangulation uses two known locations in order to plot a third. Pick two fixed features, such as the corner of a building and the corner of a fence. From each of these points, measure the distance to an unplotted element, the small tree in the illustration. Translate this distance into the scale of your base map and separate the legs of a drafting compass to this scaled distance. Place the compass spike at the building corner on your map and draw a short arc near the location of the tree. Draw a similar arc using the fence corner and the scaled measurement from that point. The point where the two arcs cross will be the location of your tree.

Scoping out slope, aspect, and elevation

Part of your analysis will require you to get a feel for the topography of the site and how that affects other factors. Topography is the configuration of the terrain, encompassing slope, aspect, and elevation.

- *Slope* refers to grades throughout the property. This is important when assessing erosion potential and looking at design elements that require flat or sloped land.
- *Aspect* refers to the direction a slope faces. This impacts how much solar gain that slope receives.
- *Elevation* refers to how far above or below sea level the site is. This impacts temperature regimes and, in some places, rainfall as well as potential for floods and tsunamis.

Slope. There are numerous ways to measure slope, some more accurate than others. One simple method requires only a straight 10-foot board, a carpenter's level, a measuring tape, and the formula slope = rise / run. If you want to get fancier, a slope can also be measured using a water level, a clinometer, a site level, or a laser level.

Dave demonstrates how to measure a slope with a few simple tools: board, level, measuring tape.

BLOOM HOMESTEAD

Base map

This is the base map for Jessi's property at the time of purchase. Note that important elements from outside her property boundaries are included.

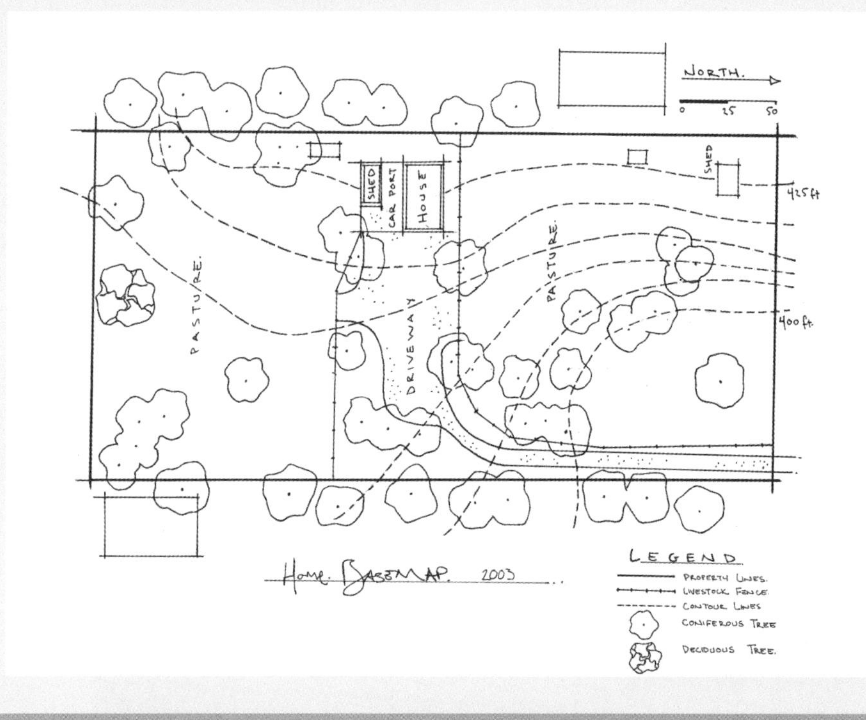

To use the simple method, start at the top of the slope and put the board on the ground heading straight down the slope. Leave one end on the ground at the very top of the slope and lift the other end off the ground with the carpenter's level on top until it indicates the board is perfectly level. Next, measure from the end of the board hanging in space to the ground directly below. Write this measurement down. Then move the high end of your board to the point you just measured on the slope and repeat. Work your way down the hill like this until you get to the bottom. Then add together all your measurements to get the rise. To get the run, note the number of measurements you made and multiply that by 10 (assuming a 10-foot board). Divide the rise by the run, and multiply by 100 to get your slope expressed as a percentage. The larger the number, the steeper the slope is. Flat sites or gentle grades of less than 10 percent are easiest to build on; for a slope of between 10 and 20 percent you will need to do some grading; for a slope of more than about 20 percent, you may need to do extensive work to stabilize soil and keep it from eroding.

Aspect. When you assess your site's aspect, you are looking particularly at how much solar gain you can expect on different parts of the site. Solar gain refers to the increase in temperature caused by solar radiation. Land, buildings, or other objects struck by the sun's rays directly will have greater solar gain than objects struck by the sun obliquely or only indirectly. Therefore, slopes that angle toward the sun's path will be warmer than those angling away from the sun's path. Similarly, windows oriented toward the sun will often result in net gain in heat whereas windows oriented away from the sun will generally cause a net heat loss.

Elevation. Through careful measurements it is possible to chart the topography of a site via contour lines on a map. Basically, each line depicts a particular elevation (generally labeled on the line) as it crosses over the land. Then the next line is another elevation (either higher or lower), and so on. Where the lines are wider apart, there is more space in between the two elevations and the land is flatter. Where the land is steeper, they are closer together. Contour lines can either be shown on the base map (if there aren't too many of them) or on an overlay (if there are a lot of them) to keep the base map from getting too cluttered.

Sector analysis

Now we will look at two methods of analysis to generate information to add as overlays to your base map. Sector analysis, the first method, looks at elemental energy flows, known as *sectors* in permaculture lingo, moving through a site. Inherently, each of these flows has a directional component. Sectors include wind and air, fire, wildlife, sunlight, and drainage. Here are some questions to ask about each sector:

- Wind and air: Which direction does the prevailing wind come from? Is it different for storm winds? Are there soft breezes that help cool things down in hot weather? If so, where do they tend to come from? Does cold air drain through the site? Does it pool anywhere?
- Fire: If there were to be a fire, which direction would it likely come from? This may relate to wind, but it could also relate to potential ignition points (for example, the gas station next door, the neighbor's teenage boys) on or off the property.
- Wildlife: Does wildlife pass through the property? If so, what path does it generally take? Do different types of wildlife take different paths? Does this change seasonally?
- Sunlight: This aspect gets more important as you move away from the equator. What does the sun's path look like in the summer? How does it differ in the winter? Are there obstructions where you don't want them?
- Drainage: How does water move through the site? Does it move differently after a heavy rain?

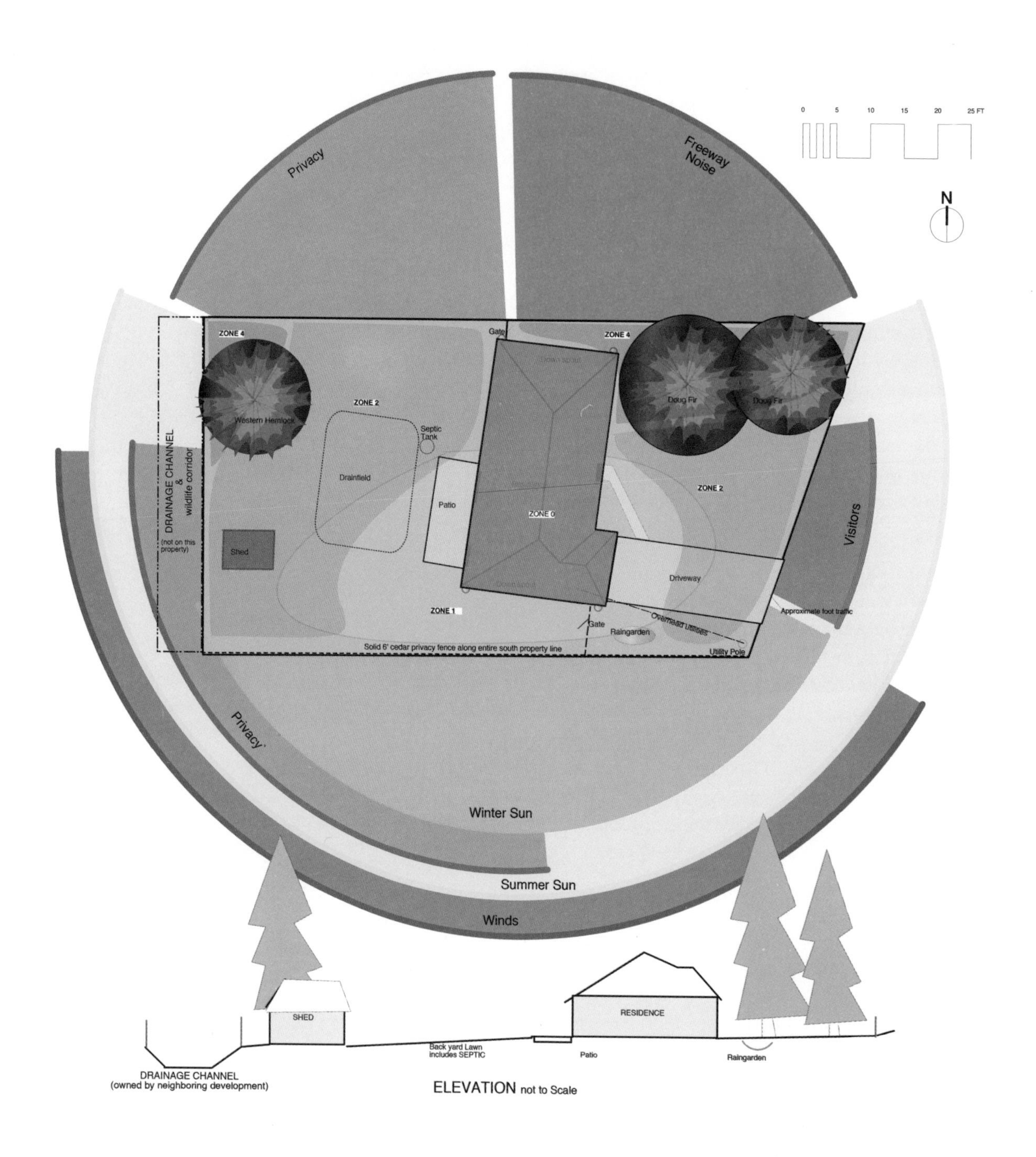

This computer-generated sector analysis drawing that Jessi created for a client represents each elemental energy coming from off-site as a large wedge-shaped sector. At the bottom of the image, the elevation drawing shows a cross-section of the relation between slope and elements.

Those natural flows will likely be present on any sector map you create. We also have energy flows or sectors that relate specifically to human-centered concerns, especially important in urban and suburban areas:

- Views and privacy: Are there any particularly spectacular views that should be preserved or enhanced? Are there places where view lines should be screened for privacy?
- Crime: Is there a particular place where vandals, thieves, or other unwelcome folks tend to come from?
- Noise: Are there sources of unpleasant noise nearby (for example, busy streets, flight paths, industrial zones)? How about sources of pleasant noise (such as rapids or bamboo groves)?
- Pollution: Often air pollution travels with the wind and water pollution follows drainage routes. Where are the potential and actual pollution sources?
- Access and human activity: Do people travel through the site? Is there a history of people moving through? Are there easements? What is the nature of the human activity moving through the site? Do vehicles move through or just people on foot?
- Other neighboring features: Is there anything else that moves through the property but has its source elsewhere (for example, the shade from large trees on a neighboring property, the odors from a tannery, eroding soil that washes onto your property)?

Representing these various flows as sectors on a map overlay can provide a good starting point for your site analysis. Make sure to label everything so your maps are clear. In addition to showing the path and direction for each sector on the overlay, you will also want to note in the margin details such as average and extreme wind speeds, easement details, and what the various views are showcasing.

Sector analysis is useful when you don't have time to look at the land in as much detail as you'd like. It can also help you to quickly understand the context of larger properties and set the stage for more detailed analysis later. If this is the only analysis method you will use, it will be important to follow up with further and more detailed research.

BLOOM HOMESTEAD

Sector analysis

The prevailing wind is strong from the southwest. There is little fire danger. Coniferous trees, both on and off the property, cast considerable shade throughout the year. Wildlife is present, but mostly winged and in the treetops, as the property's fencing limits access for ground-dwelling creatures to use it as a corridor. Although there is seldom overland flow, surface water moves through the property from west to east. The elevation on the north half of the property allows runoff to drain to a low spot in the pasture that is wetter than the surrounding area. Because the property is secluded and set back from the road, there are no privacy issues and no off-site views to maintain.

Scale of Permanence analysis

Scale of Permanence analysis was first presented by P. A. Yeomans in his 1958 book *The Challenge of Landscape*, one of a handful of works that spurred Bill Mollison and David Holmgren to come up with the permaculture design system, and has been further developed by permaculturists since. The Scale of Permanence is essentially a list of landscape features in order from most constraining to least constraining. Climate, at the top of the Scale of Permanence, is extremely difficult for you to change through actions on your property. The best you can do is to build a greenhouse to create a warmer microclimate (lower on the Scale of Permanence) in a specific area on your site. Aesthetics, at the bottom of the scale, can be changed relatively easily.

Organizing site information according to the Scale of Permanence helps you figure out where you can make the biggest change for the least effort. Focusing on one level at a time, collect as much information as possible. Some of the information you choose to collect will be based specifically on your goals. You can print off individual copies of your base map to represent each level and then mark them up with notes, or you can do this on overlays. Consolidating some of these layers also works.

Analysis via the Scale of Permanence is particularly useful for complex projects (in terms of either land or vision). It is the method of analysis we fall back on when time allows because it provides an abundance of information.

THINGS TO BRING FOR DATA COLLECTION

notebook (waterproof)

base map, possibly several copies to record data

tape measure, 100- or 300-foot

camera, to snap photos you can refer to later

shovel, for exploring soils

clinometer or site level, to help with slope measurements

compass, to measure angles and directions

recording thermometers, to explore cold and warm microclimates; check after high or low temperature events

binoculars, for observing large properties

soil probe, for checking out soil horizons without digging a big hole

field guides, to help identify local plants and wildlife

THE SCALE OF PERMANENCE AND A SAMPLING OF WHAT TO NOTE AT EACH LEVEL

Scale of Permanence	What to note
climate	high and low temperatures rainfall seasonality USDA hardiness zone
landform	topography and slopes underlying geology and depth to bedrock
water	sources changes in watershed poorly drained areas points of entry and departure downspout locations
legal issues	zoning CC&Rs and HOA rules setbacks easements wetland buffers
access and circulation	roads paths and trails points of entry and departure (for materials, people, vehicles)
vegetation and wildlife	existing and expected plant communities successional stage habitat quality species present in area
microclimate	microclimates present in area aspect of slopes areas of shade frost pockets
buildings and infrastructure	size, shape, and location paved areas utilities and services (above and below ground) septic system
zones of use	areas of property divided by character of use
soil (fertility and management)	soil types fertility pollutants
aesthetics	various features lending to or detracting from the character of the property

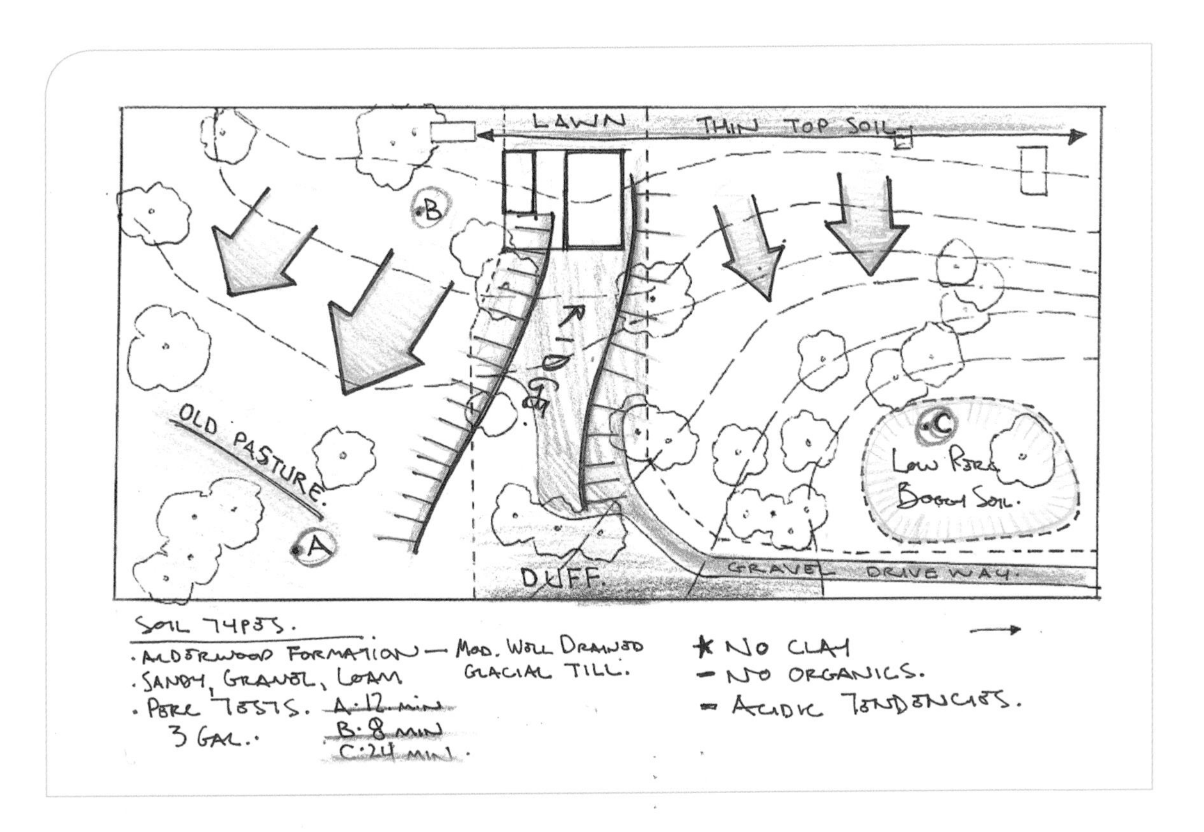

This map conveys information collected in a site soil analysis.

Assessing your data

Once you've completed your site analysis and have a large quantity of data, it's time to start drawing some conclusions about what all those clues are telling you. During the analysis process, you have likely started to form some conclusions. This is the time to spell those out and make sure your data supports the conclusions.

We like to start by identifying the drivers for the design. Some of the data you've collected will tell you what's important and what's not. For instance, if the site is in a protected valley, wind will not likely be a driving factor for your design. However, if your site is out in the open and consistently experiences gusts of up to 30 mph, wind will be a major driver for your design decisions.

The product of your assessment should be a summary of the site's constraints and opportunities. If you've used the sector analysis method, you can create a single map showing which sectors are going to have serious impacts on the design (whether positive or negative). If you've used the Scale of Permanence analysis method, you can create a map or overlay that sums up the most meaningful data you've collected for each level.

SUMMARY OF CONSTRAINTS AND OPPORTUNITIES BASED ON SCALE OF PERMANENCE ANALYSIS

landforms and soils

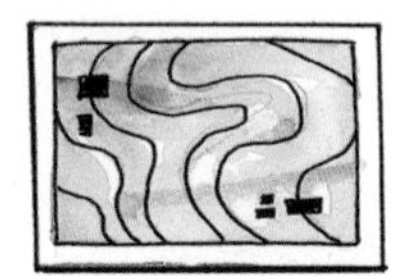

water

vegetation and wildlife

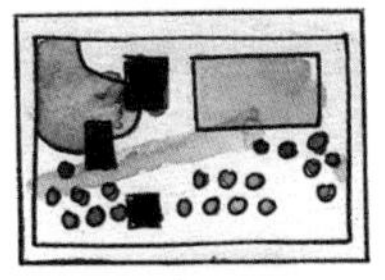

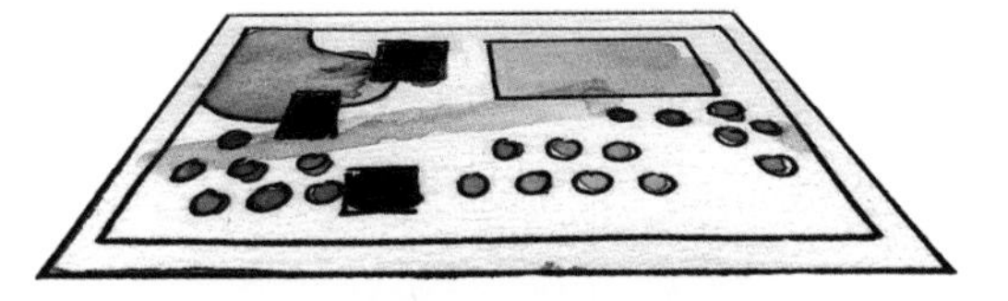

access and circulation, buildings and infrastructure

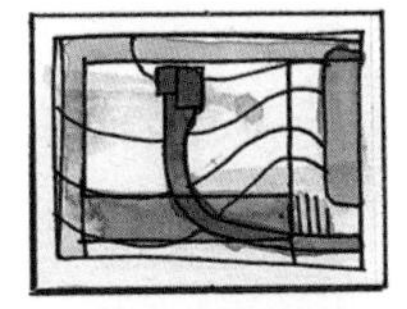

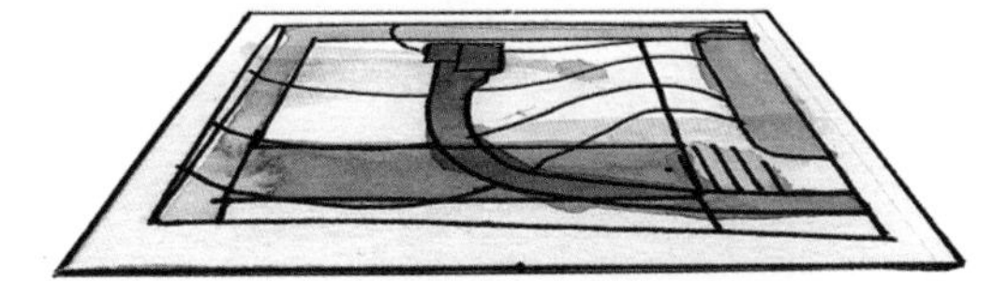

zones of use

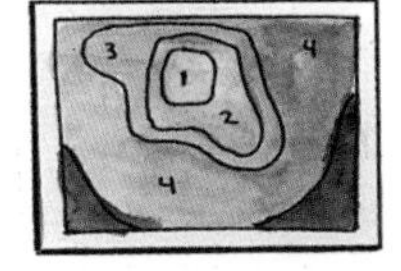

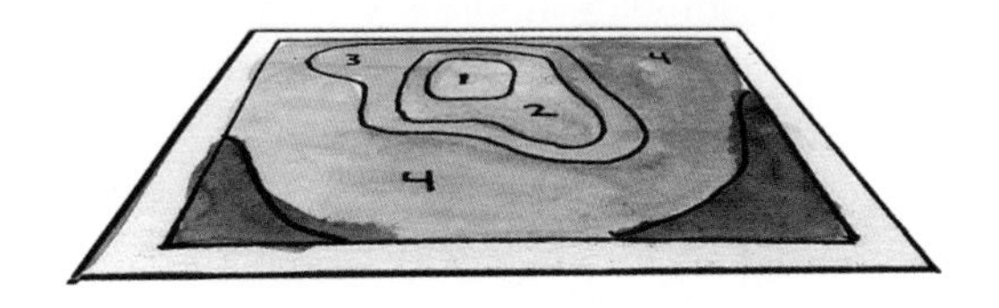

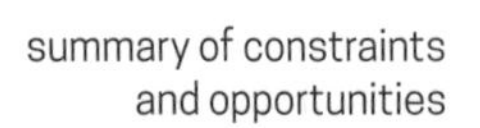

summary of constraints and opportunities

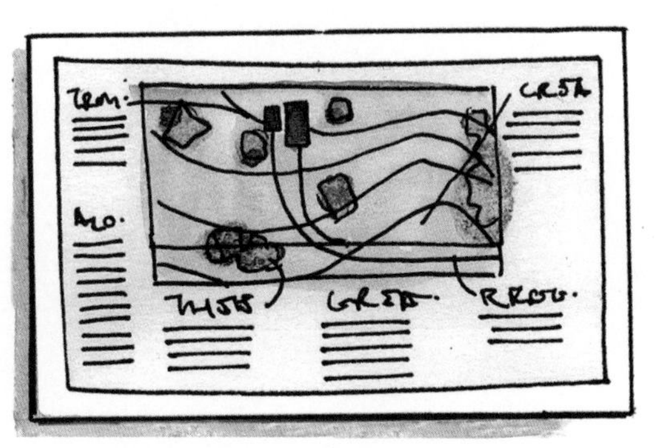

BLOOM HOMESTEAD

Scale of Permanence analysis and summary of constraints and opportunities

Scale of Permanence analysis:

Looking at the terrain from the watershed context, the site sits high on the ridge or edge of a watershed and gently slopes to the east, with the north end of the property having more of a drop into the lowest spot on the site. Rainwater naturally sheets from west to east, collecting only in the northeastern area. The rest of the site has a gentle slope.

Access into the site is via a gravel road along the northeast property line into the middle, where it curves west toward the house. Access from the street is via an easement along an existing driveway that is legally shared with two other property owners.

The plant communities consist of pasture grasses, many large and healthy native conifers, and one mature walnut tree. There is no shrub layer.

The existing utilities consist of a septic tank and drain field located in the middle of the property, with all electrical utilities following the existing driveway into the site. A well is located at the highest point in the landscape, in the northwest of the largest pasture with the largest elevation drop.

The soils are all gravelly loam, which drains so quickly that retaining moisture is a problem. Soils have a wonderful mineral component but lack organic matter and fertility.

Summary of constraints and opportunities:

The property has its downsides, one of which is major: limited solar access due to the north-south orientation of the property and conifers on neighboring properties. The existing conifers also provide a great deal of shade, making the decision to allocate land to pasture or gardening difficult. The site has very few biomass producers or deciduous plants, making those a design priority. The soil's mineral content and ability to drain on most of the site is great. The access is direct and easy into the center of the property; the minimal setbacks and utility lines also leave the site fairly open. The gentle slope to the east makes it possible to store water at the higher elevation and use it lower on the site.

Putting the Design Together

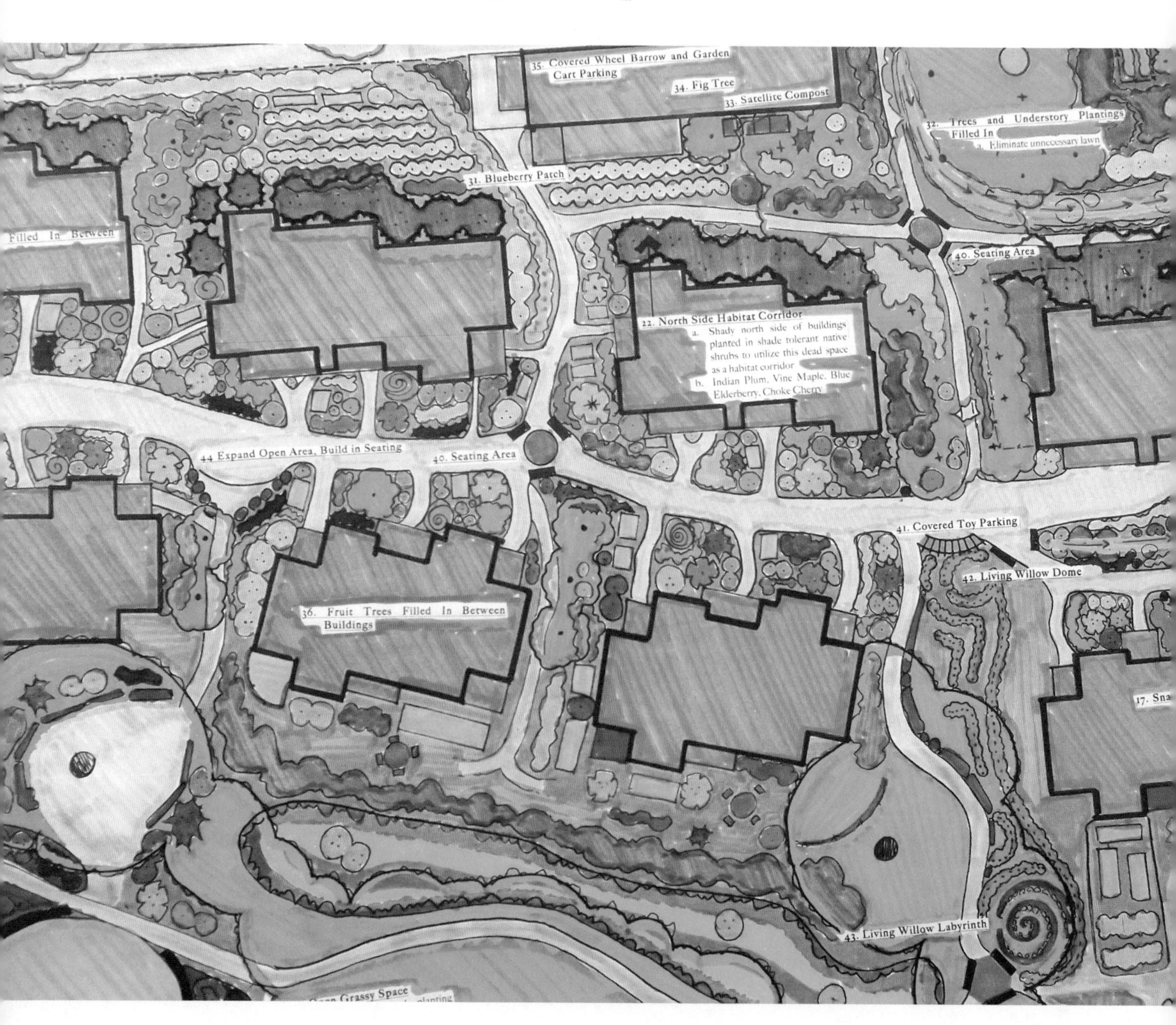

A section of the master plan for CoHo Ecovillage in Corvallis, Oregon.

After the thorough site A + A, you can sit down at the drawing board and start to develop ideas and concepts as you carefully explore all the possibilities of your project. The process we teach here is geared toward the beginning permaculture designer. It begins with conceptual design and then moves though iterations of schematic drawings. From there, you will be able to come up with an implementation plan that will guide you through the installation process. Plans and drawings can all be done all by hand or all by computer-aided drafting, or a combination of the two.

The journey from early conceptual ideas to working construction drawings can be long and require many iterations. Every project is different in the scope and level of detail needed on paper. The design will evolve as people change their minds, new information surfaces, and life gets in the way of your deadlines. If you can, don't rush it and remember to remain flexible. Let the process conform to the needs of the project.

Here are the steps and the products you'll generate in part 2 of the design process:

Step 4. Conceptual design—notes from research, analyses of elements, random assemblies, flow diagrams, bubble diagrams

Step 5. Schematic design—zone maps, several iterations of your master plan, final master plan to scale

Before we describe those steps, let's briefly consider a couple of ideas to keep in mind as you proceed through the design steps: planning as a process and efficient energy planning.

PLANNING AS A PROCESS

A successful design endeavor, much like a car, needs both the gas (let's do it!) and the brakes (let's first make sure we have a good plan). Moving forward with project details without going through the design process can mean losing opportunities and making costly errors. On the other hand, hemming and hawing over details, never feeling like you have enough information to actually get started—analysis paralysis—is counterproductive, too. Your problems may not be able to wait that long, and some opportunities will pass you by.

It may help to remember that planning is a process that involves much iteration. You can always go back and change things later. In fact, the likelihood that the plan you create up front will look exactly like what happens on the ground is pretty near zero. If you don't know how best to handle some aspect of the design, you can take your best guess, run it past someone with more experience, and be ready to flex as new information is gained. In fact, that new information may never come if you don't get started in the first place.

Also, every good engineer knows that the first time you try something it probably isn't going to work perfectly. The process of refinement over several iterations

ONE EXAMPLE OF THE DESIGN PROCESS

Author Dave Boehnlein, along with Jeremy O'Leary, Patrick Loderhose, and John Coghlan, produced a series of design drawings during a forest gardening workshop. The workshop focused on creating an ecologically functional design that met the needs of the residents of a house in Sherwood, Oregon. These drawings illustrate one possible path through the design process—the path that was appropriate for this particular project.

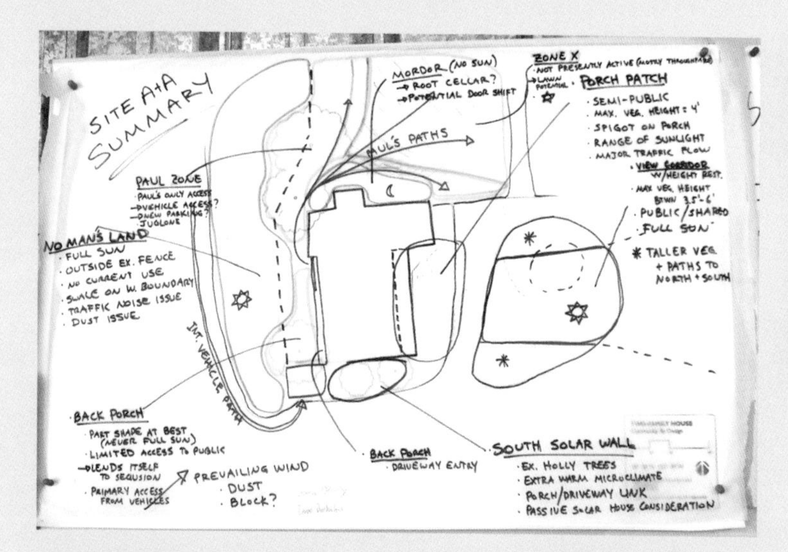

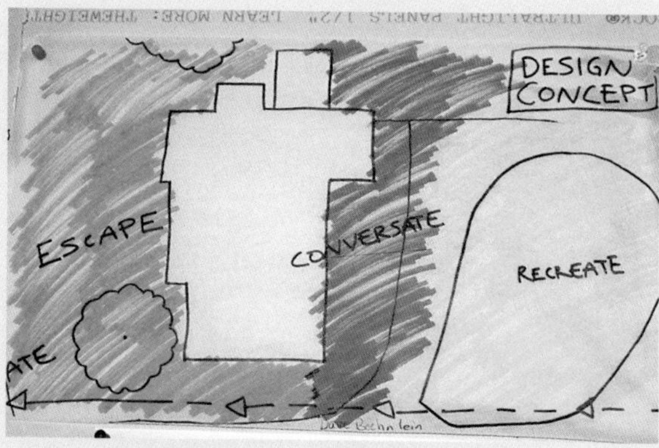

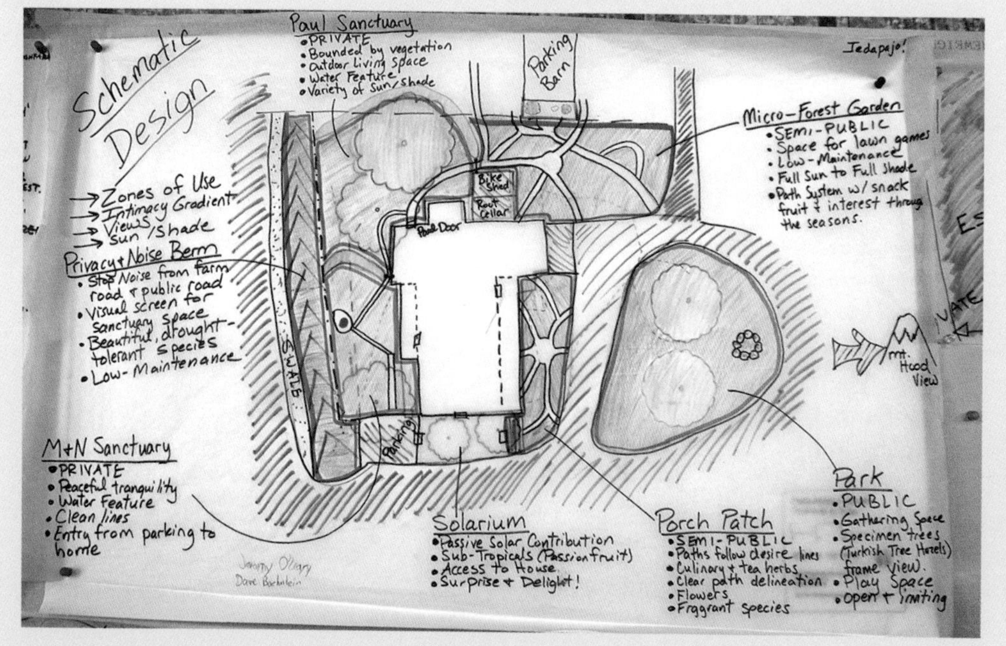

(above left) This site analysis and assessment summary lays out the major drivers for the design.

(above right) The team worked out a conceptual design based on one of the patterns found in *A Pattern Language* called an intimacy gradient, allowing for both private and public space.

(left) From this concept, the schematic design broke the landscape up into different areas with different uses and character. From here, detailed drawings were made.

and tests is how engineers create things that actually do work. Working out the kinks is part of the process. For permaculture designers, fear of making mistakes can lead to the biggest mistake of all, doing nothing. Of course, we'll want to minimize the mistakes we make but not let that stop us from trying things out. Remember to start small and have fun!

And remember that design projects done in collaboration with multiple designers can bring out the best of ideas. For instance, Dave almost always works with a team on design projects. This allows each team member to fill a specific niche. Just like in nature, the cooperative relationship leads to greater productivity. Putting more heads together to bounce ideas off of one another is one of the best ways to flush out all options and identify issues or situations that can be avoided. However, even a lone wolf designer can ask for simple feedback and have someone else look over his or her work. Having peers review your work can definitely help you see things you may have missed. If you do this, your products are likely to improve noticeably over time.

EFFICIENT ENERGY PLANNING

Paying keen attention to efficient energy planning is extremely important from the very beginning of the design process. In this context we're not talking about energy in the form of electricity, but rather natural energies in the broadest possible context. The broad patterns that we design on the landscape must make sense in the context of slope, aspect, elevation, microclimate, sectors, and regional human context. If any of these are ignored early on, bad design will follow and result in a finished product with some major functional hurdles.

In the permaculture community, we often refer to the types of mistakes from which we can't bounce back without great cost or huge amounts of work (thus meaning most people never do) as type 1 errors. Examples include buying land that can't support our goals (trying to farm coconuts in Iowa), building our cold climate home where it receives no solar gain, or taking on crippling debt with which we can't keep pace (sinking so much money into a piece of land that we have to work 60 hour weeks away from our site just to pay the mortgage).

Efficient energy planning will play out differently in different scenarios. For instance, imagine a homesteader in Scotland who decides to put her home on a gentle south-facing slope to maximize her access to low-angle winter sun (when it shines). She puts the house about halfway up the slope, avoiding the harsh winds at the top of the hill and the frost pocket that occurs at the bottom. Perhaps she chooses to move the house slightly toward a southwest aspect rather than southeast so that the hill effectively blocks traffic noise from a nearby freeway. On the other hand, for someone living in the California desert, placing a house on a southern slope would likely make it function like an oven during extremely hot summer days. Thus, permaculture design gives us a perspective, not a prescription.

In the end, we all have to work with what we've got. What if you own a property and the house (or tree or patio or garage or any other fairly permanent element) is already in a less-than-optimal spot? In this case you must figure out the biggest issues you're facing and, starting with the most negative impact, find ways to mitigate them. For instance, if you've got a home on a south-facing slope in southern California, overheating and fire danger will likely be your biggest issues. Shade trees, arbors, and overhangs are retrofits that can address the overheating issue. Fire-resistant landscaping, well-placed ponds, and firebreaks can mitigate the fire danger.

Questions to ask:

- Is my design responsive to *all* of the site information I uncovered in my site analysis and assessment?
- Have I overlooked any major drivers to the design?
- Have I rushed to the details without having a clear idea of the big picture first?

STEP 4: CONCEPTUAL DESIGN

During the early stages of conceptual design, you identify all of the elements you need to place and think about the relationships between elements independent of where they fall in the landscape. Elements are often depicted as loose forms or shapes located roughly in the locations being considered. Traffic flow through a site is often represented by dotted lines or arrows, which will later be turned into pathways and access points. At this stage you don't attempt to define exact details or specify materials. You simply get ideas on paper in a form that often looks like a treasure map.

Analyzing elements to determine the best relative locations

Every element in your design should be analyzed in order to figure out the best relative location to create beneficial relationships with other elements. First consider what inputs or needs your element will require. Bill Mollison's classic example is a chicken. Its needs are simple: food, water, grit, shelter, other chickens. In contrast, what are the outputs or yields of this element? For a chicken, the obvious meat and eggs, along with the services of scratching and aerating the soil, controlling pests, and turning biomass into manure. Another thing we look at is the intrinsic characteristics or inherent qualities of an element. In the case of a chicken, these are breed, color, and levels of tolerance toward aspects of its environment.

A good way to organize all of this information is to write out an index card for each element listing its inputs, outputs, and intrinsic characteristics. You can easily use the cards to match up the needs of certain elements with the yields of other elements. For instance, an apple tree has a need for pest control and fertilization, while a chicken needs food and shelter from predators. By placing them in good relative location to each other, both elements will get their needs met. We can do this matching process for all of the elements in our design.

A natural building sits at the edge between the native forest and the cultivated landscape at the Regenerative Design Institute, Bolinas, California.

One way to generate ideas about the best relationships possible between elements is a simple exercise called random assembly. Divide a sheet of paper into three columns and in the middle column write a list of relationship words or phrases such as *next to*, *on top of*, *below*, *in*, *on*, *under*, *around*, and *containing*. Shuffle your index cards of elements and lay them out in the columns on either side of your list. Reading across each column, you may find that some of the relationships make sense and others are silly. Chickens don't go *under* ponds, but maybe ducks go *next to* the pond. Some of the combinations will inspire creative ideas and build connections or systems between multiple elements. Dave once did this exercise and the relationship that randomly popped up was sauna on top of pond. This may seem silly at first but if you think about it, a sauna on a floating dock in a pond would be fire safe and you wouldn't have to go far to take a cold plunge.

Generating design ideas

The main goal during conceptual design is to explore a variety of scenarios and possibilities for your project. Here we describe a few methods you can use to generate ideas. While we recommend using as many of these methods as possible, not all will be applicable to your project.

Unlike most of Japan, the island of Shodoshima has a Mediterranean climate, so this island has become an olive-producing region.

Do an analogue climate assessment. Analogue climate assessment is finding other climates in the world that are similar to your own in terms of temperature ranges, precipitation, and patterns of seasonality, and gathering time-tested ideas from those places about dwelling sustainably. In each of those analogue climates you can explore a wide variety of questions, including these:

- What are the principal agricultural crops of those regions?
- How did traditional people build their houses in those places?
- What were the traditional healing systems of those locations and what plants did they rely upon?
- What were the daily activity patterns of people in those areas?

As an example, author Dave Boehnlein and Doug Bullock did analogue climate research for a project in northern California and discovered that one analogue climate was a region in Iran where the chief agricultural products are saffron and barberries. Another analogue climate, in Tajikistan, had native forests where the bulk of trees are pistachios, pomegranates, and almonds. This provided huge clues as to what productive species might fit into the landscape they were designing.

Look to traditional and indigenous cultures. The Industrial Revolution signified a major change in how people related to their environment. Much more energy became available for human use, and the knowledge of conservation and efficiency that had been developed by people to make sure they always had enough to meet their needs began to fall by the wayside. As permaculture designers, we have a lot to learn from cultures with components still intact from a pre-industrial time. These include traditional cultures such as the Amish in North America as well as modern cultures that retain vestiges of their long-standing traditions, such as the Japanese. We can look at how these people live and what wisdom has worked for thousands of years that we might be able to integrate into our own strategies.

Look at local precedents. One of the best places to gather information to use in your design is other projects in your area. Although you may not find another full-on permaculture site to check out, there may be many precedents for individual pieces of your design in your local area. For instance, if you want to build a straw bale house, go see other examples in the area and ask the owners about their experiences. If you intend to grow fruit trees, go see what local farmers are growing and ask them about potential pest and disease issues. If we pay attention to local precedents, we can save ourselves a lot of headaches.

Look at global precedents. We can also use the Internet to find inspiring examples of permaculture work being done around the globe. Whether you are looking for big picture conceptual ideas or the details of a particular project, you can collect what you find online in a scrapbook or on an idea board. Many tools and websites online such as Pinterest let you organize ideas on boards. Or simply cut out photos or words from print resources and attach them to your drawing with strings or arrows.

Listen to intuition. In the modern world, most of us are not terribly comfortable making decisions based on our instincts. In fact, in both subtle and overt ways we are often encouraged to gather all data beforehand, quantify the situation, and then make a decision based on the numbers. However, in some situations you might have a gut feeling about some design decision that has nothing to do with numbers or scientific analysis. Don't ignore it! You may have a great idea. However, at the same time, don't just create a whole design based solely on your intuition. When you have a gut feeling (intuition), note it and research it using the best scientific and analytical approach available to see if it really is a great idea. You may find that the more experience you have in sustainable design and land management, the better your intuition becomes.

Mimic nature. When we can mimic what's happening in nature or apply the patterns we see there, we cut some of the guesswork out of our designs. It's important, however, that we mimic nature in function as well as form. When you have a problem to solve, always ask yourself, What would nature do? For example, if lots of pine trees grow in a certain soil type in your region, it would make sense to explore planting nut pines that occupy a similar niche in their native climate. Similarly, when trying to manage weedy pioneer species you can observe that as an ecosystem moves forward in the process of succession, these species usually get shaded out. Thus, you can plan ways to create the shade they can't tolerate.

Creating flow diagrams

Flow diagrams depict how resources and materials move through a space. To make a simple one, take a copy of your base map and with a pencil that you keep on the paper, trace your steps through the property as if you were walking or driving from one space to another on an average day. What you end up with is a web of paths taken that shows where the most traffic is happening. Flow diagrams can be used to depict routes for transporting goods through communities, walking paths through a rural plot, wheelbarrow routes through a garden, or your functional patterns in the kitchen.

You can make a new flow diagram each time you work though a different area in your design and figure out details. This will help you identify any obstacles impeding flow and figure out what would make the flow more efficient, such as moving an element to make a path more direct. Another way to evaluate flows generated by your placement of elements is to mock up paths and obstacles on the land with tape, a line of string, or a garden hose and walk through this imaginary landscape as if you were planning to use it. Think of a livestock barn you would need to visit every day to care for the animals, check and refill water and food supplies, access grooming supplies, and such. By laying out a mock-up of the barn and walking through the spaces, you can get a clear idea of whether your layout will work.

BLOOM HOMESTEAD

Conceptual samples

During the conceptual design process, Jessi used bubble diagrams to draw different scenarios on multiple copies of her base map. This allowed her to make sure she had laid out good broad patterns for her landscape before she got down to the details.

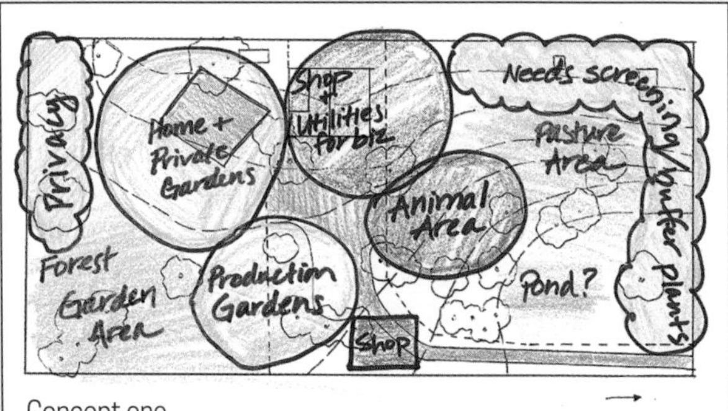

Concept one

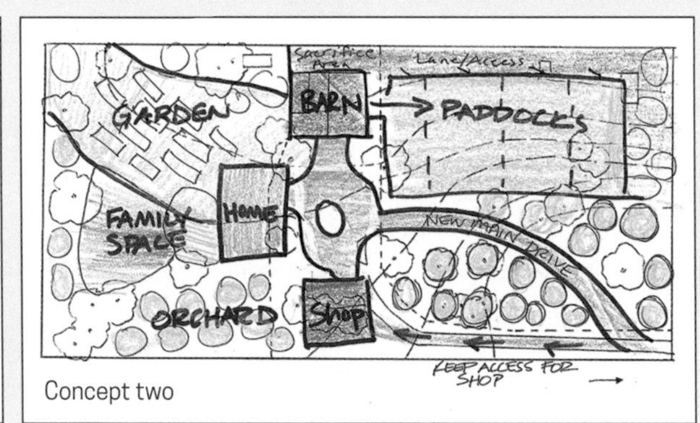

Concept two

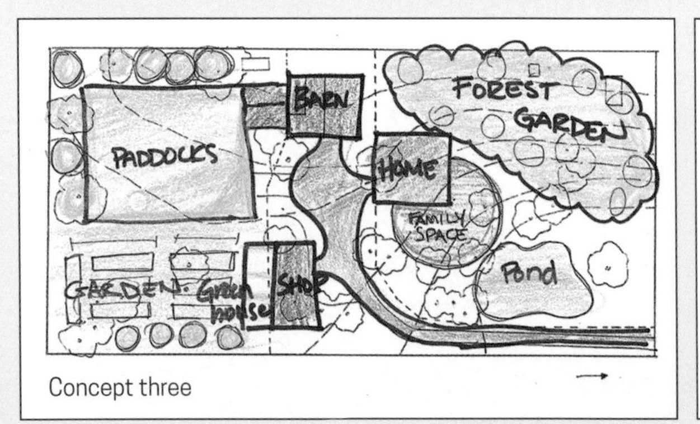

Concept three

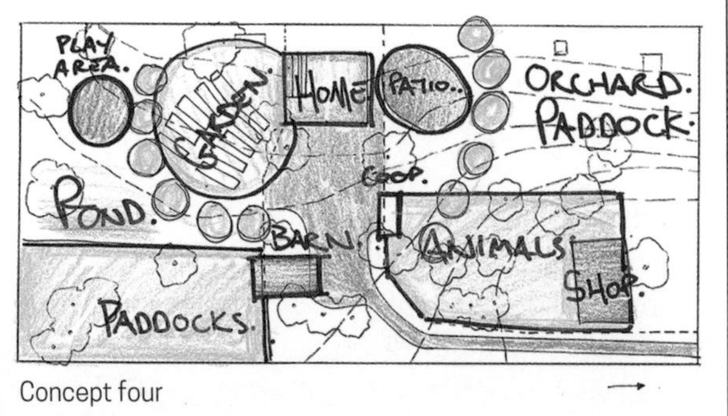

Concept four

STEP 5: SCHEMATIC DESIGN

During the schematic design step, your goal is to land individual elements on the base map. This means you draw all buildings, roads and paths, water bodies, and other major site features to scale where you intend them to be installed. During this step, it is important to create multiple iterations of your layouts so you can have several options to choose from (or hybridize).

Take tracing paper and lay it over your base map to make quick sketches of all the possibilities. These sketches don't need to be complete or neat; they are meant to get you thinking about all the options available and how the elements all relate to each other in your design. Your first iterations should be quick and dirty. Use fat markers, not mechanical pencils, so you can't get into too much detail. Alternatively, it can be helpful to create scale cutouts of your various elements and move them around on your base map to quickly test different configurations; you can snap pictures of these. As you find some things you like and start to refine them, you can add in more detail. At the end of this step, you will have a master plan from which to work.

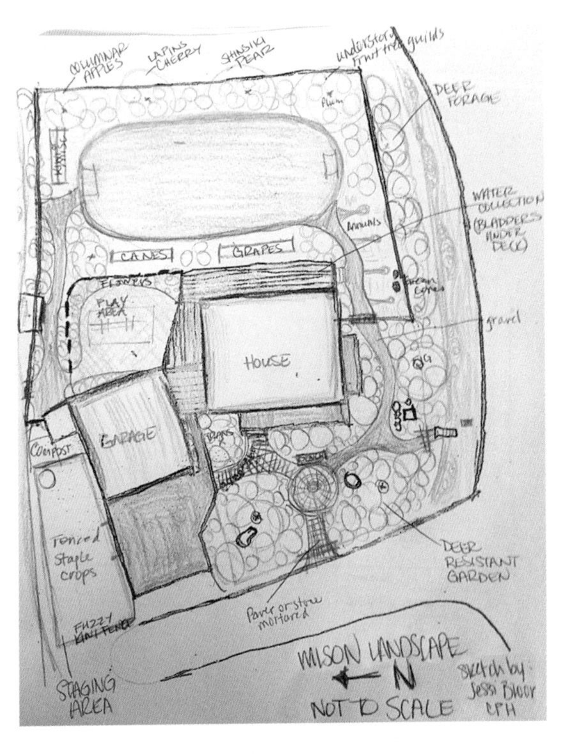

This simple schematic drawing by Jessi gave her clients a quick, low-cost look at how to get started on implementing their goals.

A master plan depicts your long-term vision for your site in a visual format. Not all of the details of your master plan need to be figured out during this phase, but it should be comprehensive enough to guide you toward your goals. Fully developing a permaculture project can take a long time, during which it is important to keep the big picture in mind. A master plan gives you something to refer to at any time to see how the current task at hand fits into the big picture, and it's also great for communicating the project vision to everyone involved in the development process.

Your master plan will be to scale and will have labeling that identifies areas, buildings, types of plantings. Depending on the size of the property, trees may be drawn individually (for smaller landscapes) or depicted en masse (for larger landscapes). In some cases, it may help to create section drawings in certain areas to illustrate how they will work. For example, this could depict the side of a house, showing where the rain cisterns are located and their approximate height or width in relation to the structure, which will be important in cases where you rely on gravity to move water. Remember, however, that the master plan and accompanying drawings are still meant to provide a big-picture look at the property and are not going to indicate where every violet should be planted.

Most of the time it's best to do several schematic versions of your master plan to get all the possibilities out on paper so that you can come to some decisions. For example, if you have a space that is 20 feet by 20 feet and you need to locate a structure that is 8 feet by 4 feet, you can draw it out to scale and visualize the space left over when that structure is placed, while considering the many ways of orienting it within the space. Rotating it in one direction over another may shade an area

This master plan shows a mosaic of land uses for the site of the Our Table Cooperative in Sherwood, Oregon. This is an innovative project using permaculture design thinking to create opportunities for farm enterprises that generate a livelihood for the growers. The plan was produced collaboratively by Communitecture Planning and Architecture, Terra Phoenix Design, permaculturist Jenny Pell, small farm advisor Josh Volk, and animal agriculture expert Nita Wilton.

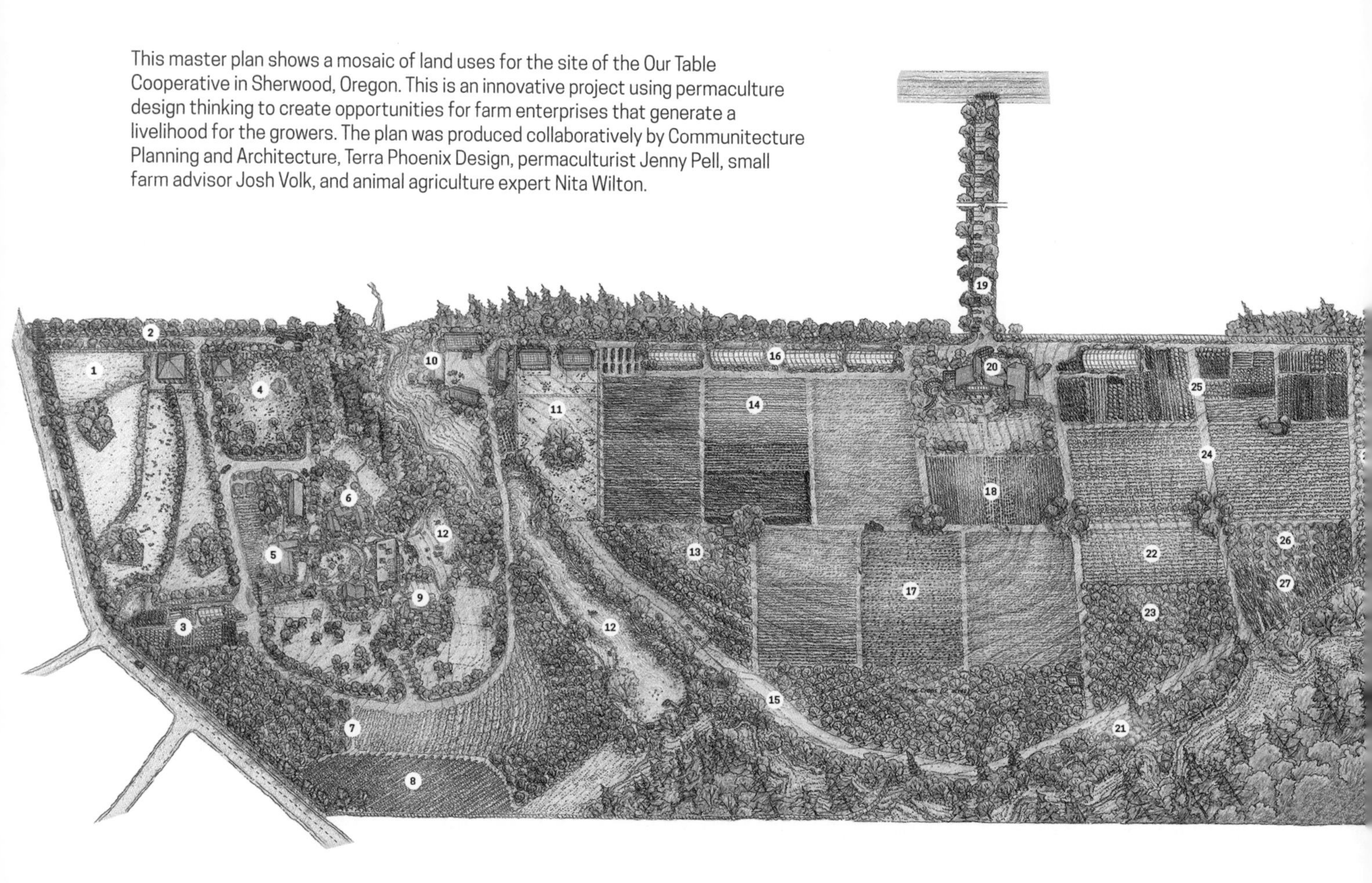

KEY

1. small livestock (3.5 acres)
2. multifunctional hedgerow (along edge of property)
3. nursery (0.5 acre)
4. poultry and forage crops (1 acre)
5. house 1 (existing)
6. house 2
7. orchard and soft fruit (4.5 acres)
8. trellised fruit
9. house 3
10. workshops, sheds, and storage
11. livestock and poultry
12. pond
13. tree crops (1.5 acres)
14. annuals rotation (5 acres; vegetables, grains, beans, cover crops)
15. rotational grazing pasture
16. greenhouses
17. annuals rotation (5 acres; vegetables, grains, beans, cover crops)
18. flowers, herbs, medicinals (0.75 acre)
19. parking and nut and fruit trees (0.5 acre)
20. farmstand, cold storage, packing and washing shed, food processing facility
21. mushrooms and understory medicinals
22. interplanted berries and medicinals (0.8 acre)
23. tree crops (1.25 acre)
24. soft fruit U-pick (2.2 acres)
25. nursery (2.2 acres)
26. tree crops (0.75 acre)
27. bamboo (0.5 acre)
28. nut and coppice tree buffer (0.5 acre)

where you want sunlight, or it may not line up well with the terrain, making one part of the system less functional.

We will outline several design methods that can help you make concrete placements. However, before we present those, let's look at how we think about siting elements in a landscape.

The critical importance of siting elements

The placement of elements in relationship to each other is critical to creating a functional permaculture design. The summary of constraints and opportunities you prepared in part 1 of the design process is your starting point for placing elements in your plan. Consider microclimates and sectors as you look for good locations for your various elements. On a windy site, you may consider placing a wind turbine and/or designing a windbreak to protect your sensitive elements. In an area prone to specific disasters such as wildfire, earthquake, flooding, or tornadoes, you will want to look at the best design strategies for that situation.

Then consider the lay of the land, aspect, and slope. A shady slope that is dark all winter long is probably not a good place for a solar array but may be a great place for a root cellar. If you are on a slope and want to have a garden and a pond, it may seem logical to have the pond at the bottom of the slope where water would naturally end up, but for irrigation it is better to have that water storage at a higher elevation than the garden so watering can happen via gravity. Perhaps in your situation it might make sense to have a pond at each location. Gravity never fails, so consider gravity-fed water systems whenever possible.

At garden blogger Stacy Brewer's house, access from the front yard to the back was an essential component of her design. She decided that this access should not only be functional for her but also provide beautiful habitat for pollinators.

Another major consideration when siting elements is access. How will you get to those elements for maintenance and to move resources in and out? When planning for efficiency, you want the elements you need most often closest to you. But what if your site or existing structures don't allow for that? What if your only sunny spot for a garden is farthest away from your house? You can design the highest priority features that require sun in that spot and make sure the spot is highly accessible. Could you gain more sun close to the house by removing a tree?

When placing elements, it's a good idea to start with those that have more restrictions in terms of where they can be sited, and work in the easier elements afterward. Remember, the Scale of Permanence will help you to identify the most constraining factors in your design. Narrow down your placement options based on landform before vegetation and based on legal considerations before aesthetic ones. Often once the most restrictive elements are locked in, the rest fall into place fairly easily.

Planning with permaculture zones

One useful way to think about the placement of elements is in terms of zones of use. Typically, a permaculture design identifies five zones, with zone 1 at the center of activity and zone 5 on the farthest reaches. Besides frequency of use, each zone is characterized by typical kinds of infrastructure, plantings, and animals placed there. Here is how each zone is described:

Zone 1. This is intensive-use space, the space closest to your center of activity, and often includes herb gardens, flower beds, and annual food crops that need attention multiple times daily. Utilities that are used daily, such as rainwater catchment systems off your house and food digesters to compost kitchen waste, are often located here. Toolsheds, greenhouses, and enclosures for small animals, such as rabbits or pigeons, are often in this zone as well.

Zone 2. This is where elements are often placed that require attention between once per day and a couple times per week. Semi-dwarf fruit orchards, perennial vegetables, and small livestock such as chickens and ducks are often found in zone 2. You may also find compost bins, beehives, and workshops here. Barns, stables, and coops are often placed on the border with zone 1 so the humans can easily feed and care for the animals, while the animals themselves have access to zones farther out in their pastures and runs.

Zone 3. This is often called the farm zone, and you might frequent it a couple of times a week to a couple of times a month, depending on the season. It is commonly the zone for staple crops, large fruit trees, pastureland, and produce grown for market sale. Large water storage is often placed here, and soil fertility is often managed through the use of green manures and animal manures since it tends to be too expansive to fertilize with a home compost system. Cattle, horses, sheep, and goats might be pastured here.

Zone 4. The least managed of the zones, often bordering wilderness, zone 4 often encompasses woodlots, pasture, or wildlife areas that may be visited as little as a couple times a year. Sometimes it makes sense to place low-maintenance nut crops such as chestnuts or acorn-producing oaks here. The pastures tend to serve larger range animals such as cattle; the shelters and watering troughs they require are often the only nonliving elements placed in this zone.

Zone 5. Originally, zone 5 was considered wilderness, where people go only to observe. However, as our understanding of the traditional land management practices of indigenous people has increased, it has become apparent that many landscapes around the world previously thought to be wild were actually managed in a low-intensity way that guided their development. For instance, many parts of North America were frequently burned as a way to set back succession and keep oaks as the dominant species. The resulting oak savannas provided acorns for direct consumption and supported wildlife for hunting. Thus, zone 5 may not need to be completely hands-off. Either way, zone 5 requires little, if any, attention other than proper conservation efforts and occasional foraging, recreation, or hunting. Bill Mollison says that "this is where we learn the rules that we try to apply elsewhere."

PERMACULTURE ZONES

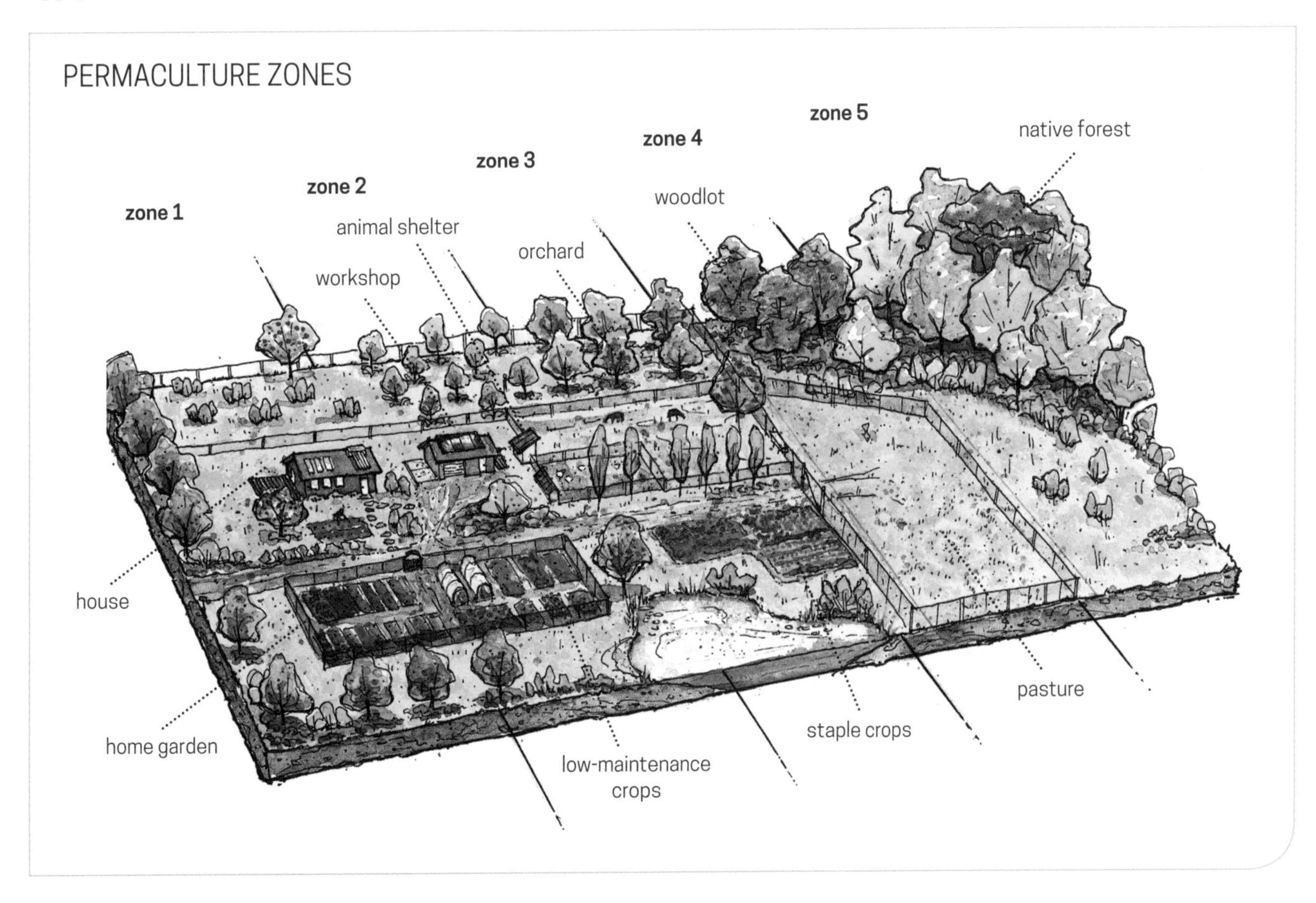

These zones not identified in the original permaculture literature have found their way onto the list over the years, representing additional areas to which permaculture design principles can be applied:

Zone 0. This zone encompasses indoor spaces, usually referring to our homes. In warm climates, the lines between inside and outside are often blurred. For instance, when you sit on the veranda under the grape arbor for a cool glass of lemonade, are you inside or outside? In cold climates, we close the doors in the winter and are either distinctly inside or out, so thinking of the inside as a different zone with extra-intense usage for at least part of the year makes sense. Zone 0 typically includes a kitchen and living spaces used almost constantly. We can apply permaculture design thinking to these indoor spaces by looking at ways to incorporate living systems into the home space—by using foliage for shade, for example, and growing houseplants to cleanse the air. This is also the space where animals such as cats, dogs, and bins full of compost-processing worms may live as a part of our indoor-outdoor systems.

Zone 00. Some permaculturists think of this as the zone of self-care, the place where we care for our mental, physical, spiritual, emotional, and social well-being. Because we spend 100 percent of our time in this zone and carry it with us wherever we may be, some figure it's worth planning and designing. There is some controversy about including this zone in permaculture design since it deviates from the original intent of the zone concept and risks losing the integrity of zone planning by trying to stretch it to mean too many things. However, the ethics and principles of permaculture can certainly apply to our inner spaces, so use this if you find it helpful.

Zone 6. Some permaculturists consider zone 6 to represent the broader community, both local and global. It encompasses how our property and our choices tie into the greater world around us in terms of interaction and impact. We can apply the ethics and principles of permaculture to all relationships we have in the world. This zone can be where we make the largest impact, as our actions and energy can reach larger numbers of people and often inspire others to make changes in their lives.

There are two ways to use the zone concept. For sites that already have established patterns of use, you can determine the zones that already exist and then think about whether to place elements in ways that work with these zones or create a design that modifies the existing zones. For instance, putting a beautiful vegetable garden with a hammock and picnic table in a little-used part of the yard can essentially change its zone by enlivening it.

You can also use the zone concept to test your various design solutions during the schematic design step. Take one of your maps with all the elements placed and draw where the zones would be if the design were implemented that way. Do you see any elements that have found their way into zones where they shouldn't be? Do

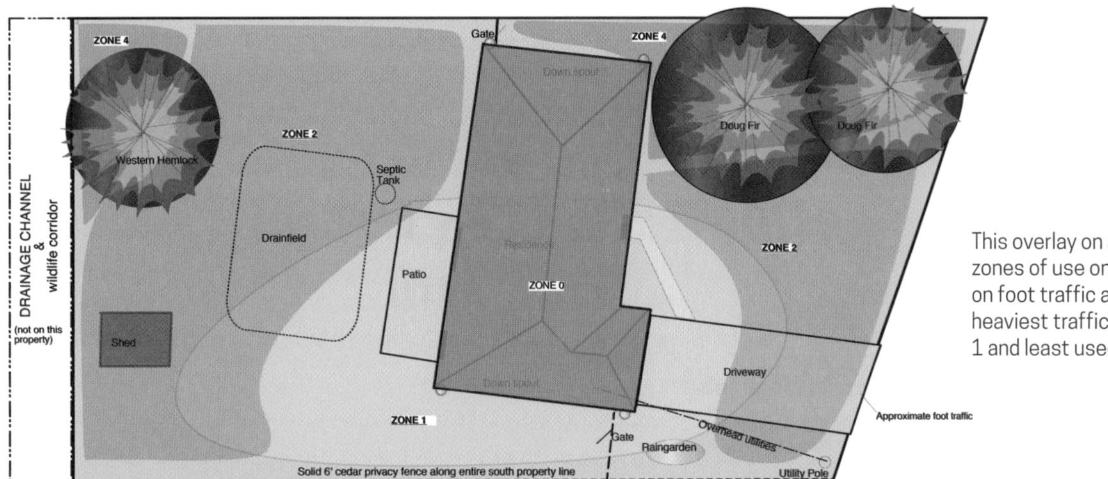

This overlay on a base plan shows the zones of use on this property based on foot traffic around the site. The heaviest traffic is indicated with zone 1 and least used space zone 4.

things seem to lay out in a way that helps with efficiency? If not, maybe you can make some minor changes for big benefits in efficient design.

There is no hard-and-fast rule about whether a given element should be in one zone or another. Remember, this is a tool to help you create an efficient design, so define things in a way that makes sense for you. Zones are frequently illustrated by concentric circles, but this is hardly ever how they lay out in reality, as topography and other limitations or parameters often modify their shape and size.

Here are a few notes to keep in mind as you lay out your zones:

- You can have several of the same zones spread throughout a site but not connected. Also, there are often no hard boundaries between zones, although they may change at fence lines, pathways, and other physical borders.
- In most cases, nonliving systems such as structures and energy infrastructure—homes, greenhouses, photovoltaic panels—are larger and more numerous in zones 1 and 2. These nonliving systems tend to get smaller, less numerous, and more spread out as you move toward zones 3 and 4. In zone 4 you might not have any energy infrastructure and the only buildings might be a small animal shelter or hunting blind.
- Conversely, living systems tend to be smaller in all aspects in zones 1 and 2. In zones 3 and 4, living systems tend to consist of larger individuals in larger quantities covering larger amounts of land. For instance, a zone 1 apple tree might be a dwarf because you don't have room for a big one with all the other things going on. It will likely be a good variety for fresh eating so you can snack on it as you do your

It's important to plan for areas to rest within your garden where you can stay connected to nature. In many climates a beautiful cob bench like this would last longer if it were covered by a roof.

yard work. A zone 3 apple might be a standard-size tree in the midst of an orchard block of fifty apple trees intended for commercial production. These apples might be keeper apples that you never snack on but instead harvest all at once for winter storage.

- If you have multiple points of intense use, it's possible to have multiple zones 1 on a property. For instance, in a village, each home would be a zone 1 for its residents, as would the community gathering space. Many small farms these days have not only the farmer's home but also a processing facility and possibly points of sale; each of these could be a different zone 1 with different use characteristics.
- Zones do not have to be contiguous. In other words, it is possible for a patch of zone 5 to be in direct contact with your zone 1. In fact, you may design with this in mind if you like to watch wildlife from your front porch.

URBAN ZONES

What if you live in a city? With more concrete and asphalt than exposed soil, cities are often considered food deserts. But cities also offer opportunities to gather resources from the waste stream and use public land or even rooftops for growing food. Being near larger communities from diverse cultures can offer many benefits. In addition, you can start to think about the broader context when defining your zones.

For instance, you may not really have any zone 4 areas on your ⅛-acre city lot. But what zone 4 do you have in your neighborhood? Perhaps a public park could serve as your zone 4. Wouldn't it be great if someone made a living grazing sheep in public parks to replace lawn mowers? How about planting nut trees along the edges of our sports fields? What could make for an easier walnut harvest than just raking them up off the turf? And keep in mind that there are always opportunities for small patches of zone 5 anywhere. That overgrown thicket in the corner of the backyard is where the most wildlife is likely to be.

At the Picardo P-Patch community garden in Seattle, some gardeners grow low-maintenance staple crops such as potatoes and winter squash. Crops needed more often, such as salad greens and culinary herbs, may be grown at home in the yard or even in containers on the deck.

CORE PERMACULTURE ZONES

	Zone 1	Zone 2	Zone 3	Zone 4	Zone 5
Frequency of interaction	multiple times per day	once per day to two or three times per week	two or three times per week or per month	two or three times per month or less	occasional
Infrastructure	house, outdoor kitchen, potable water, energy infrastructure, toolshed, greenhouse	barn, coop, workshop	animal shelters	animal feeders	None
Plantings	greens, herbs, snack fruits, berries, flowers	some staple crops, home orchards	commercial plots, large fruit trees, animal forage	firewood, nuts, bamboo, pasture, timber	native ecology
Animals	pets, small livestock	poultry, fish, fowl	cattle, horses, sheep, goats	cattle, large animals	wildlife

ADDITIONAL PERMACULTURE ZONES

	Zone 00	Zone 0	Zone 6
Frequency of interaction	constant	many times daily	varies
Infrastructure	your heart, your mind, your body, your spirit, your relationships	bedroom, kitchen, bathroom, living room	community center, businesses, other communities, institutions, allied organizations
Plantings	medicinal plants	houseplants, window boxes, containers	local farms, city and national parks, nature reserves
Animals	symbiotic microorganisms	pets	wildlife, strays, neighbor's animals, interaction with other humans

DESIGN BY EXCLUSION

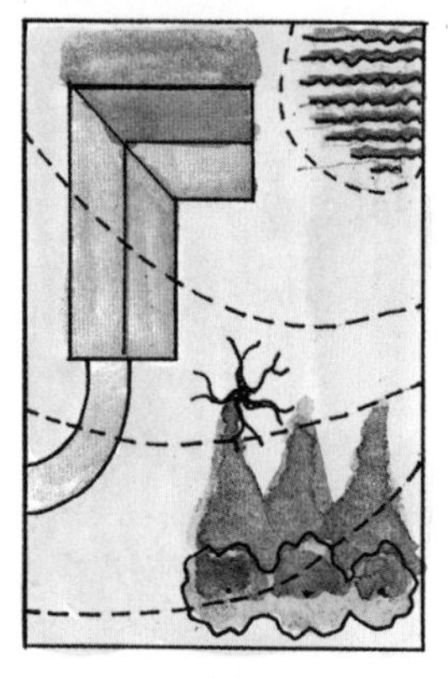

winter
heavy shade
high water table

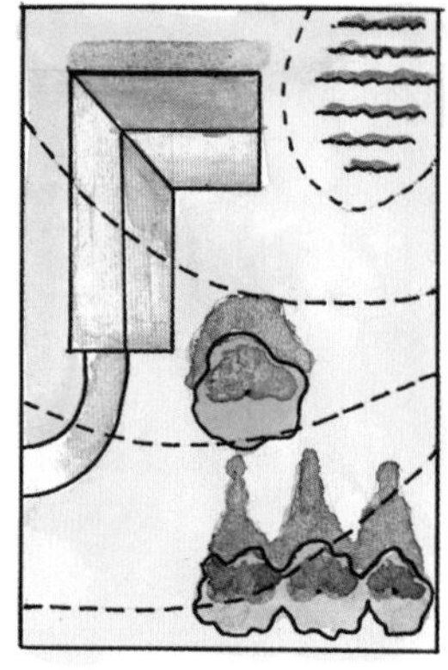

spring and fall
moderate shade
moderate water table

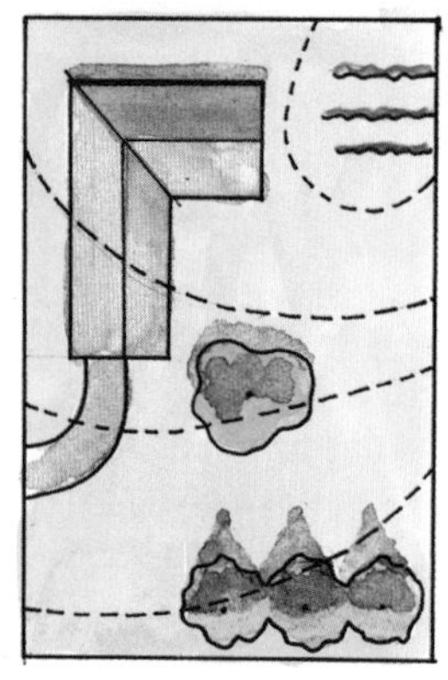

summer
light shade
low water table

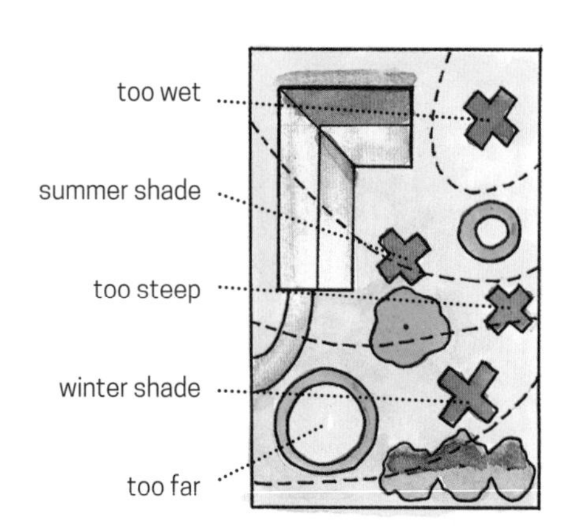

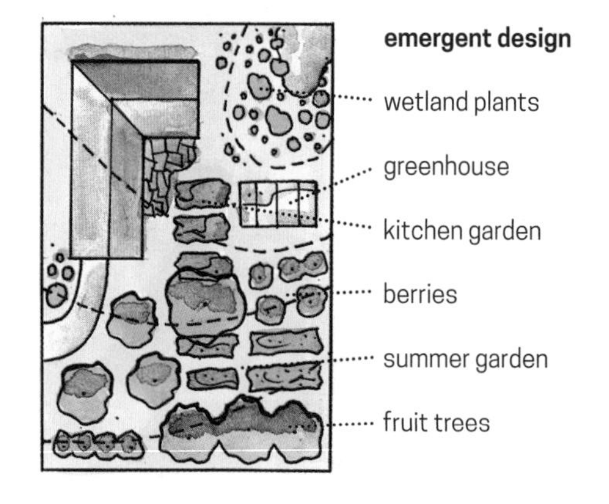

Using design by exclusion

In situations with no clear indication of where to place an element, you can use design by exclusion. This technique was first suggested by landscape architect and planner Ian McHarg in his book *Design with Nature*. It consists of posing the question, Where should this element *not* be placed? After you rule out all the worst options, you can see which ones are left standing.

To begin, create several overlays on your base map, each showing a different constraining factor (such as soil instability, soil moisture, or aesthetic value). If you want to get really elaborate, you can create an overlay for each season showing the constraining factors during that season. Then consider which locations are not suitable for the element you're trying to place. Start with elements that have very specific requirements for siting.

For instance, if you are trying to figure out where to put your greenhouse, you can create a map layer for each season's constraining factors showing the locations that would be inappropriate for a greenhouse due to those factors. The illustration of design by exclusion shows that in the winter there are areas with too much shade and areas that flood. In the spring and fall, the shade and water problems have shrunk but still exist. In the summer, shade is minimal in most cases (except where the deciduous tree sits) and the wet spot is dry. By looking at all of these together, you can figure out the best place for the greenhouse to get year-round sun. Similarly, if you were trying to place a greenhouse geared for use in a particular season, you might make a different placement choice.

Planning for access

Access to elements on your site is another very important layer to work through in your design during the schematic step. Every element in your design should be easy to service—to harvest from, monitor, clean out, or just visit. Your pathway system can be organized in a hierarchical branching pattern, with the most heavily used, widest routes narrowing to smaller arterial pathways that carry comparatively little traffic. While the shortest distance between point A and point B will be the most obvious route to connect elements, there are many other things to consider, including the widths of paths, the materials used to surface them, and the maintenance needed for upkeep.

Consider what and who will be traveling through these spaces to get an idea of how to design them.

Vehicular access. Roadways should be at least 10 feet wide for cars and trucks, with consideration for turnaround space and parking. Surfaces can be anything from gravel to asphalt or grass strips. For other traffic with wheels—bicycles, wheelbarrows, wheelchairs, strollers—it is important to think about providing a firm rolling surface and a comfortable width.

Two-person walkways. Main entry points and pathways for people to walk side by side should be at least 4 feet wide. If these paths are used by the public, they should have a durable surface that can be used by people of all ages, agile or not.

One-person paths. For any paths where access is minimal and people walk single file, you may only need to make your pathway 18 inches wide, but any width up to 3 feet would be more comfortable. If you need to maximize soil space—say in a garden setting—you may only need stepping-stones that allow someone to stand in an area and reach different points.

Elevation also comes into play when you are designing access. Both high-elevation and low-elevation access to a site can be important. If you need to bring materials onto your site, it is best to do that at a higher elevation so the materials can be distributed downhill, working with gravity. If you are transporting materials off-site, it is best to do that at a lower elevation on the property because it is easier to harvest and move product downhill than to carry it uphill.

In one of Jessi's annual gardens, she uses salmon-shaped stepping-stones that can be moved around in between growing beds as the plantings change from season to season.

Garden blogger Stacy Brewer uses bottles to define path edges and burlap to keep weeds down in the paths.

It's important to allow for access around your site's perimeter and along all fence lines. For fenced areas this will allow you to do routine checks because damaged fences could lead to predation and loss of valuable commodities. If a livestock fence fails, can you easily access it for repair? Many elements may not need routine care, but you should plan for access to do routine monitoring for any repairs just in case something happens.

Designing utility spaces and boneyards

While we're considering final placement of elements, we need to remind you to incorporate utility spaces and boneyards into your design so that they don't have a negative aesthetic impact (imagine old car parts on the front lawn). Everyone implementing permaculture systems will likely have a large collection of materials that have been gathered and saved for later use—stuff like salvaged lumber, pipes, and fencing—as well as a bunch of tools. These things all need a home. An area for such a purpose doesn't have to be ugly and can be quite organized with some thought put into it.

These spaces generally don't need sunlight or a certain soil type, but they do need to be accessible. They also need to be protected from the weather. A great place for them is often on the backside of a shed or accessory building where they are easily accessible but can be visually screened without much trouble. It is fairly easy to build shelving on the outside of a building with a small shed roof to protect the materials from the weather.

These bulk materials bays provide an organized way to manage wood chips, compost, topsoil, gravel, and other materials used frequently in a permaculture landscape.

Master plan

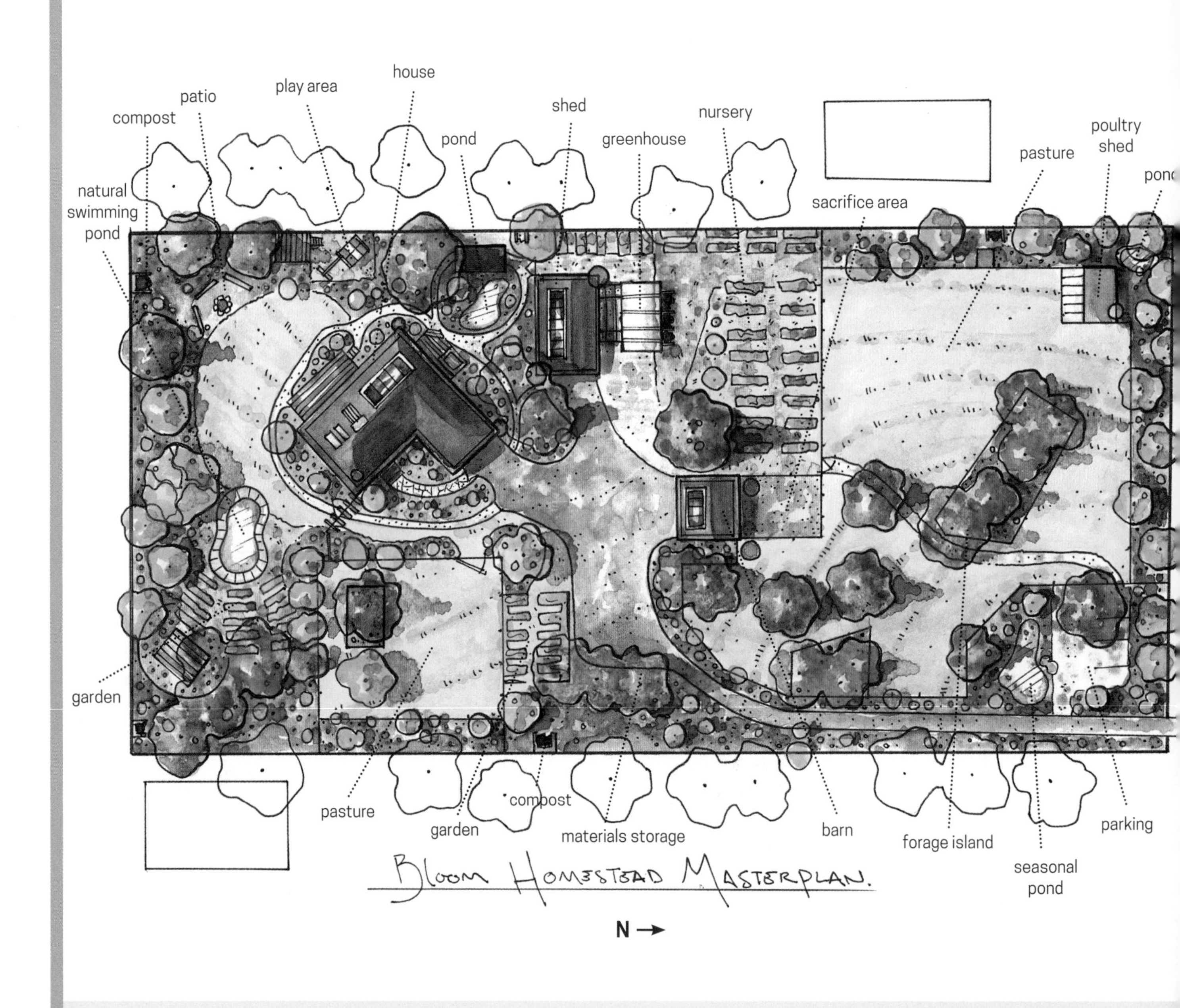
house
play area
patio
compost
natural swimming pond
pond
shed
greenhouse
nursery
sacrifice area
pasture
poultry shed
garden
pasture
garden
compost
materials storage
barn
forage island
seasonal pond
parking
BLOOM HOMESTEAD MASTERPLAN.
N →

In the master plan for Jessi's property, gardens are placed where soil and sun exposure are ideal. Along the perimeter of the property and along fence lines closest to the house, fruit and nut trees are placed in a diverse food forest border. Annual gardens are placed with ideal sun exposure in the growing season and in areas that are easily fenced off. Compost systems are placed at different locations throughout the site near the garden areas that they service.

For animals, structures and forage areas are placed for good access and management along with multiple compost areas for the bedding and manure shoveled from their housing. There are two main grazing areas for the larger animals (horses, pigs, goats) that can be divided further but that allow for pasture rotation. A sacrifice area (a small gravel paddock where animals can be kept to allow the pasture to rest) is placed behind the barn for good drainage during the rainiest season. The smaller animals are accommodated with a mixture of free-range (for the working nursery ducks and chickens) and confined-range systems (for the turkeys). The larger utility shed that was already on-site is used for seasonal meat animals and has access to four different paddock options for grazing, depending on the time of year.

The temporary dwelling is replaced by a residence that can house both the family and an office for the business. Multiple outdoor rooms and private gardens nearby include play space, herbs and annual beds, and areas for outdoor cooking and eating. The deck has a small lean-to greenhouse for citrus and other tender plants as well as containers to grow many plants used daily that benefit from the hot microclimate. The home uses solar hot water and a woodstove with many functions in addition to a heat pump for energy and heat. Solar orientation and roof light comes through via skylights and solar tubes. Drying racks and storage space are under cover on the deck, and a basement provides space for stored produce and the laundry.

The business office in the daylight basement of the home has a large, open floor plan and natural light. The easily accessible space where the temporary dwelling and attached structures stood becomes the greenhouse and shop area with a nursery. Parking for business vehicles, employee vehicles, and customers is placed in a few areas throughout the property based on who will be parking there. Equipment and biodiesel fueling is in a central location for easy access.

Some lessons were learned from mistakes made in preparing this master plan. The placement of the greenhouse was rushed, simply using space that needed the least site prep. That means the greenhouse doesn't have adequate sunlight for all the things Jessi wants to grow. The greenhouse would have been better placed in the paddock east of the house despite the extra site prep needed. Besides that, having public access to the business through private space is not ideal. Jessi would have preferred a clear boundary between public and private space, but without a separate structure, the shared access arrangement is the best option.

Figuring Out the Details

(left) Jonathan Brandt gives a design presentation to his classmates at a forest gardening workshop, showing the different phases of permaculture design from conceptual drawings to schematics to details and elevation drawings.

(right) Permaculturist Marisha Auerbach's backyard was all lawn when she moved in. In mere months, she was able to transform the space into a productive edible landscape using sheet mulching and fast-growing annuals. With a good plan you can implement your design very quickly.

Now that you have a master plan, you can begin thinking about how to tackle the work and create the detailed working documents you will need to get the job done. Different aspects of your design will require different amounts of detailing. For instance, if one of your tasks involves planting an area with habitat improvement species, the only detailing you may require before sticking your shovel in the ground is to figure out how many of each plant you need. Conversely, in order to build your house you will need to create detailed blueprints, identify and obtain materials, and make sure you have all the proper permits. Other tasks involved in implementation planning include budgeting and planning for maintenance.

Here are the steps and the products you'll generate in part 3 of the design process:

Step 6. Implementation planning—phased implementation plan, timeline, budget

Step 7. Detailing and working documents—blueprints, detailed landscape drawings, planting plans, water and energy system layouts, rotational grazing schemes, resource lists, and so forth

Step 8. Maintenance planning—maintenance calendar

STEP 6: IMPLEMENTATION PLANNING

Once you have your master plan in hand, you're ready to get to work on the land. However, most people quickly feel overwhelmed because there are so many things to do. Where do you start? This is where some basic project management skills come in handy. You need to think about your project's overall phasing and scope, timeline, and budget to determine what to implement when.

Creating a phased implementation plan and identifying scope

First, it is helpful to identify all the major tasks that need to happen to build out the master plan. This includes things like "install water lines to western field" or "plant orchard." Put each major task on an index card and then sort them based on a sensible order of operations. For instance, it may be critical to get water lines to the western field before you plant the orchard there so that your trees don't die of drought before you get the water system in.

The tasks that need to happen first become phase 1. The rest lay out as phases 2, 3, 4, and so on. Ideally, you won't have more than four or five phases. If there are places on the site that are actively degrading (such as major erosion or roads at risk of washing out), they should be addressed in phase 1. Another rule of thumb is to renovate and care for what you have before installing new elements. The last phase can contain any optional tasks. Tasks that can be done anytime can have their own special category. A major benefit of this kind of phased plan is that even if you never make it to the last phase, you're always improving your property.

With your phased plan in place, you can begin narrowing your scope from the big picture to the details of the task at hand. Earlier in the design process you may have identified the scope of your design work from a big-picture perspective (for example, are you designing the front yard, the whole yard, or the neighborhood?). At this point when we talk about scope we are referring to identifying the extent of each given task you listed in your phased implementation plan. For instance, if one of the projects on your list is "Install orchard," does that mean just planting the trees or does it also include installing irrigation, improving the soil, and sheet mulching? Or are some of those tasks separate items on your list? It's important to identify the scope for each item on your list so you are able to make accurate materials lists and estimates of time required. This approach will allow you to move forward with your implementation with a minimum of hiccups.

Timeline

In addition to identifying the scope of any given project, it is helpful to create some sort of timeline so you can make the best use of an always-limited resource and keep the momentum going. Those projects that drag on for years often never get completed. As a general rule, you should always allow for extra time. If you think a task will take six days, give yourself seven or eight. It is always better to be ahead of schedule than behind. Scheduling work days in advance on your calendar can prevent the "I don't have time" excuse because you won't overbook yourself.

(*This page and opposite*) This phased implementation plan, part of the master plan for the Chaikuni Institute in the Peruvian Amazon, defines the scope of each phase and the steps to take.

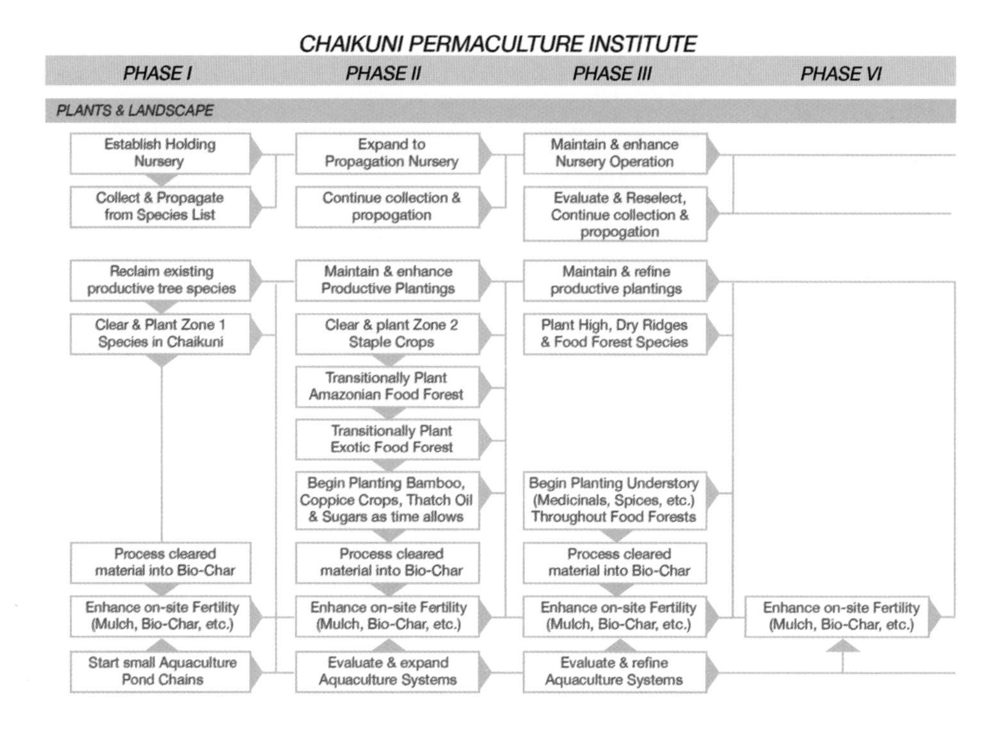

	Monday	Tuesday	Wednesday	Thursday	Friday	Saturday	Sunday	Monday
Day 1	Remove existing plants and lawn							
Days 2-3		Excavate and grade patio site to proper depth and slope						
Days 3-4			Lay and compact gravel layer for base					
Day 4				Lay down fine rock dust or sand for setting stone				
Days 5-7					Lay down stone material			
Day 8								Complete joint work and clean up

A Gantt chart may help you see how all of the tasks needed to complete a project lay out and where there is overlap (indicating you may be able to work on multiple stages at once). This sample Gantt chart details the steps to complete a stone patio project.

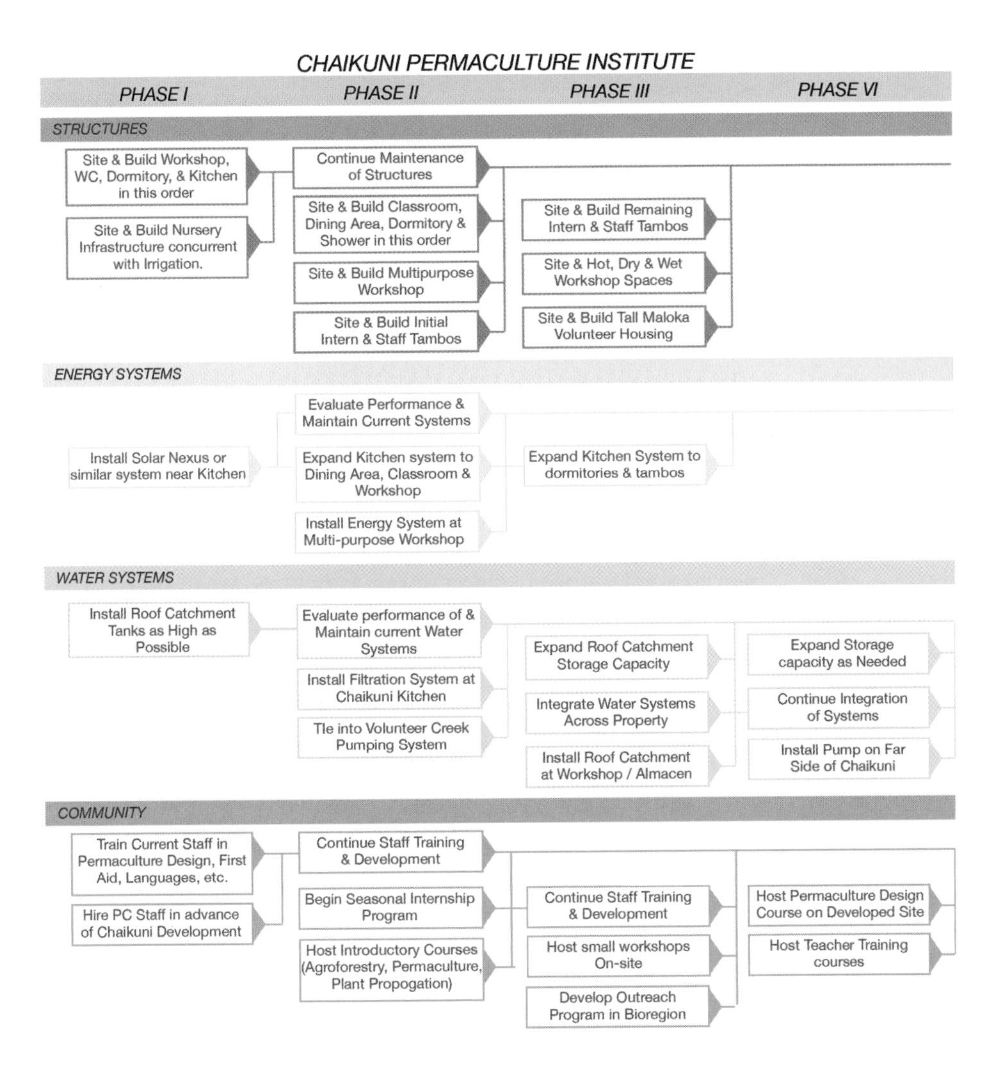

Budget

Depending on the project's scope and the resources that are readily available, a huge range of costs can be associated with it. For people who are resourceful and handy, a project can be fairly low cost or even free. For others, loans may be involved or a fairly large amount of money. Often people don't know how much to budget for materials or services because they've never dived into a similar project. If this is the case for you, it can be helpful to have guidance from professionals who know about all of the options for your project. You can also research the cost of certain aspects of your project on your own.

Creating new permaculture systems can often cost more up front but end up saving money in the long term. For example, Jessi's solar hot water system wasn't cheap to plumb through the house and tie into existing plumbing, but over several years the system has paid for itself in energy savings. Spending up front can set you up for resilience and cut your resource use over time. Look at implementing your permaculture design as an investment in your future. Sure, anything can be done on a shoestring, but if and when spending a little more will mean you get quadruple the longevity or the maximum efficiency out of an element—whether for water storage or energy production—then it's probably worth it.

Consider these costs when budgeting for your project:

Expertise. Hiring professionals can save you time and potentially costly mistakes. This could mean hiring a hydrologist for consultation in the early phases of design to help you sort out drainage issues or a plumber to help with renovating your conventional plumbing system to use rainwater or create an up-to-code greywater system.

Materials. You have a lot of options when it comes to procuring materials. You may have everything you need on-site for a particular project, especially if you're using natural materials. For example, rock found on-site is a great option for paths with heavier traffic; if wood chips from a fallen tree are available, they can be used for paths with lighter traffic. You may end up buying some materials and also sourcing some from salvage yards. You may have to do some research to find special components. Locally available materials should be your first choice from an embodied energy and sustainability standpoint. Look at material options not only from the financial and ecological points of view, but also in terms of their maintenance needs; some lower-cost or lower-quality materials may require more upkeep, which costs more in the long run. For example, wood chips must be replenished as they break down over time, and rock may need touching up.

Tools. Most projects require tools of some sort, whether digging tools, measuring devices, or even heavy equipment. Ideally you have access to tools through friends, family, or even a tool library in your community. Renting tools is another option. And if you know you will use tools repeatedly, it may be worth the investment to purchase them. You can also buy used tools; checking with a rental yard is a good place to start.

Some structures in the garden can be built from recycled materials, like these simple trellises supporting edibles at Bullock's Permaculture Homestead.

Labor. If you are not physically able to do labor yourself, this is an expense to consider, assuming you aren't counting on volunteers for everything. Knowing your physical limitations is important so you don't end up injured from this endeavor. For most homeowners, a simple potluck work party of neighbors is a great way to get work done and build community, provided the gathering is well organized, with a clear objective, a clear process, a known number of people you can count on, enough tools and materials, and enough people to guide the process. You do need to be selective about the tasks you get volunteer help with; removing blackberries is a great project for a work party, while wiring your photovoltaic system, not so much.

Teamwork and professional partners

It is entirely possible to complete a project from start to finish by yourself. But if you feel uncomfortable with some aspect of the work because of your skill level, the technical specifications of the project, or limited access to or experience with certain tools, you may want to use the expertise of people who specialize in their field. Relying on the knowledge of professionals or experienced helpers could save you a lot of time, money, and headaches.

There are several ways to go about finding good people to work with. Asking around is one place to start; nothing beats a personal recommendation from someone who has already worked with the person you are considering. Searching the Internet or local directories is another route. Although you may find a lot of information about someone on their website, it's a good idea to set up interviews or meet-n-greets before hiring anyone. Have a few questions in mind. Do they work hourly or by the job? What does the contract cover and when is payment of the invoice due? How long have they been working in their field? Is it possible to see their work or talk to past clients? You can also ask about their ethics or how they handle difficult situations. Always do your due diligence when bringing someone else on board to make sure their approach is a match with your sustainability goals.

STEP 7: DETAILING AND WORKING DOCUMENTS

Most projects require working documents to move forward, but the level of detail can vary. To get permits, you may need to present your local government with construction drawings with the nuts and bolts specified and the stamp of approval from an engineer or architect. On the other hand, a planting plan for a forest garden can often be done by hand and labeled with botanical names in a key. Most working documents will have a detailed material list, which can serve as a shopping list for material sourcing. The purpose of these documents is to have a detailed road map thought out in advance before starting a new project.

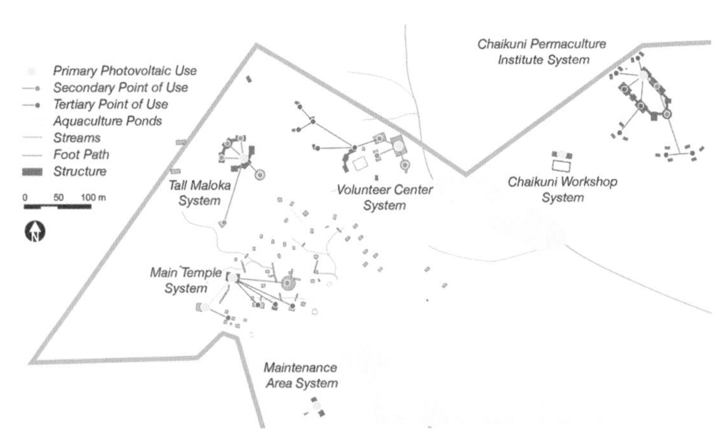

This Terra Phoenix Design drawing goes into details of the energy system for one of their projects.

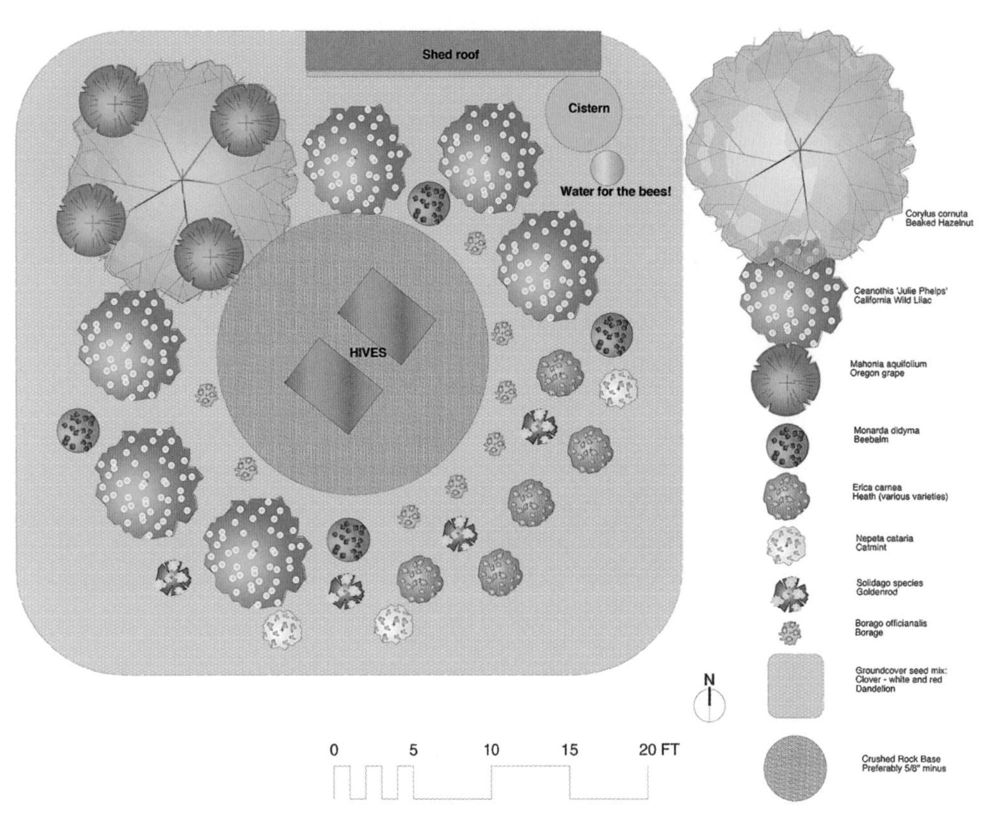

This CAD drawing is Jessi's design for a bee guild with specific vegetation to support the hive with both shelter and insectary plants.

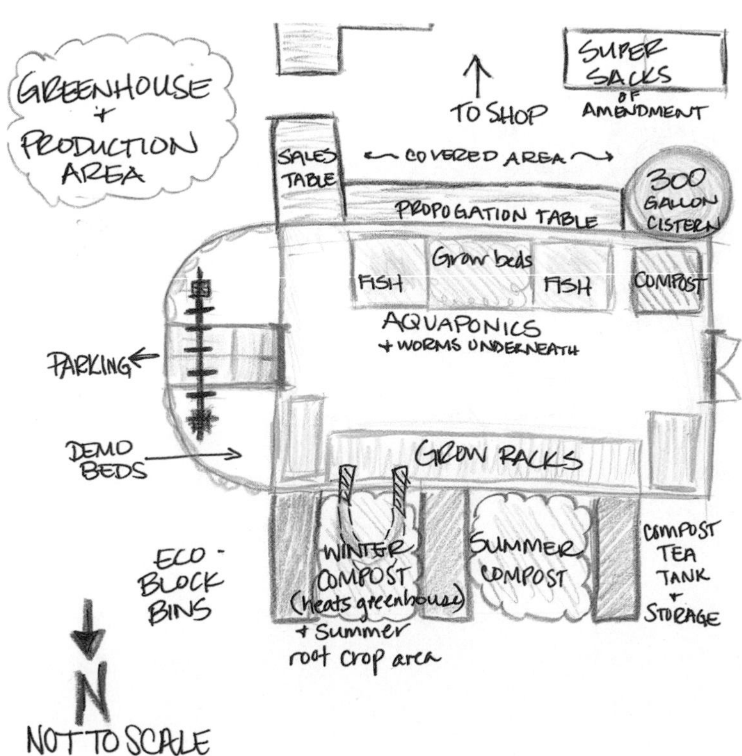

These details of Jessi's greenhouse show where large elements will be placed in relationship to each other.

Thinking about aesthetics

Although you've probably been thinking about aesthetics throughout the design process, this is the time when you can figure out the details that really make a design look and feel spectacular. According to the Scale of Permanence, aesthetics are the easiest thing to change, so we have a lot of control over this aspect. In their zeal for functional systems, some permaculturists have let aesthetics fall by the wayside; some even pooh-pooh ornamentals and aesthetic considerations. This is a response to a world where aesthetics are often the only drivers of design decisions, where images of digitally enhanced or touched-up bodies and lawns have resulted in damaged environments and human psyches. However, nature is inherently beautiful, and in our enthusiasm for function we need not let form fall by the wayside.

People are drawn to beauty, and our goal is to make permaculture so beautiful that everyone wants it in their lives. *Biophilia* is the understanding that humans possess a biological inclination to affiliate and connect with natural systems and processes. Any design we create can honor that attachment and provide connection to those things. Spaces designed to incorporate light, color, natural shapes, and plants have been shown to increase the health and well-being of humans using those spaces. Such spaces can increase workplace productivity and get sick patients out of hospitals sooner in comparison to spaces without those features. The healing power of nature is also evident in the example of therapy animals; many studies have documented the way in which people are aided in their healing process by spending time with dogs, cats, and horses. The same also goes for gardening, which has led to an entire field known as therapeutic gardening.

(left) In permaculturist Marisha Auerbach's productive landscape, this simple straw bale adds rustic beauty and a place to sit.

(right) Functional plants can be downright beautiful. This perennial kale provides great variegated foliage and can be grown in partial shade.

This beautiful native vine maple (*Acer circinatum*) outside of Jessi's office was placed and carefully pruned to provide a soothing view enlivened by visits from birds, squirrels, and chipmunks while she works.

(bottom left) This bed in Jessi's garden emphasizes color and repetition. The yellow and purplish blue flowers complement each other in a pattern punctuated with border grasses that keep the eye flowing through the landscape. Yarrow and hardy geranium, both insectary plants, are tucked in among berry shrubs.

(bottom right) The ground cover and shrub layer in this bed in Jude Hobbs's garden is full of productive plants, rich with textures, giving it a lush feel.

People vary in their ideas of what's beautiful, but many ideas can be accommodated in permaculture design. For some, a beautiful landscape consists of a manicured lawn and neatly trimmed hedges; for others, beauty is found in the eroding banks of a stream or in a wild prairie scrubland. The latter folks are often easier to bring on board the permaculture bandwagon because many of the examples they might run into are pretty wild and shaggy. But by focusing on the aesthetics of our designs, we can also attract people who like a tidier environment. For these people, we can incorporate features such as tidy borders and manicured edges to contain their ecologically functional garden. Those who prefer a wild garden can also use traditional landscape design techniques to give their space more aesthetic appeal.

Traditional design schooling emphasizes aesthetic principles such as repetition, unity, color, scale, and balance. These principles are not commonly considered in permaculture design, but there's no reason not to consider them. These aesthetic principles can easily be incorporated into your detailed design.

Themes and repetition. Repeating a theme across a space creates a feeling of unity or cohesion. This can be done with color, texture, or materials. For example, you can use a color repeatedly in flowers, containers, and maybe even paint choices on structures. You can repeat a plant with a texture that contrasts with the plants around it to tie a space together.

Color. Consciously using color in your design can make it more attractive and also give it a certain feel. For example, if you want a place for people to feel relaxed, you may choose violets and blues, while in an area where you want the user experience to be more upbeat with higher energy, you may choose reds and yellows. Our eyes are naturally drawn to lighter colors, which is important to pay attention to if you are using white; darker colors get noticed less and tend to hide imperfections.

Balance and scale. Balance and scale apply to both spaces and the elements included in those spaces. For example, if you have a space that is only 10 feet by 10 feet, it would make no sense to plant a tree that reaches 40 feet tall and wide at maturity. Instead, you want a tree that fits nicely within that space without underwhelming or overwhelming the area. Then you can use plants of an appropriate size at each layer underneath that tree to fill the space. To achieve balance, you can use matching plants on either side.

Creating connection

A landscape that helps you to be more resilient also creates a deep sense of connectedness. A design can help people make positive physical and emotional connections to spaces by incorporating features that provide a sense of safety and shelter. This can simply be protection from the elements by a surrounding or enclosure of natural materials; it can also include providing nourishment from food grown in the vicinity.

A permaculture design can also promote emotional connectedness by incorporating plants or elements that have historical meaning and evoke positive

This pergola in Jessi's garden provides a soothing sense of enclosure, plus kiwis and shade from fruiting vines in the summer. The materials were salvaged from a nearby barn, which adds another layer of meaning.

memories. When each element in your landscape—from the plants to the materials—has a story behind it, it becomes more interesting and friendly. For instance, if the rock border for one of your garden beds is made of stones taken from the foundation of the house you grew up in, it evokes memories and connection. A fruit tree you plant as a memorial to your grandmother has sentimental value. A tool rack built from pallets acquired at your favorite brewery moves beyond the ordinary. The more stories we can tell about the elements in our landscapes, the more personal those landscapes become and the stronger our connection to them.

STEP 8: MAINTENANCE PLANNING

There's a myth that permaculture means no maintenance or less work. While it's true that a permaculture approach will pay off down the road when your established systems really do require less energy from you, everything in our lives requires a certain amount of maintenance. Particularly during the demanding establishment period, a permaculture landscape requires ongoing care, which ebbs and flows as the seasons change.

It's fairly common for projects to get implemented well and then end up not being cared for properly, resulting in a waste of resources. All of the systems you design and implement will require ongoing maintenance. Research this during the design process so you know what you're in for, and plan maintenance schedules in advance to ensure that things are not forgotten. The maintenance schedule will look different for every homestead; it can be as simple as a chore list or as elaborate as a spreadsheet. Breaking maintenance chores down by zone and writing them on a calendar can be helpful.

Home systems that need to be maintained may include everything from energy systems to water systems to food production systems and more. For heating, maintenance could mean managing your fuel supply and making sure the chimney of your woodstove is inspected and cleaned annually. For water catchment systems, it means making sure gutters and their filters are clean before the rainy season and after big storms. Maintaining your pantry or root cellar may mean an annual cleaning out and/or reorganizing, depending on crops and storage time; it may mean planning for simple preservation methods, such as dehydration.

Planning ahead for irrigation is important, whether you implement a drip system or water by hand, especially in the early years when your landscape is getting established. Timers and other water management devices are handy but also need to be checked and updated depending on the time of the year and the seasonal moisture levels in the soil. Watering by hand is a great way to get to know what is happening with your plants and be able to monitor their growth.

Zone 1 gardens are the most labor-intensive area in terms of maintenance, from starting seeds in the off season to consistently checking in on the plants during the growing season and making sure they have what they need—water, nutrients, mulch, harvesting—to saving seed at the end of the harvest season. Perennial food plants such as fruit and nut trees may require less maintenance, but it is important to plan for pruning, checking on pest damage, thinning crops, and harvesting and preserving the food grown. These plants will most likely require some cleanup, but that depends on the systems you have in place and the aesthetics you have in mind.

If you have small animals such as bees and poultry, you will need to check on and inspect hives and coops fairly often. Have a routine thought out and schedule larger maintenance tasks so they are not neglected. Bees can be looked at once a week in the growing season or much less, depending on the style of management you choose. Egg collection is a daily chore, as is filling water buckets unless you've set up an automated watering system to handle this need.

Large animals require a certain amount of daily maintenance as well. Routine grooming tasks such as hoof trims can be done every few months, but milk must be collected daily from dairy animals. Breeding will require another layer of care to be planned for well in advance.

RESPONDING TO FEEDBACK AND LEARNING FROM MISTAKES

We need to be open to feedback during the design and implementation process. When we get feedback that something isn't working—for example, we've tried planting cherry trees somewhere three times and they keep dying—we shouldn't just give up; we should return to our design process and walk through it again for that specific element or system so we can come up with a better solution. If we quit and never try again, we have short-circuited the feedback loop and missed the opportunity to develop skills, get smarter, and grow our skill sets. Only by accepting feedback and working with it will we grow as designers and land stewards.

Also, it's important to remember that things change whether we'd like them to or not. In spite of having done a thorough job of site A + A, it's possible that something in your environment or life will supersede the data you've collected. Maybe the shade you thought was restricting your garden placement goes away because a tree blows down. This means rethinking your design and deciding if it's worth changing based on this new information. Either way, that change in conditions means an opportunity to fine-tune or wholly reorient the direction of your design, so try to see the positive side of whatever changes come along and work with them.

Implementation planning

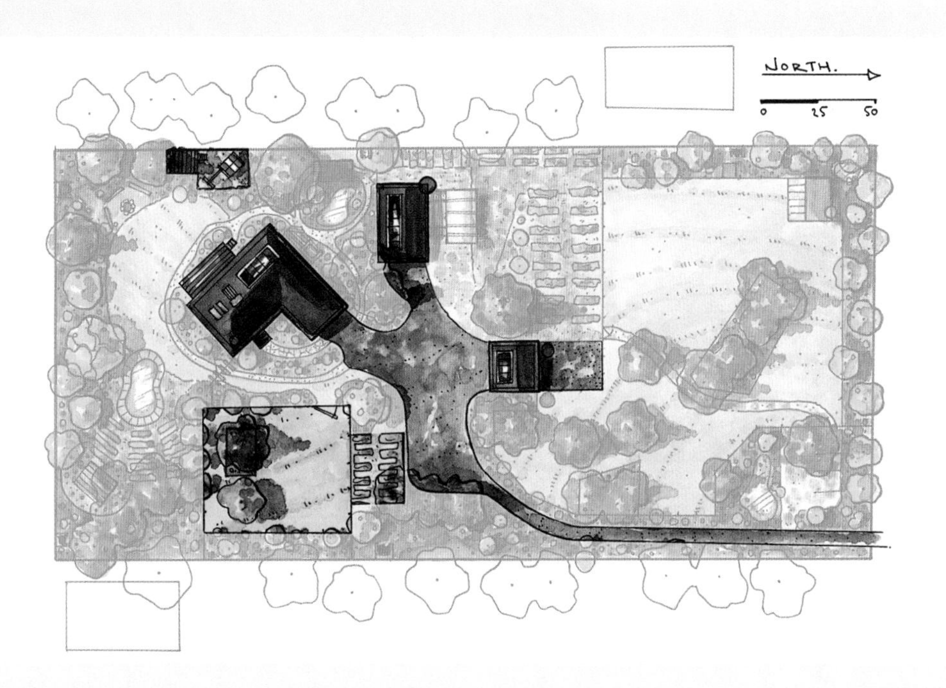

PHASE 1

Soil fertility—build compost bins, plant biomass-producing plants

Build annual food garden area

Infrastructure—construct buildings (barn, home)

Build play area for children

Fence secondary pasture for grazing animal rotation

Soil fertility development is a priority for this project, so fast-growing plants will be planted immediately and animal manure composting systems will be developed. The barn and animal housing will be the first element to be designed in detail and implemented.

(*continued*)

Implementation planning *(continued)*

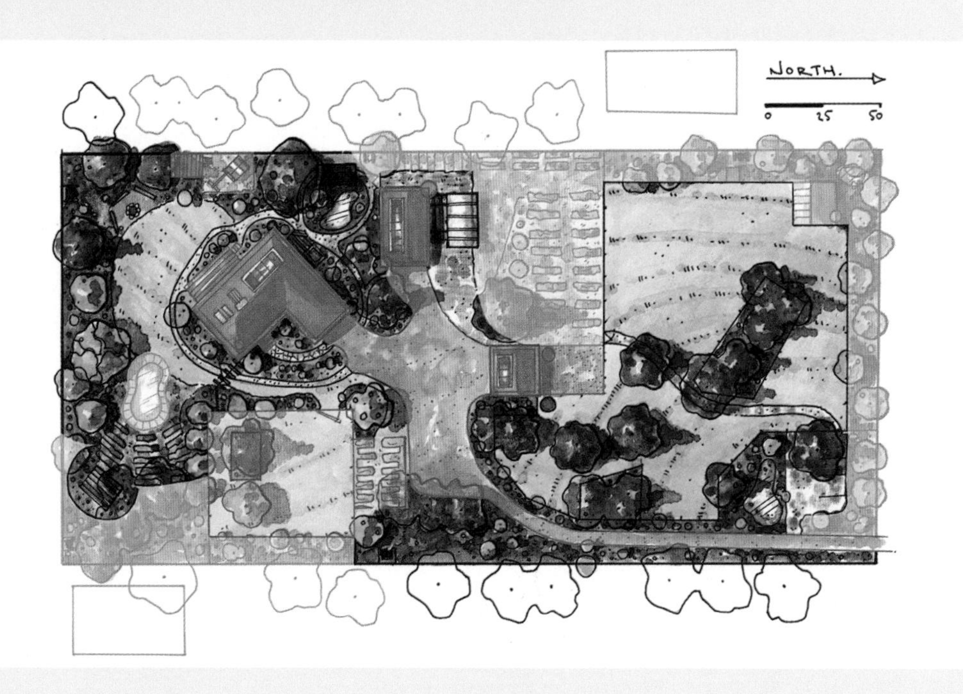

PHASE 2

Plant southern property lines for perennial food production and privacy

Water catchment systems—attach cisterns to all buildings and dig pond

Build greenhouse with aquaponics system

Develop more annual food garden space

Develop pasture and forage systems

Water catchment systems are another priority to take advantage of the rainfall, so a collection tank will be placed with each rooftop as it is built. With this rainfall harvest, more plantings can be put in. The greenhouse supports more growing activity and helps to extend the season so seeds can be started earlier in the spring.

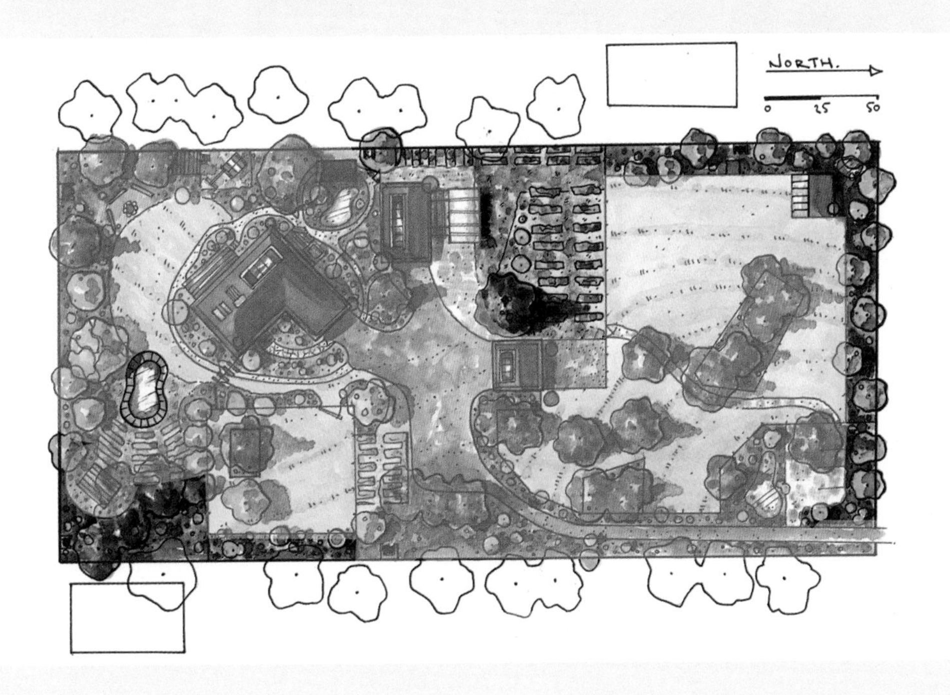

PHASE 3

Install energy systems such as solar technology

Develop nursery

Remodel shed

Develop forage zones for poultry and wood production

Develop forage zones in pastures

Install natural swimming pool

All of these tasks are last in the phasing because the foundation needs to be ready for them. The construction of buildings comes before the installation of solar. The shed and additional forage zones are lower in priority due to their proximity to the home but also to allow time to figure out what types of animals are easiest to care for on the site.

ANYTIME

Perennial food plants will be incorporated every fall, while in the springtime annual plantings will be done with a consistent but minimal effort due to lack of time in the busy growing season.

PERMACULTURE SYSTEMS

Permaculture is a process of holistic design that forces us to think in systems. A system assembles a number of different elements to accomplish some essential function and ideally creates a closed loop. This section of the book provides a closer look at the various systems that every permaculture design should encompass. When we create a master plan, it should lay out systems for managing soil fertility, water, energy, waste, shelter, food production, and animals and wildlife. Each of these systems is a major piece of the sustainability puzzle. As you read about these systems in detail, remember that permaculture design gives us a perspective, not a prescription. Use the examples to better understand the underlying issues, not as cut-and-paste solutions that are sure to work on every site.

Several lists of plants we find particularly useful or worth trying for various purposes appear in this section. These plant lists include this information, as well as additional notes where appropriate:

Scientific name. We encourage you to become familiar with the scientific names of these plants as it will make for clearer communication.

Common name(s). We give one or two of the most frequently used common names.

USDA hardiness zones. The United States Department of Agriculture (USDA) has looked at average annual minimum temperatures and used that data to divide the United States into thirteen different zones. In this system, the lower the number, the colder the temperatures tend to get. The hardiness zone range associated with each plant tells you the zones in which it is likely to survive and thrive. Bear in mind, however, this is only one factor in figuring out if a plant will succeed in your area.

Soil Fertility

Improving Tilth and Nutrients

Soil fertility is the basis of a thriving permaculture landscape. Thus, managing soil nutrients and soil life is a key piece of working with your landscape. And the good news is that no matter what state your soils are in, you can work with what you have in order to improve tilth and soil fertility. It starts with assessing what you have and then moves to thinking about crops and systems for soil improvement.

ASSESSING WHAT YOU HAVE

Testing your soil is a great way to get a feel for what's going on. It's wise to take several samples in an area; aim to get a cross-section of soil from the top 6 inches, not including organic matter that hasn't yet decomposed. Samples can be mixed together for a composite sample or analyzed individually. A micronutrient analysis should be fairly inexpensive and provide a lot of information. It generally gives the basics, such as soil pH and percentage organic material, and also tells you how well equipped your soils are in terms of the nutrients essential for plant growth. This allows you to come up with a fertility plan to supply the nutrients that are lacking.

Soil pH

The pH of your soil tells you a good deal about which crops will do well and what you might want to add by way of amendments. If you haven't had a soil test and you don't have any litmus paper or a pH tester, you may be able to get a feel for your soil's pH through taste or smell. Acidic soils tend to taste and smell sour, while alkaline soils tend to taste and smell sweet. Soil pH matters because extreme acidity or alkalinity causes certain nutrients to get locked up in the soil. Most nutrients are available at a pH between 6.3 and 7.0. Most plants like soils with a pH between 5.5 and 7.0.

Macronutrients

Several nutrients are used by plants in fairly large quantities. These include nitrogen, phosphorous, potassium, sulfur, calcium, and magnesium. Nitrogen (N) is what fuels growth and leaf production in plants. Phosphorous (P) helps plants to flower and produce fruit and seeds. Potassium (K) helps to make plants more resistant to environmental damage and diseases. If you purchase a box or bag of fertilizer (chemical or organic), it usually has an analysis of the N-P-K content. That allows you to see proportionally how much of these "big three" you are adding to your soil. If you are going to use amendments in the soil, match the content of the amendment with the nutrient levels found in the soil test.

Micronutrients

Micronutrients, or trace elements, are the nutrients that plants need to have in smaller quantities to stay healthy. Essential micronutrients include boron, copper, iron, chloride, manganese, molybdenum, and zinc. Even though the actual amounts needed are extremely small, plants will not be healthy without adequate quantities of these. If your plants don't seem to be healthy, you can look up the symptoms on a plant nutrient deficiency chart. In this way, plants can serve as indicators of soil health.

Soil life

As important as it is to understand the nonliving part of the soil, the living (or biological) part of the soil is what helps nutrients to cycle. In a broad sense, when you manage soil fertility you can think of all the techniques you apply (such as mulching and composting) as ways to feed the critters living in the soil. Some even go so far as to say, "Feed your soil, not your crops."

The living part of the soil is composed of a vast cast of characters doing a wide variety of jobs, such as buffering soil pH in the root zone of plants. Primary consumers break down organic matter directly. These include fungi, bacteria, earthworms, gophers, and so on. Secondary consumers eat the primary consumers and include mites, some beetles, springtails, protozoa, and amoebas. Tertiary consumers are the equivalent of top predators in a macro-ecosystem. Tertiary consumers include centipedes, moles, snakes, and lizards. Just as we can tell something about the health of a macro-ecosystem by the presence or absence of top predators, we can tell something about the health of the soil's micro-ecosystem by the presence or absence of these characters.

To create the conditions that will support a diverse soil food web, we want to start by having diverse food sources. That means diverse plants in our landscapes. In this case, diversity above begets diversity below because diverse plants means a diversity of critters are needed to break down the organic material produced.

Early in succession, ecosystems tend to have bacterially dominated soils. This occurs because bacteria are particularly well suited to breaking down nonwoody material. In the cultivated landscape, annual gardens and pastures emulate this successional stage.

One particularly important type of bacteria found in these early successional soils is symbiotic nitrogen-fixing bacteria. These bacteria have the incredible ability to take nitrogen from the air and turn it into a form usable by plants. They live on the roots of certain plants, many of which are in the legume family. The plants are able to get nitrogen from the bacteria living on their roots, while the bacteria receive some carbon compounds from the plant in exchange.

Later in succession, ecosystems tend to have more woody plant matter that needs to be broken down to allow nutrients to cycle. Fungi are specialists in breaking down complex lignin

Earthworms are the great aerators and organic material processers of the soil community, yet they are just one of the millions of microorganisms that live under your feet.

found in wood. Therefore, as ecosystems move farther along in succession, the soils tend to move from being bacterially dominated to being fungally dominated. In the cultivated landscape, these are the orchards, food forests, and woodlots.

One particularly important class of fungus is called mycorrhizal fungus. These fungi grow in webby strands underground and develop relationships with plants much like rhizobial bacteria. Some mycorrhizal fungi grow in close proximity to plant roots, while others actually grow inside the root tissues themselves. Luckily, this is another symbiotic relationship. The mycorrhizal fungi spread throughout the area and, in a way, connect it all together like a giant network. They can help move nutrients and water to where they are needed most in the ecosystem. Mycorrhizal fungi act like extensions of the roots of plants, transporting water or nutrients from far away to the plants that need them. In exchange, the fungi receive carbohydrates from the plants. Some mycorrhizal fungi also exude antibiotic substances that help to prevent plants from picking up soil-borne diseases. On sites where soils have been disturbed to the extreme (for example, through scraping or fungicide use), very few mycorrhizal fungi may be left. When planting into these desolate landscapes, you can use mycorrhizal inoculants, which come in the form of tablets, powders, and gels. Make sure you use the right kind of mycorrhizae for the crops you are planning to grow.

To support the soil food web, gear your management activities to support the successional situation you are trying to create. For your annual gardens and pastures, use compost and mulches made from nonwoody vegetation and manures to support the bacteria that naturally dominate there. For your orchards, forest gardens, and woodlots, use composts and mulches made from woody vegetation to support the fungi that naturally dominate there.

If you understand the baseline defined by your soil's physical properties, you can figure out where you might need to work to bring that soil to a state of productivity. The soil life is what will help you make that happen. In fact, most soil fertility strategies work with the living (or once-living) part of the soil to improve fertility, both in the immediate sense and for the long term. Therefore, understanding and managing the soil food web well will allow you to create thriving, productive landscapes even in challenging environments.

Contaminated soils

It's possible that your soils have contamination problems, ranging from petrochemicals to radioactive material. Some of the most common contaminants you might run into are heavy metals. For instance, the area around an old house may well be contaminated with lead, which was used as an ingredient in paints for many years. If a property has a history that includes orcharding, sugarcane production, or other crops, it could be contaminated with lead and arsenic, both of which were prominent ingredients in pesticides widely used for many years in commercial agriculture. In the Puget Sound region of Washington, some areas are contaminated with cadmium because of a smelter that operated for many years in Tacoma, belching out toxic materials that blanketed anyplace downwind.

Since testing can be fairly expensive, we recommend it only if you have reason to suspect contamination. In that case, the RCRA-8 test is worth the cost and will tell you the levels of eight different metals in your soil. You can compare these numbers to the background numbers for your area to see if anything looks out of sorts. Sometimes interpreting the test results can be difficult, in which case we recommend that you get help from your local extension service or department of ecology.

If you do find some sort of contamination, there are several ways to handle it:

- Replace the topsoil. This is pretty energy intensive and doesn't really solve the problem (where does that toxic soil go?), but it may be the best option in certain circumstances.
- Plant food crops above contaminated soils. Whether you use containers, boxed beds, or just put down a barrier between the contaminated soil and your imported garden soil, you can always choose to raise your crops above the toxins. If no food crops are actually rooted in the contaminated soil, you are less likely to come into contact with the metals through working the soil or eating the produce.
- Add lots of organic material. By increasing the organic content of your soil, you can increase the number of bonding sites available to hold onto heavy metals. If heavy metals are held by the organic matter in your soil, much less of them will make their way into the plants.
- Avoid root crops and leafy vegetables. When plants take up heavy metals, they tend to do so in their leaves. Roots are directly in contact with the soil and even the best washing may not remove all heavy metals. Crops that are not in contact with the soil are safer. Best of all are plants where you are eating the fruit or seeds. Heavy metals in a plant often won't transfer into the fruit and seeds.
- Use plants or fungi to extract heavy metals from your soils and concentrate them in their tissues, in a process known as phyto- or mycoremediation. These tissues can then be removed from the site and disposed of. Over time, you may be able to meaningfully affect the level of contamination on your site. You shouldn't eat the plants or mushrooms used in this process.

FERTILITY MANAGEMENT

Fertility management is among the most important things to think about as you develop a design for your landscape. If you aren't cycling and recycling nutrients within your system, you will need to bring in fertility from elsewhere. During the establishment period when your landscape is young and getting started, this will likely be the case. However, if you do a good job of managing your nutrients, you should need fewer and fewer outside inputs over time.

Balancing nutrients is another important aspect of fertility management. For instance, too much nitrogen can make plants grow too fast. When plants don't have time to harden off cell walls, they are more susceptible to disease and damage from cold weather. Conversely, lack of nutrients can mean lackluster growth, production, or nutrient content in food crops. If the minerals aren't in your soils, they won't be in the food produced by those soils. Therefore, paying close attention to nutrient balancing is critical to your health as well as plant growth.

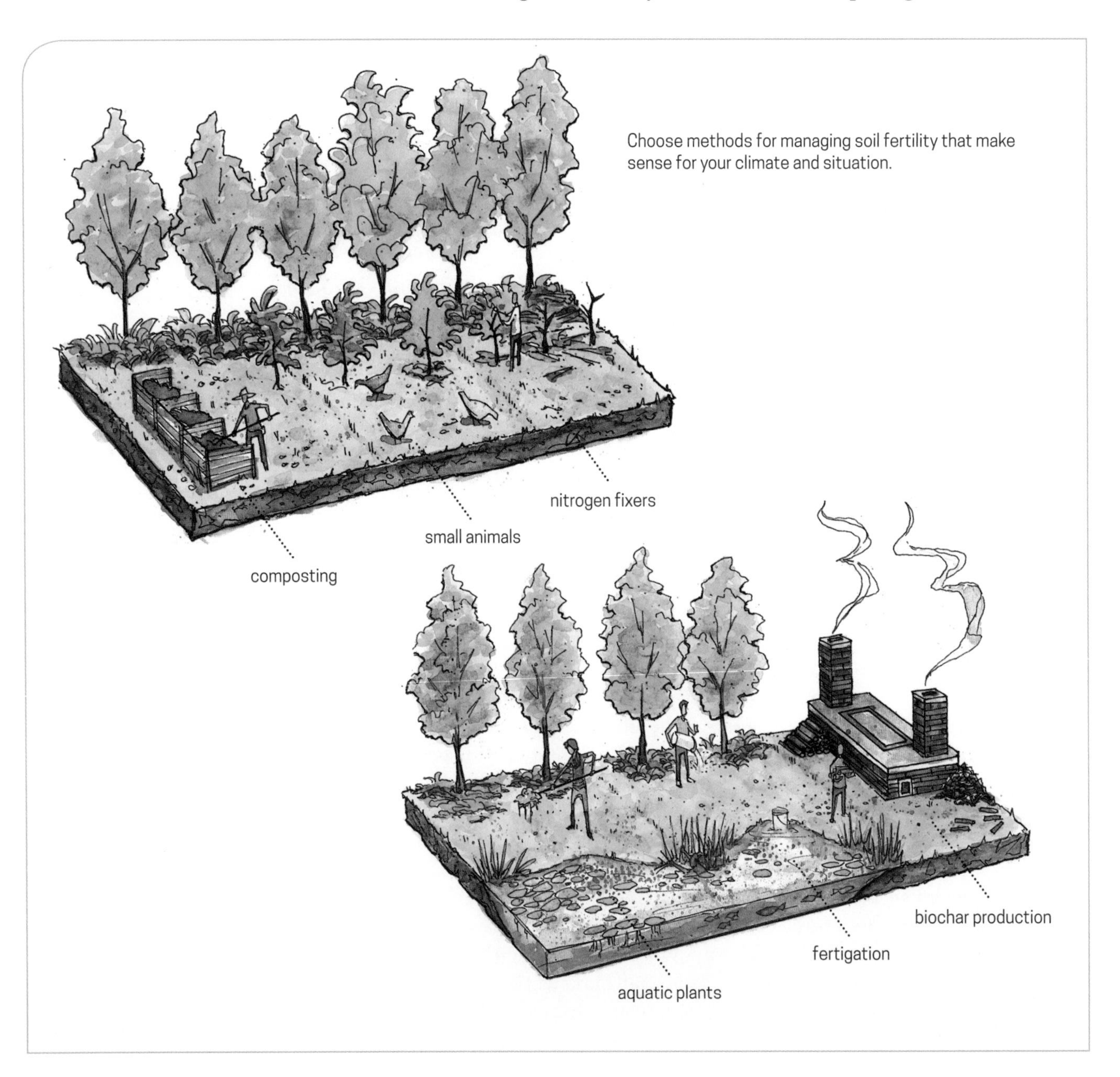

Choose methods for managing soil fertility that make sense for your climate and situation.

Composting

Composting is a great way to cycle organic material back into the landscape and take advantage of the nutrients tied up in food scraps, weeds, and brush. It can be done by anyone, even apartment dwellers with minimal space. Many books are available on the details of composting; here we offer several ways to take your organic wastes and roll them back into your soil.

Three-bin composting system. In the simple three-bin system, you set up three bins that are easy to access yet rodent-proof. When you have yard waste or food scraps, you throw them into the first bin. When the first is full, you use a pitchfork to turn the pile into the second bin and start filling the first again. When the first is full again, you move everything down the line and keep filling the first. By the time the first is full again, the third should have compost that is ready to use in your garden. This is a simple, fairly low-maintenance way to do compost, but it may not always get hot enough to kill weed seeds due to its piecemeal construction.

Worm bins. Worms are among the most valuable organisms on earth. Digesting organic matter in the top layers of the soil and creating a rich food for plants is an important job. We can get worms' rich manure by providing bins of bedding for them to live in and biomass to digest. Worm composting is often done in a two- or three-bin system with the compartments stacked on top of each other. The bins can be simple wooden boxes or even old Tupperware bins, or you can buy worm bins that can be easier to harvest from. Worms like a temperature between 55 and 75°F and are best kept inside during the winter if you live in a cold climate.

In its simplest form, a worm bin is a box with some holes in the bottom for drainage. This box is filled partially with some sort of moist bedding material (Dave likes to use black-and-white junk mail run through a paper shredder). Bury some food scraps below the bedding at one end of the bin and release some red wiggler worms. They'll turn your food scraps into beautiful vermicompost or worm castings, an excellent fertilizer for the garden. You can just keep adding more food scraps and bedding until the bin is full, then you can harvest your nutrients and start over. Don't forget to catch whatever drains out the bottom; this "worm tea" is an excellent garden amendment in its own right as it is very high in humic acids, an excellent way to build up your soil's nutrient-holding capacity.

Food waste digester. A food waste digester is a great, low-input way to deal with food scraps. You can make your own or buy one called a Green Cone. A Green Cone consists of a plastic basket that you bury in the ground and a big plastic cone that attaches to this and extends aboveground. You just throw your food scraps in the top and they slowly decompose. The underground basket allows all kinds of decomposers living in the soil to have access to the compost, but it keeps rodents and other pests out. You can just keep throwing food scraps into the Green Cone until it is full. Then it is best to have another one to switch to while the first cooks down. Once the second is full, you can harvest the rich compost from the first and use it in the garden. If you only throw food scraps from a small household in there, it can take quite a while for it to fill. That means minimal work.

(top left) This three-bin composting system fits nicely in an alley behind a small urban garden.

(top middle) Yuko is adding food scraps to the worm bin she made in her apartment. She uses shredded junk mail as bedding to keep her worms happy.

(top right) These Green Cone food waste digesters in a backyard in Houston, Texas, are used to process food scraps. When one is full, the other is used while the contents of the first decompose.

(bottom) At Bullock's Permaculture Homestead, residents make high-quality hot compost using a ring of wire mesh. This compost is used in annual garden beds and for aerobic compost tea.

Hot and cold composts. Hot composts are generally those that heat up a lot in the process of decomposition. Manures and food scraps help to make this happen. Hot compost piles are also generally made all at once. The benefit of hot compost is that those high temperatures will usually kill weed seeds, resulting in a better product to use in annual beds and areas for direct seeding. The disadvantage is that they are often more work to build and turn and they tend to lose more of their nitrogen to the atmosphere during the composting process.

Cold composts do not achieve high temperatures during the composting process. Brush piles and/or compost piles that you add to slowly over time are of this type. They are pretty low-maintenance but can take quite a while to break down. They are not terribly effective at killing weed seeds, but they retain more of their nitrogen.

From a design perspective, hot compost piles are best placed in zone 1 as they require more management and are more likely to be used in annual gardens. Cold compost can be placed in zones farther out as they may require almost no management, are more likely to be used around tree crops, and may be unsightly.

Small animals

Small animals can be allowed to forage through the landscape to meet their food needs while simultaneously providing a variety of services. Such services include controlling pests, controlling disease through cleaning up fallen fruit, and fertilizing through direct application of manure.

Allowing animals such as chickens, sheep, and pigs to forage through your perennial landscape means less work spreading fertility for you. In Ohio, permaculturist Chris Chmiel even uses goats and chickens to manage weeds and fertility in his pawpaw (*Asimina triloba*) patch. The elegance of this system is based on the understanding that goats don't like to eat pawpaws, but they eat almost everything else. The animals conveniently spread manure so that he doesn't have to apply any nitrogen fertilizer.

Depending on which animal species you choose to have on your site, they can be managed in a number of different ways to increase fertilization where you want it. Fencing or confined-range systems let the animals live as naturally as possible while you benefit from one of nature's best designs for soil fertility. For instance, a chicken tractor (a movable floorless pen that can be any shape or size) is one smart way to manage chickens to work on the soil in small areas such as raised beds or a backyard lawn.

Soil-building plants

If we think about the process of succession starting with bare ground, highly successful and adaptive pioneer species usually move in first and create better growing conditions for plants that will eventually take their place. These plants have some basic characteristics we can look for and use to create and maintain our own fertile ground.

Nitrogen fixers have bacteria living on their roots that can take nitrogen from the air and convert it to a form accessible to plants. This is a great attribute to look

Cytisus 'Zeelandia' is a nitrogen-fixing shrub that is helping this cardoon grow in poor soils. *Cytisus* can be opportunistic in some places, but this cultivar is a sterile hybrid that produces little, if any, viable seed.

for in plants where we have nitrogen-deficient soils or plants that require a lot of it to thrive. For example, beans, alders, acacias, and clovers all fix nitrogen. In nature, most nitrogen fixers are pioneer species adapted to living in an early successional environment.

Dynamic accumulators are especially good at growing roots that mine for nutrients and minerals and concentrate them within their tissues. Some plants are known for accumulating specific minerals that could be beneficial for many uses. For example, dandelions are dynamic accumulators of potassium, phosphorous, calcium, copper, and iron. These plants can be used in composting and to make specific fertilizers. These can also be used for our own consumption, whether that be as nutrient-dense food or medicine.

Biomass producers are especially good at creating large amounts of biomass. The foliage from these plants can be great for regenerating organic matter in soils or compost systems and also good for creating green mulch for easy application. Some of the most prolific biomass-producing plants include comfrey, miscanthus, bananas, sugarcane, and bamboo.

SOIL-BUILDING PLANTS

Achillea millefolium
yarrow
zones 3–9
accumulates phosphorus, potassium, copper

Brownea macrophylla
Panama flame tree
zones 11–13
nitrogen fixer

Cajanus cajan
pigeon pea
zones 9–13
nitrogen fixer

Elaeagnus multiflora
goumi
zones 5–8
nitrogen fixer

Elaeagnus umbellata
autumn olive
zones 4–8
nitrogen fixer

***Equisetum* spp.**
horsetail
zones vary
accumulates calcium, cobalt, iron, magnesium

Gliricidia sepium
madero negro
zones 10–13
nitrogen fixer

***Lupinus* spp.**
lupine
zones vary
nitrogen fixer
accumulates phosphorus

Medicago sativa
alfalfa
zones 3–8
nitrogen fixer
accumulates iron

Nasturtium officinale
watercress
zones 2–10
accumulates potassium, phosphorus, calcium, sulfur, iron, magnesium, sodium

Petroselinum crispum
parsley
zones 5–9

Schizolobium parahyba
Brazilian firetree
zone 10 and warmer
nitrogen fixer

Symphytum ×uplandicum
Russian comfrey
zones 3–9
accumulates potassium, phosphorus, calcium, copper, iron, magnesium

Trifolium pratense
red clover
zones 3–9
nitrogen fixer
accumulates phosphorus

Urtica dioica
stinging nettle
zones 3–10
accumulates potassium, calcium, sulfur, copper, iron, sodium

This aquaculture tank at a permaculture farm on the Big Island of Hawaii is being used to produce massive quantities of biomass, which is used as mulch in the landscape.

Aquatic plants

Many species of floating aquatic plants, such as duckweed, reproduce incredibly quickly. You can use that characteristic to your advantage by harvesting those plants and using them to mulch your crops. This means that you can use ponds to generate fertility even in extremely nutrient-poor conditions. In warm climates, it is possible to harvest and apply these aquatic mulches every couple months. In cold climates, you may just skim this material off once or twice a year. Always leave enough behind to repopulate your pond.

Fertigation

Fertigation means applying water-soluble liquid fertilizers to your crops via the irrigation system. It works great, as liquid amendments are immediately available to the plants. Usually, fertigation involves using a chemical fertilizer solution because liquid organic fertilizers tend to clog irrigation systems. However, doing this organically just involves designing an appropriate delivery system. If you have aquaculture ponds or tanks, you have a great source for nutrient-infused water on hand. You can just apply this directly to your plants via such means as buckets, sprinklers, or flood irrigation, and they will receive nutrient benefits that you won't get from watering with well or municipal water.

Biochar

Biochar is a form of charcoal. Special stoves are used to turn organic material (such as wood and agricultural wastes) into biochar. This process is called pyrolosis, and it involves burning everything in the organic material except the carbon under high-heat, low-oxygen conditions. Biochar itself is just the pure carbon left over after the process is finished. It contains no nutrients for plants. However, it has many, many bonding sites for nutrients to attach to it. Therefore, while the presence of biochar in your soils will not directly improve fertility, it will improve your soil's ability to hold nutrients. That can mean a lot in depleted soils. It also provides habitat for soil microbes in its pore spaces.

The best way to use biochar is to charge it with some sort of nutrients first. For instance, you may choose to soak your newly made biochar in a mixture of fish emulsion (hydrolysate) and liquid seaweed extract. This will charge it full of nutrients, which you can then apply to your landscape. In some poor soils if you just applied those liquid products directly to the soil, they would likely just wash away. So ultimately, biochar provides a way to condition soil so that the nutrients you apply aren't lost from the system, but instead are held for the long term so plants can access them as needed.

Compost tea

Compost tea is just what it sounds like, compost steeped in water to achieve specific goals. Compost tea can be aerobic or anaerobic; each has a different purpose. If you want to try this method of nutrient management, do in-depth research first.

Aerobic compost tea does not apply a significant amount of nutrients directly to your land. Instead, its purpose is to extract and multiply the population of beneficial organisms found in a compost pile and inoculate your land with them. Using compost tea helps to diversify the soil food web and make the positive impacts of compost go farther, and it may help prevent pests and diseases. One way to make aerobic compost tea involves placing high-quality compost into a filter bag and submerging it in a compost tea brewer (homemade or purchased). Next, fill the brewer up with clean (nonchlorinated or nonchloramined) water and begin to aerate it by using an appropriately sized aquarium pump to generate a roiling of air bubbles on the surface. Finish by adding a conservative amount of easily digestible food (flours, fish or seaweed extract, and such) for the microorganisms to feed on during their procreation and let it brew for several hours. From there, apply it directly to the soil or cover entire plants with the beneficial tea.

With anaerobic teas, you use a fermentation process to extract nutrients from organic material that can then be applied to the land. The fermentation process tends to stink a lot, so this is probably not the best idea for your suburban yard. In fact, there is some question about efficacy and possible dangers of anaerobic compost teas (such as pathogens like *E. coli* or salmonella).

(left) In the forest garden at Florida Gulf Coast University, sweet potatoes create an effective living mulch.

(right) At this residence in Edmonds, Washington, burlap coffee sacks are being used as a weed barrier at the beginning of a sheet mulching project. This is a common technique used by Jessi's company, NW Bloom Ecological Landscapes.

Mulch

Mulching—applying a layer of material to the surface of the soil—can make a big difference in the success of your landscape. This is because mulch

- conserves soil moisture by preventing the desiccating effects of sun and wind,
- helps with weed suppression,
- can reduce runoff by building up the spongy organic material in the soil,
- prevents raindrops from pounding the surface of the soil so hard and thus prevents erosion and compaction,
- insulates the soil surface, keeping your plant roots warmer in the winter and cooler in the summer, and
- can look attractive.

There are many different kinds of mulches, each with its own benefits. Nonliving mulches such as gravel, rock, and black plastic do a great job of weed suppression, but they don't add any organic material to the soil, so they are not often the best option. We prefer to use compost or mulches with more organic matter because they add nutrients to the soil. Dead mulches—plant material that was once alive, such as wood chips, compost, and rice hulls—require occasional reapplication and do best as weed suppressants when they are layered 2 to 6 inches deep. Living mulches are plants that cover the ground such as clover, vetch, or orchard grass. Depending on your goals and the crops in question, you may cut back living mulch crops occasionally to allow them to renew themselves.

Sheet mulching is a technique that can be used during establishment to essentially hit the ecological reset button. Let's say you want to convert your lawn into an annual garden. You can use a layer of biodegradable material such as cardboard or newspaper to smother the grass, while simultaneously enriching the soil by covering this barrier layer with organic material, such as wood chips or compost, to weigh it down. Once the vegetation underneath has died over time and the barrier has mostly broken down, you can plant in this rich new soil. In some cases you can also wait for a shorter period of time and perforate the barrier in places to plant directly. Just make sure you look carefully at what you are trying to smother as some weeds, such as morning glory and horsetail, can actually thrive under sheet mulch.

Cover cropping

Cover cropping is a practice for protecting your soil and helping to replenish its nutrients. It involves planting a living mulch that gets turned into the soil. We often call crops grown for this reason *green manures.* In temperate climates, crops are often harvested in the fall, leaving the soil bare. During the cold months, it's ideal to grow soil-building ground covers to prevent erosion and to feed the soil. In any climate, you can also grow cover crops anytime you don't have a vegetable crop on the land. This can help to break disease and pest cycles and recharge soils. If you garden year-round, one of the design challenges is to find space and time to do cover cropping. Fava beans and buckwheat are both interesting cover crop options because they do their job as cover crops and also provide food.

The basics of cover cropping are simple:

- Select and plant your cover crop at an appropriate time based on your climate and needs.
- Make sure the cover crop has proper care. This depends on the species selected, but some might need to be mowed or watered. For the most part, you should find that cover crops are very low maintenance.
- Have a plan to kill your cover crop before it goes to seed or the vegetative growth gets out of hand. Depending on the species, this could mean chopping it and letting it lie, tilling it under, or harvesting the seeds and letting it die on its own.

Common temperate cover crop plants are rye, legumes, buckwheat, oats, mustards, and sorghum. It's important to plant the right crop or crops for your situation. Each has different attributes such as frost and heat tolerance, nitrogen fixing,

In this New Hampshire garden, buckwheat is grown as a cover crop on the beds that aren't in active production. This protects the soil and prevents weeds from becoming established.

and/or weed-suppressing effects; and different cover crops are good for certain times of the year or climates. One thing to note is that rye and sorghum are known to suppress the growth and germination of other plants, which is great for weed control but not for growing seeds.

Cover cropping can be used in conjunction with fallowing, or letting your land rest. By planting a cover crop and letting part of your land sit for one or more seasons, you can break pest and disease life cycles and help to build soil organic material.

SOIL FERTILITY STARTING POINTS

	Easiest	More involved	Most involved
Recycle yard and kitchen waste.	Compost your food scraps.	Build a worm bin.	Make your own biochar.
Use dead or living mulch.	Mulch your garden. Collect biomass from neighbors.	Plant green manures (cover crops)	
Use fertigation or compost tea.	Make plant tea to fertilize your garden.	Spray aerated compost tea.	

Water

Making the Most of a Limited Resource

Water is a precious commodity that is vital to life. It is easily the most important natural resource we rely upon on a daily basis. Our lives are dependent on water for drinking, food production, transportation, and many other uses. But the supply of fresh potable water available on this planet is limited, and as the population grows, the supply is threatened by contamination via chemicals or other toxins and mismanagement or overuse. In many places, rainwater is our primary renewable resource, and by managing it wisely we can gain many advantages as well as support thriving ecosystems. In this chapter we explore how to capture this resource and put it to use in our lives.

We never know the worth of water until the well is dry.

—THOMAS FULLER

(above) This large galvanized cistern collects rainwater from the barn at Seattle's Picardo P-Patch community garden.

THINKING LIKE A WATERSHED

Understanding the hydrologic cycles of your landscape is critical to your permaculture design. Depending on climate and site specifics, water is something you may have too much of, not enough of, or both at different times. Because water is a beneficial and perhaps scarce resource, you should cycle and recycle it through your landscape as much as possible.

Here are some important questions to ask about water on your land:

- Where does rain fall and collect during a storm event?
- Do bodies of water nearby influence my site's hydrology?
- Where do I get my drinking water?
- Where does water go after it is used in my toilet and my sink?
- What activities occur above me in the watershed that may have effects on the water in question?
- How can I put water to work on my property in a way that makes it cleaner when it leaves?
- What is below me in the watershed that will be impacted by my actions?

HIERARCHY OF WATER USE

You can use water in many different ways in your permaculture design, and ideally you want to match the quality of the water to the use. Most of our municipal water systems use high-quality potable water for flushing toilets or watering the garden. Although many homes are plumbed that way, changes can be made. You may have other forms of water that are lower in quality that could do those jobs in a less wasteful way.

Start by thinking about the sources of water available on your land. These may include the following:

- rain and snow
- rivers, streams, and creeks
- ponds and lakes
- springs
- wells
- runoff from adjoining land
- municipal tap

Here is a generalized list of water uses in order based on the water quality required, from highest to lowest:

- drinking and bathing
- cleaning

This shallow well sits in a front yard in Shodoshima, Japan. Water is accessed via simple jack pump, which is a great nonelectric way to pump water.

- irrigation and agricultural use (including aquaculture)
- recreation
- fire fighting
- outdoor pond and lake systems
- infiltration into the ground
- wetlands

Looking at this list, you can decide the most appropriate use for the different water sources on your site. For instance, municipal tap water can be reserved for drinking and bathing. Water from a creek shouldn't be consumed without filtering, but it may be perfectly suitable for an irrigation system (assuming it isn't polluted). Sediment-loaded runoff that comes onto your property from a neighbor's fields isn't appropriate for either of these uses, but it may be fine to let it infiltrate into the ground via water-harvesting earthworks. After filtering through the soil, that water will likely make its way to a wetland. Ideally, before the water leaves your property you will have systems in place to ensure that it is as clean as, or cleaner than, when it entered your land.

For this small Arizona landscape, permaculturist Andrew Millison designed a water system to collect rainwater in a cistern that's ready for irrigation use. This cistern overflows into the landscape, which has been designed to retain as much of that water as possible so that it infiltrates into the ground.

TAKING ADVANTAGE OF AMBIENT WATER IN THE LANDSCAPE

For folks in arid climates, water conservation in the landscape is probably already on the radar. However, this applies to all of us regardless of climate. We need to do what we can to make use of the ambient water in the landscape so we can minimize the use of more expensive and energy-intensive municipal water. For many, this means catching water in rain barrels and water tanks. That's a great idea if you want water on hand for specific uses, but it usually isn't enough. If you live in an arid climate that averages 15 inches of rain per year and you have a 10,000-square-foot lot, more than 85,000 gallons of rainwater will hit the property. You would be hard pressed to put enough rain barrels or tanks on your property to catch that much water, so to make the most of it, you need other strategies.

You can store water in many ways, differing in cost, permanence, and capacity. When you think about how to store water, also consider how easy it will be to access for the use you have in mind; for instance, taking a shower with water from a cistern is easier than drawing it from the soil. Storing water in ponds is often the best choice on all counts. However, even if you have tanks and ponds, you also want to consider infiltration strategies to store water directly in the largest storage reservoir you have, the ground.

COMPARISON OF STRATEGIES FOR STORING WATER

tanks or cisterns	ponds	soil
more expensive per gallon stored		less expensive per gallon stored
less permanent		more permanent
less total rainfall capture		more total rainfall capture
easier to access for use		harder to access for use

Priorities for rainwater management

The order of priorities for managing rainwater in the landscape is as follows:

1. Slow the water down to diminish its erosive potential and drop out sediment.
2. Spread the water across the landscape so there is more surface area to absorb it.
3. Sink it into the soil and store what you need for later.

Allowing water to infiltrate into our landscape has a number of positive impacts:

- Overland flow and downstream flooding are minimized. Letting the water sink into the landscape slows it down enough that rivers and stormwater systems don't have to deal with it all at once.

- A healthy groundwater resource is maintained and the aquifer is recharged. The water table can actually be prevented from dropping in dry areas if enough people do it.
- Erosion is prevented. If we slow water, spread it out, and sink it in, we stop the water from flowing forcefully over the surface of the land and eroding it.

Developing the soil sponge

We've all seen how much water a sponge can hold compared to something like a piece of concrete. In our soils, organic material is what helps the ground to act like a sponge. Therefore, building up the organic content of your soil is one of the best strategies for increasing infiltration and helping the soil do a better job of harvesting water.

How much water can soils hold? Generally more than we think, although it varies from place to place and with different soil conditions. As a rule of thumb, each percentage point of organic material can absorb 6/10 inch of rain (during a single rain event). For instance, a native Iowa prairie with its rich topsoil (up to 10 percent organic material) can absorb up to 6 inches of rain in a single event; in other words, no overland flow or flooding occurs unless the single rain event is greater than 6 inches. By contrast, a conventional Iowa cornfield has less than 2 percent organic material and thus sees overland flow in rain events as small as 1.25 inches, resulting in flooding, erosion, and sedimentation.

Keyline design

Keyline design is a revolutionary system of managing water and soil on a site. Developed by the innovative Australian mining engineer and farmer P. A. Yeomans, it was a major source of inspiration for Bill Mollison and David Holmgren when they conceived the permaculture design system. Keyline design principles are applicable to almost any terrain but are best applied in broadscale landscapes with rolling hills. Applying keyline design principles to a landscape can help you to improve water infiltration over time and lay out a site that uses water effectively.

Design using keyline principles starts with identifying the major keypoint of your slope. The keypoint is the point where a hill goes from convex to concave. Above the keypoint is a zone characterized by erosion. Even in an intact ecosystem, some soil and organic matter naturally moves downhill. Below the keypoint is a zone of collection. This is where all that material from above ends up. Above the keyline, soils tend to be poorer; below the keyline, they tend to get richer over time.

The keyline is a contour line on the landscape that goes through the keypoint. Knowing where the keyline falls can help you design the rest of your landscape. Because land above the keyline tends to erode, you should maintain vegetation cover—preferably forest—on this part of the slope. Below the keyline is the most appropriate place for the bulk of your productive landscape. The flat land at the bottom can support annual agriculture without eroding. The steeper areas below the keyline can support orchards and other tree crops. The keyline itself is generally a good place for human settlement. It's high enough to avoid flooding and

(above) At New Forest Farm in Viola, Wisconsin, permaculturist Mark Shepard has used keyline design principles to maximize use of water on his site. Earthworks and plantings help to slow, spread, and sink water on his site, which helps maintain a healthy water table and provide for the needs of his crops.

KEYPOINT OF A SLOPE

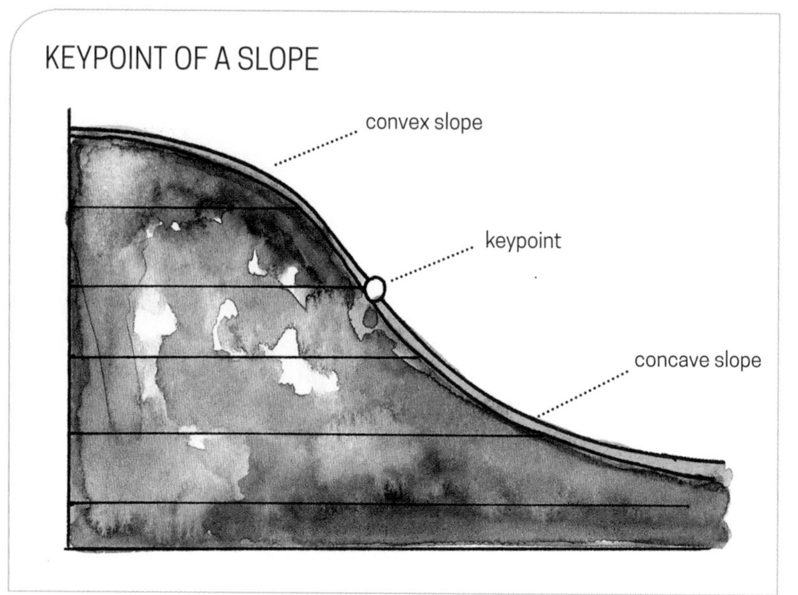

KEYLINE DESIGN CONCEPTS AND TERMS

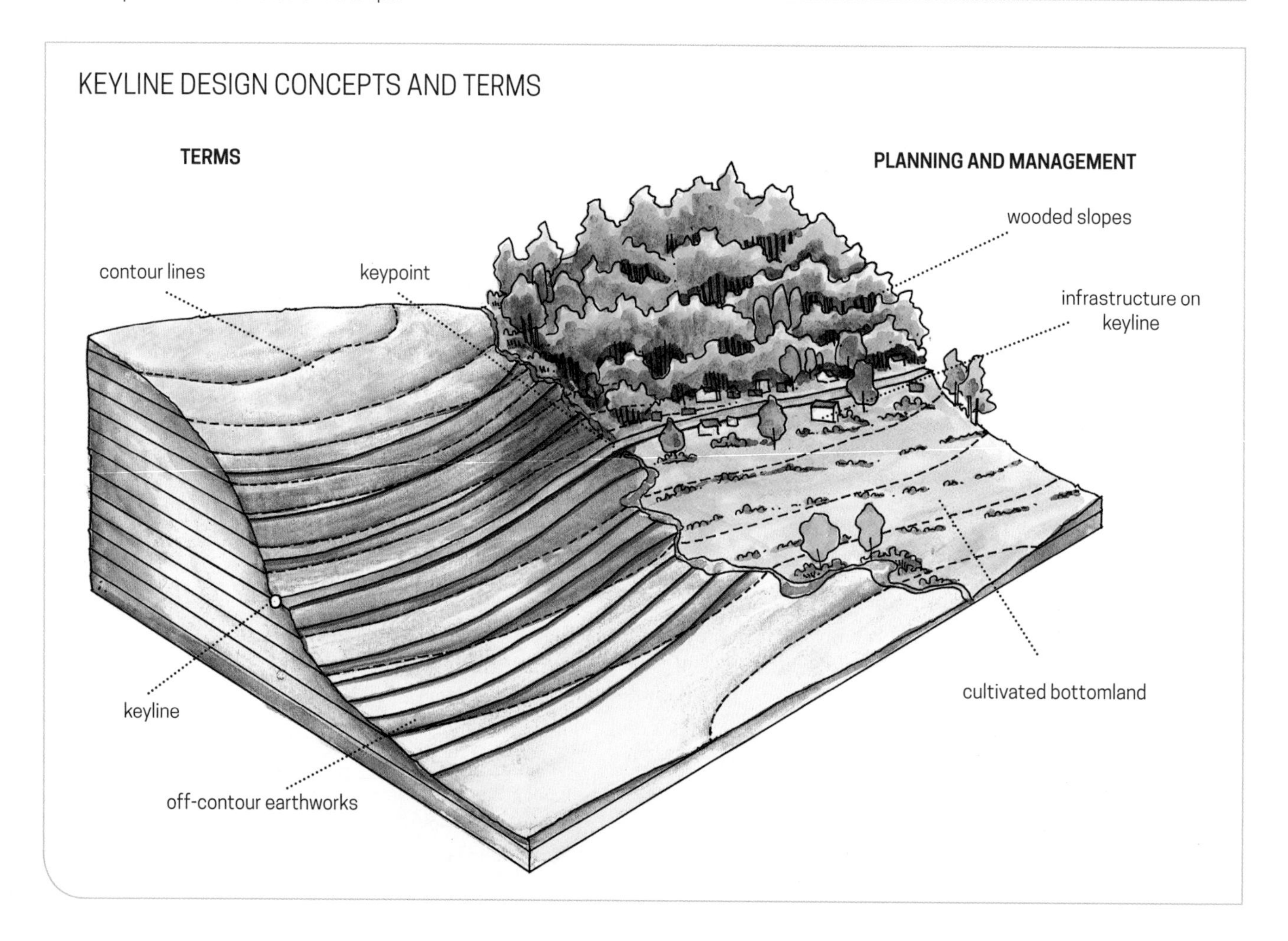

cold air drainage and low enough to avoid steep, erosive slopes and greater fire danger. The keyline is also a great place for access because one can easily access the upper, lower, and settled parts of the landscape from a relatively flat road or footpath. And the keyline is usually the highest point in the landscape where you can situate large water storage ponds without danger of instability.

The keyline design system also offers strategies to help spread water across agricultural landscapes. In the natural landscape, water tends to move toward valleys and away from ridges. If you use a tractor to rip lines parallel to your keyline, you will find that they eventually start to go off contour because landscapes have variation. (Ripping, unlike plowing, opens up furrows in the soil without turning it.) As water flows down toward the valleys, some of it will actually drop into the riplines and move in the other direction, toward the ridges instead. This is an excellent way to spread water out on your land. More spreading means more potential for infiltration.

At Occidental Arts and Ecology Center in Occidental, California, plantings are on contour to allow water to infiltrate and to prevent erosion.

Ripping lines down to the subsoil parallel to the keyline will improve infiltration and root penetration over time. A specialized ripping device called a keyline plow was invented by P. A. Yeomans for this purpose. Compacted soils can be hard or impossible to rip, but if you can rip just a 1-inch-deep line, the soil will begin to allow water infiltration and root penetration. As water moves down, some organic material will likely fall in as well. The next year you might find that you can go a little deeper because the soil has been softened by water and root development. Eventually, you should be able to rip as deep as your shank goes. The whole time, topsoil will be developing.

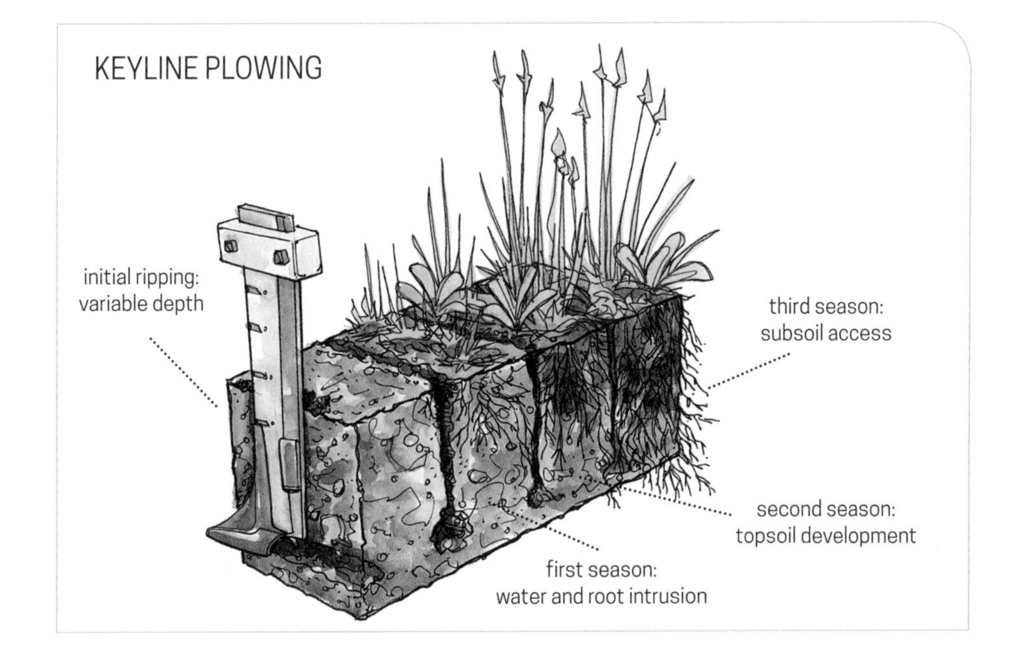

WATER-HARVESTING EARTHWORKS

In addition to building the soil sponge, which could take years in some situations, and using keyline design principles, you can also install earthworks for situations where you want to keep more water on your site. A variety of kinds of earthworks can help slow, spread, and sink water into your landscape. These are particularly useful for managing too little or too much water on your site.

For instance, if your soils are saturated and you want to dry an area out, you can put in a curtain drain. A curtain drain is a trench dug slightly off contour to capture water moving across the landscape and route it somewhere else. The trench usually has perforated pipe buried in the bottom and is backfilled with gravel. This can be helpful around foundations that are too wet.

Another way to deal with too much water is to install a dry well. If your site doesn't drain because of a compacted layer in the soil, you can dig a hole that penetrates the compacted layer. It can be filled with gravel or just capped with a grate. Runoff flows into the dry well where it can infiltrate below the compacted layer.

In arid climates with free-draining soils where rains tend to come infrequently and in large events, you can make the most of limited water by sinking garden beds into the ground. Water collects in the beds and is allowed to infiltrate, thoroughly soaking the soil. Similarly, you can use raised beds in wet climates to allow for better drainage. Raising garden beds above the rest of the landscape not only makes them drain better but also exposes more of their surface area to sun and wind, which also help to dry them.

Other earthworks strategies include infiltration swales and diversion drains, ponds, terraces, and rain gardens.

Infiltration swales and diversion drains

Infiltration swales are long, meandering basins placed on contour lines, thus making them dead level. When rainfall causes water to flow across the surface of the land, or just under the surface, the infiltration swale captures that water so that it can slowly percolate into the ground over time. This percolation can help to create a freshwater lens below the soil surface (assuming it doesn't hit an impervious layer, like bedrock, first). Think of the freshwater lens as a giant bubble of water suspended above the water table that can be utilized by vegetation during dry periods.

By including infiltration swales in a sloped landscape, you will cause the amount of water that percolates into the soil and recharges aquifers to increase. This will lead to decreased irrigation needs and, over time, a healthy water table. Planting bunchgrasses or other dense vegetation above the swale will help to filter out sediment before it gets to the infiltration swale. Planting trees or other crops on the mound (berm) on the downhill side of the swale (or just below it) will allow them to take advantage of that soil moisture during dry periods.

Infiltration swales are appropriate earthworks for gentle to moderate slopes. They're a poor option on steep slopes, where they can be more likely to blow out in a heavy rain. In those locations, you can try creating berms out of brush laid out

EARTHWORKS FOR WATER MANAGEMENT

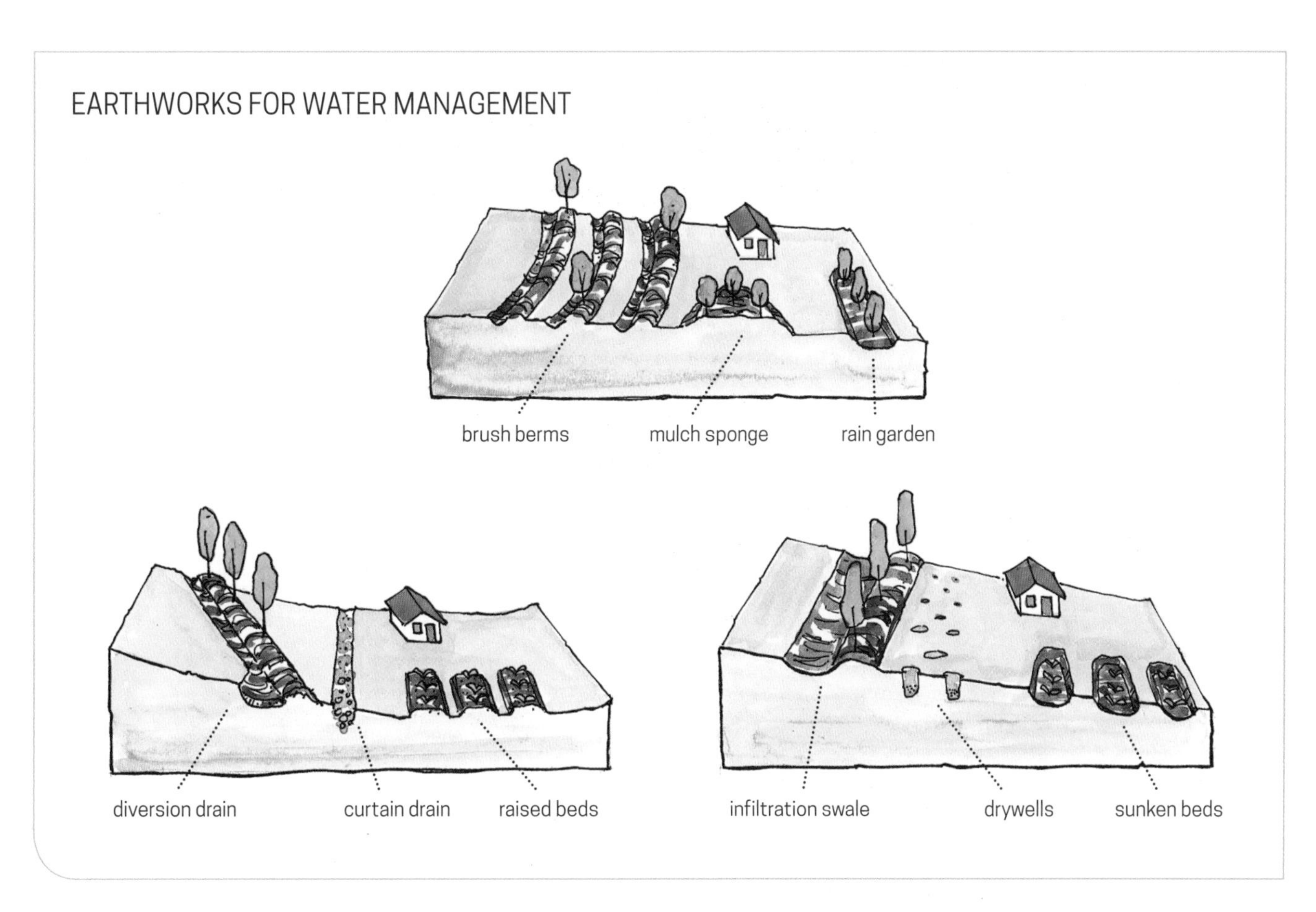

AN INFILTRATION SWALE SYSTEM

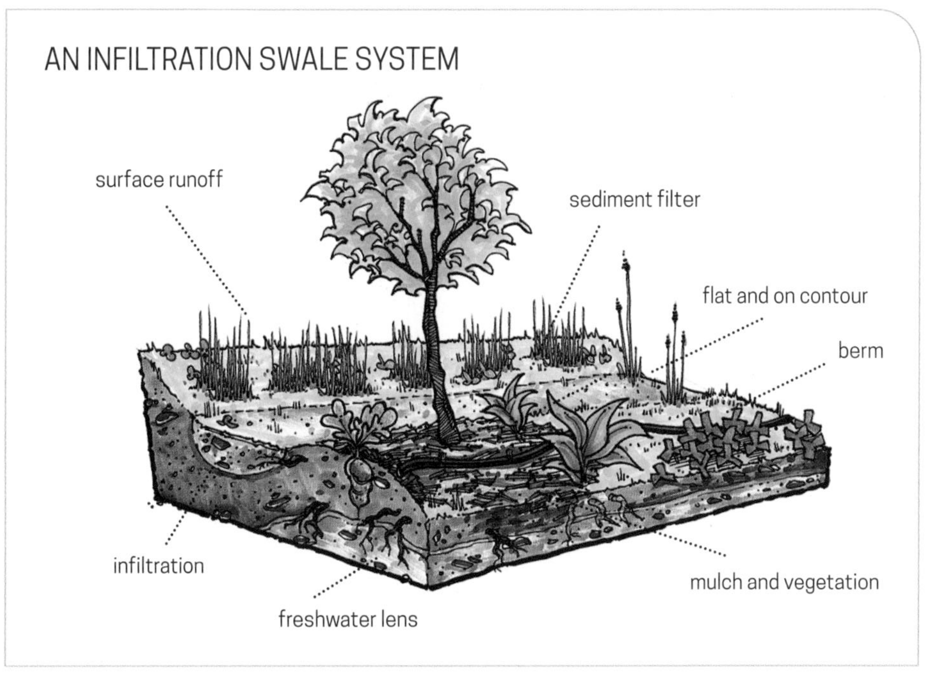

At the Costa Rica Center for Natural Living, many slopes were planted with vetiver grass (*Chrysopogon zizanioides*) on contour to prevent erosion.

and staked on contour. These will slow down water and help with some infiltration without disturbing the soil or collecting large amounts of water. Also, all swales should have an overflow strategy designed from the beginning. They can either sheet over the top evenly when full (ideally to be caught by the next infiltration swale downhill) or be connected to other water-harvesting earthworks via diversion drains.

A diversion drain is a trough in the soil that isn't dead level but takes water from one place to another. These are placed in the landscape off contour with the goal of moving water to a specific location (such as a wetland, a pond, a tank, or a creek). They need to have a gentle enough slope that a large rain event won't cause them to erode. A good rule of thumb is to have them drop 1 foot for every 200 to 300 feet of distance or more, but check with an engineer if you are unsure.

Ponds

Ponds can create an incredibly beautiful, functional addition to a landscape. They can store water for irrigation or household use (often the cheapest accessible storage), serve as wildlife habitat, and offer essential recreation in hot climates. If well placed in the landscape, ponds can also provide opportunities for aquaculture and/or crucial protection from fire.

Elevation and topography are major considerations when you are deciding where to place a pond. A pond high in the landscape can feed via gravity to any lower point on the property. Gravity-fed water systems are always the most resilient as they don't rely on electricity or pumps that can fail. Chains of ponds cascading down slopes (in existing draws if you're mimicking nature) connected via diversion drains or spillways can provide water storage that also looks quite natural.

It's important to think about how these ponds will get filled, however. If there isn't a reasonable water source high on the slope, ponds can also be filled via pumping from a location downhill. If the ponds will be harvesting overland flow, each pond moving downhill should be larger than the last to accommodate a larger potential volume of water. At the same time, it's critical not to place large ponds on steep slopes as the danger of dam blowouts during major rain events is much higher. You should always have an idea of how much water your ponds will need to accommodate (based on size of the drainage area and largest expected rainfall events) before you size them.

Here are some cardinal rules of pond construction:

- Every pond must have an overflow that goes to a known location.
- Keep woody vegetation away from pond dams so the roots don't penetrate them.
- In dry climates, try to minimize surface area and shade the water with trees to reduce evaporation.
- When you are constructing a large pond, a complex chain of ponds, or anything that makes you uncomfortable, involve an engineer and/or experienced heavy equipment operator early on.

Ponds generally love clay soils. One of the benefits of a clay-lined pond is easier establishment of an ecosystem due to the natural bottom. If your soils aren't clay or they just don't hold water (for instance, if a clay layer sits on top of a vein of sand), you have a couple of choices. You can try to line the pond with clay brought from elsewhere or else use a synthetic liner made of EPDM rubber or polyethylene. These materials do a great job of holding water, but they have a few pitfalls such as susceptibility to puncture by animals or people and a tendency to get slippery from algal growth.

Regardless of how the pond is sealed, you will want to do some terraforming to create flat benches at different depths. Not only will this provide appropriate habitat for a wider range of plants and animals, but it will also make it easier to get out if you go for a swim. Flat benches in liner ponds also provide places to set potted plants such as water lilies, lotus, and wapato.

Terraces

If you lack flat land for access roads, structures, human spaces, or agriculture, terraces may be a good option. From a water perspective, terraces prevent erosion as they slow the flow of surface water, and they function to some extent as infiltration swales as they allow some water to slowly infiltrate into the soil. Creating small terraces is pretty straightforward, but terracing large landscapes can be a huge amount of work. And terracing can cause severe land damage if it is not done properly, so it's best to proceed with caution with this form of earthworks and consult an engineer if in doubt.

Terraces can be built with or without retaining walls. As a rule of thumb, if the slope is under 25 percent it may not need a wall constructed. If it's between 25 and 45 percent, retaining walls are recommended. If the slope is above 45 percent, it's a good idea to plant dense vegetation to hold the terrace in place, and to consult an engineer.

Rain gardens

A rain garden is a shallow depression in a landscape that collects water from impervious surfaces nearby. The depression allows stormwater to infiltrate over time, much like a swale. It is filled with a special bioretention soil mix, which has both organic and mineral content. Rain gardens contain plants that are suited to the conditions of being inundated during large rain events and possibly dry for long periods, depending on the region's climate. These design elements are often included near parking lots, parking strips, and small yards where water from impervious surfaces is directed to them. In urban areas they decrease the amount of water flowing into municipal stormwater systems. Instead this water is used to replenish groundwater and minimize irrigation needs. To design a rain garden, you must determine the square footage of collection surface and compare it to the soil's percolation rates to calculate the area needed for stormwater storage.

MANAGING WETLANDS

If the property you're dealing with contains wetlands, you have both enormous opportunities and responsibilities. In terms of pure biomass, wetlands are the most productive ecosystems on earth. They serve as the kidneys of the earth, cleaning the water that passes through them. At the same time, wetlands can be severely damaged by human activities. In many parts of the world, regulatory agencies have placed extreme restrictions on what one can and cannot do in wetlands, with good reason. However, as with most legislation created as a response to damage from human activity, wetland rules seldom leave room to integrate productive use that causes no harm (or in some cases actually restores wetlands). Therefore, be aware that some of the concepts presented in this section may not be legal where you live.

(top left) This pond at Red Rabbit Farm on Orcas Island, Washington, is a source of water for a variety of purposes including fire protection, recreation, and wildlife habitat.

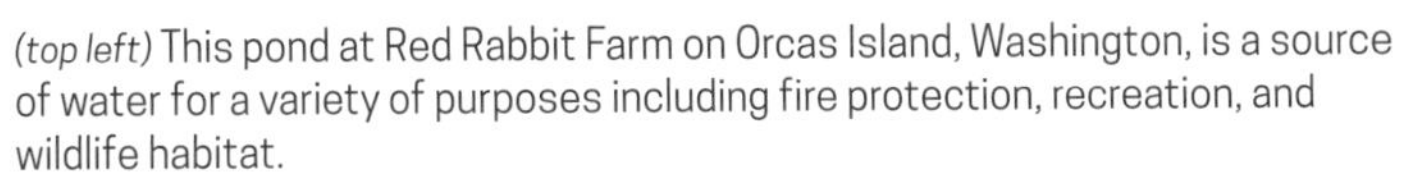

(top right) Permaculturist Sam Bullock operates a small excavator while creating a terrace on his land.

(bottom right) This rain garden at Jessi's house accepts the overflow from a rain barrel that collects water for irrigation. It infiltrates almost 15,000 gallons of water per year.

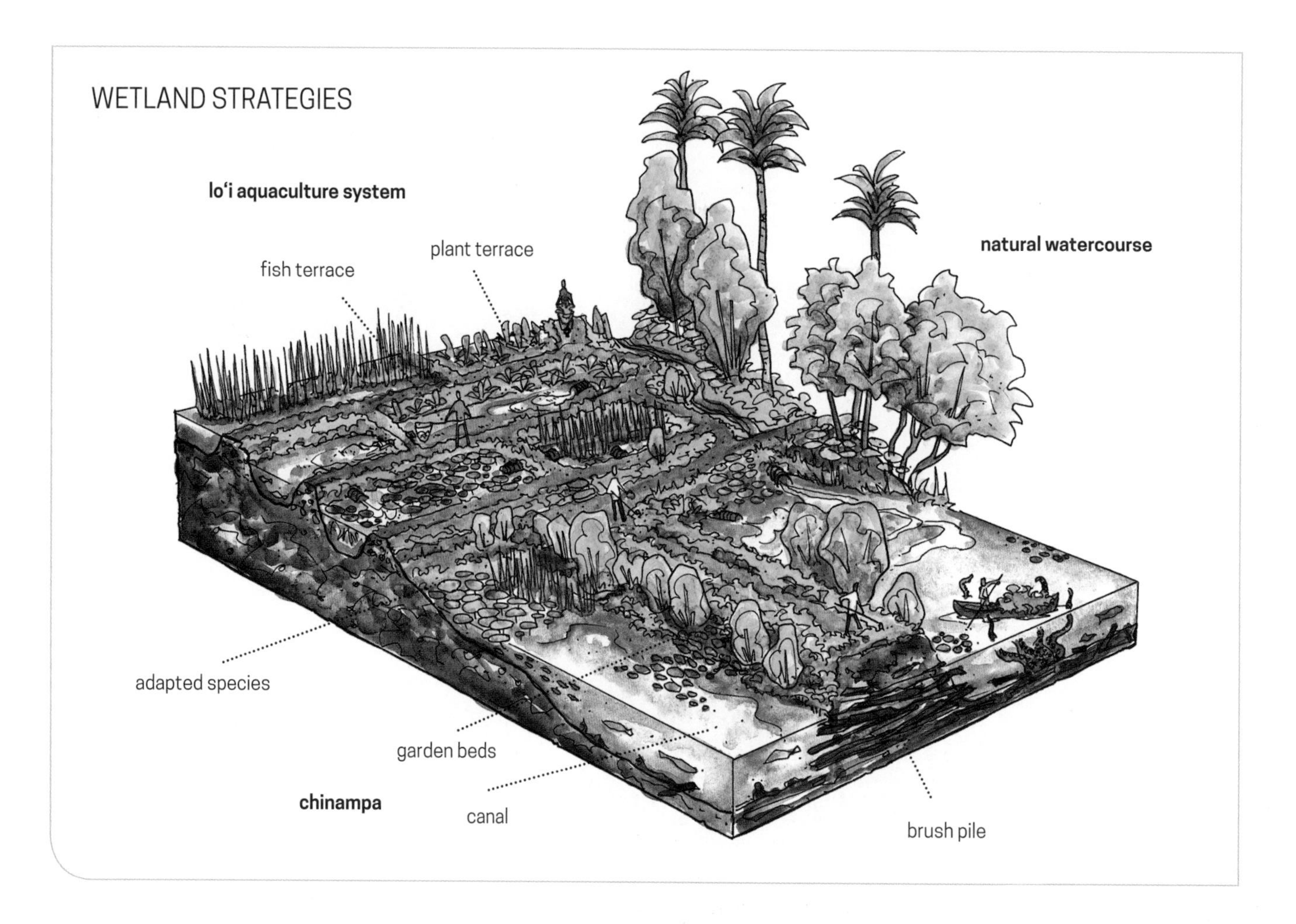

When dealing with wetlands, we must fall back on the precautionary principle: Do no harm. If the activity proposed in your wetland would lead to degradation or decrease in ecological function, you need to rethink your plan. These are a few indicators of wetland health to be aware of:

- healthy population of amphibians
- diverse population of macroinvertebrates
- good water quality as measured with a simple lab test
- lack of silt and sedimentation eroding from the nearby landscape

If the wetland on your property is already damaged in some way, it is important to restore ecological function as a priority. This may mean locating sources of sedimentation and pollution and addressing them. In both cases, vegetating wetland edges can help. Vegetation can catch soil runoff from nearby agricultural fields and eliminate surplus nutrient inflows through the use of plants that absorb the nutrients. These vegetated strips can contain productive plants for both people and wildlife. Once pollution and sediments are under control, you can reintroduce native plants, animals, and macroinvertebrates to help kickstart the return to ecological function.

Assuming you're willing to ensure wetland health as a top priority, you can apply some strategies to wetland areas to make them even more productive. These include adapted crops, chinampa agriculture, and lo'i and paddy systems.

Adapted crops

Many crops are adapted to wetland conditions. Typically these crops can withstand inundation (seasonal or permanent) or be completely aquatic. By choosing crops adapted to wetland conditions, you may be able to derive production from your wetland areas without making massive changes.

PLANTS FOR SATURATED SOILS

Aronia melanocarpa
aronia or black chokeberry
zones 3–9

Asimina triloba
pawpaw
zones 5b–7

Borassus flabellifer
toddy or Palmyra palm
zone 10 and warmer

Colocasia esculenta
taro
zones 10–13

Cornus sericea
red twig dogwood
zones 3–8

Cydonia oblonga
common quince
zones 5–9

Durio zibethinus
durian
zone 12 and warmer

Eleocharis dulcis
Chinese water chestnut
zone 9 and warmer

Nasturtium officinale
watercress
zones 2–10

Nelumbo nucifera
water lotus
zones 9–12

Sagittaria spp.
wapato
zones vary

Salacca edulis
salak
zone 10 and warmer

Salix spp.
willow
zones vary

At this permaculture farm, human-made chinampa islands and peninsulas allow for productive use of the wetland without damaging the ecology.

Chinampa agriculture

Chinampa agriculture was developed by Aztec people in lowland parts of Mexico who modified wetlands to create opportunities for production. These people would pile brush in shallow wetlands to create a matrix of coarse woody material. They would then dig rich, fertile swamp muck from around the brush pile and heap it on top, resulting in areas that were about a foot above the level of the water, encircled by canals. The people would then plant wetland tree species at the edges so that their roots would hold the newly formed "island" together. On these chinampas, as they were called, Mesoamericans were able to produce vast amounts of food with little to no irrigation, since water wicked up from the wetland. When time for harvest came, the chinampas were accessible by canoe, which eased transportation of produce to markets in the great cities.

Small-scale chinampa agriculture is still feasible today. Creating small chinampa peninsulas and islands in wetlands provides opportunities to increase edge for species that like to have access to water year-round. Good fruiting plants for these conditions include pear, quince, aronia, and hawthorn. This is also a way to produce lots of willow for basketry or wattle-and-daub construction.

Construction of these terraced rice paddies on the island of Shodoshima, Japan, began in the 1400s. For hundreds of years they have allowed people in this valley to meet their needs for food production while protecting the slopes from erosion.

Lo'i and paddy systems

Many cultures in Asia and the South Pacific used wetlands or even went to great lengths to create them for the purpose of cultivating crops. For instance, the lo'i system used by Polynesian people involved creating vast terraces cascading down valleys. Over time, these terraces would fill in with silt and water, creating the perfect conditions for the cultivation of wetland taro. Water would flow from terrace to terrace via gravity. These systems also managed both erosion and flooding.

Imagine the Hawaiian lo'i system with alternating terraces of fish and plants. The manure-rich water from the fish terraces would help to fertilize the plant terraces, while the plant terraces would clean the water for the fish terraces. If you incorporate a lo'i system, always end it with a series of plant terraces to ensure that water returning to the environment is clean.

In countries such as Japan and China, rice paddies have been maintained on slopes for thousands of years. These paddies have been around for so long that much wildlife has adapted to paddy conditions and actually relies upon the human cultivation system to meet their needs. These systems offer an amazing opportunity for the integration of production systems and ecosystem.

These photos show an area at the Chaikuni Institute in the Peruvian Amazon before, during, and after construction and planting of a series of lo'is. The area was converted from pioneer vegetation to a thriving food system in less than nine months. This lo'i is supplying huge volumes of taro and other food species to support residents, visitors, and local people.

This tank on the corner of Jessi's barn stores water for irrigation. A valve allows selection of where the overflow will go: either to a trough for animals or to a nearby pond (during the rainy season).

RAINWATER CATCHMENT AND STORAGE SYSTEMS

Rainfall is a wonderful resource if you have the infrastructure to capture it for use. In urban and suburban areas, gutters, downspouts, and storm drains carry rain away as quickly as it falls. To avoid wasting this precious resource, a key part of a permaculture design is a rainwater catchment and storage system.

In the broadest sense, such systems are composed of a number of parts, raising questions to think about when designing:

- Sources: Where does your water come from?
- Collection: Where is the water in question being collected from—a roof, driveway, city street, spring box?
- Distribution: Once the water is collected, how is it transported through the system—pipes, hoses, trenches, pumps?
- Storage: After collection, where is the water stored—barrels, tanks, ponds, cisterns?
- Overflow: When any given storage method in the system reaches capacity, where does the excess water go?
- Points of use: Where is the stored water used—sinks, spigots?
- Cleanup: How is the spent water cleaned after use?
- Recycling or return to earth: How does the water get looped back into another part of your system or returned to the earth?

For any water system design (using rainwater or not), you need to consider all of these parts. If any part of the system is missing, it won't work.

Collecting potable water from a roof

If you're collecting rain from a roof for potable use, it is especially important to consider the roofing material, how to divert the first flush, and filtration. Testing water is the best way to make sure what you're drinking is up to par.

Roofing material. Asphalt shingles are not an ideal surface for potable rainwater catchment as they are likely to exude toxic substances throughout their life to some degree. Metal roofs are among the cleanest. Painted or sealed metals are the best; galvanized metal can potentially leach toxins as the water flows over them. Wooden roof materials or thatch can be a good option, as long as they are not

(left) This home in Arizona has some of the essentials for any rainwater harvesting system: a metal roof, a first flush diverter, and an attractive storage tank.

(right) This rainwater catchment system at the Regenerative Design Institute in Bolinas, California, has a simple first flush diverter. When rain starts, the vertical pipe fills with water. Once the pipe is full with the first flush, the water begins going into the tank. A hole at the bottom of the vertical pipe lets it drain slowly, and a cap at the bottom allows cleaning out after the rain event.

treated with moss killers or other herbicidal concoctions. Slate and clay tile are both great materials for collecting rainwater.

First flush diverter. Debris such as leaf litter, bird droppings, dust, and more accumulates on surfaces your water is collected from. At the beginning of a rain event, all of that debris gets washed into your system unless you use a first flush diverter to send that first dose of water somewhere else before it goes into storage. This simple device for roof and gutter systems can take many forms; some commercial products as well as easy DIY options are available. First flush diverters can be made with spare pipe angled as a T before the water is drained to the collection unit. A ping-pong ball or similar float is used to fill the chamber and seal it so the dirty water is capped off and the cleaner water can flow over the ball and into the main collection unit.

Filtration. If you are collecting water for potable use, filtration is especially important. The wide variety of options for bringing water up to potable standards include ceramic filters, reverse osmosis, and UV light. You can also make your own slow sand filter, a low-tech biological method of water filtration, as long as you design it with care.

The slow sand filter is basically a container filled with sand ranging in size from a 5-gallon bucket to a massive community-scale system. The larger the system, the more water can move through it in a given time period. The inflow water falls onto the surface of the sand, where a layer of microorganisms that forms will do a lot of the filtering work. As the water sinks, it also gets filtered by layers of sand. Once it reaches the bottom, it can flow into the outflow tube and exit the filter clean. It

Permaculturist Penny Livingston-Stark shows the slow sand filter in her rainwater catchment system at her home in Bolinas, California.

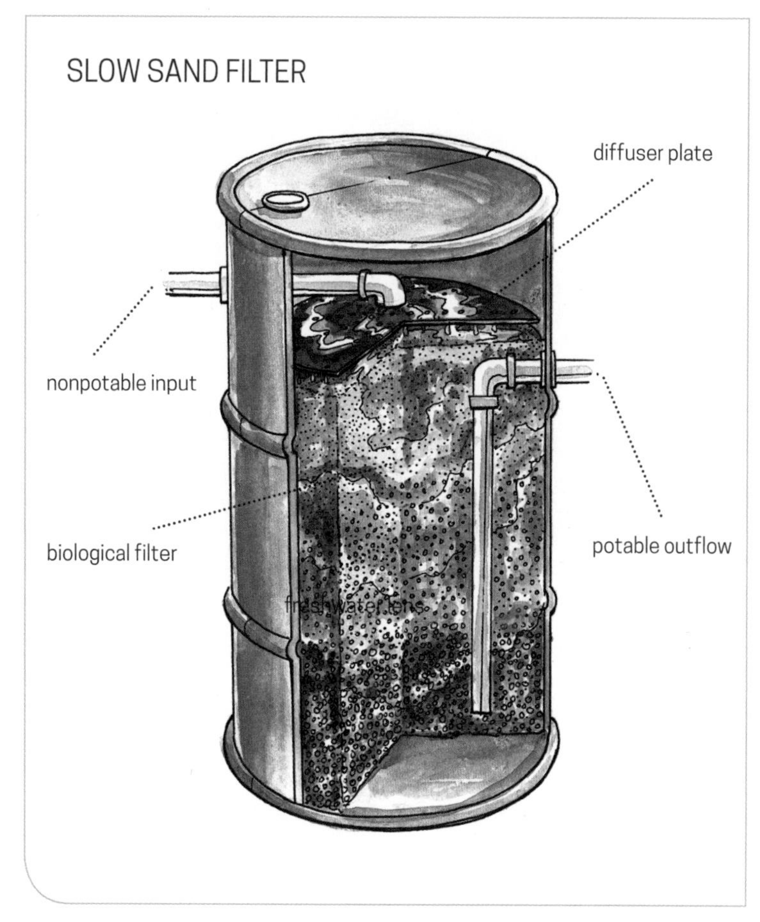

is crucial that the outflow pipe is lower than the inflow. This basic system should eliminate 99 percent of the biological agents that would cause you harm. You can also hire specialists to help figure out filtration for situations with lots of people or more complex systems.

Storage options

When it comes to storing water, options include barrels, tanks, cisterns, and ponds. All these come in different materials and sizes. Sometimes available space or budget for storage units will help narrow down your selection.

As you consider how much storage space you need, just remember that few people ever complain about having too much water storage capacity. Answering these questions will help you determine the size of the system you need for your goals:

- How much water can you possibly catch? You can figure it out by using this simple equation: (average annual rainfall in inches ÷ 12) × (square feet of collection area × 7.48) = gallons per year. For instance, if you lived somewhere with 50 inches of rain per year and you had a roof of 100 square feet, you could catch up to 3,100 gallons of rain in an average year.

* How much water do you use? You'll have to do some estimating to figure this out, and you might want to separate potable use from irrigation.
* What is the longest period you are likely to go without rain? In other words, what is the longest period you need to get by without your rainwater tanks getting recharged?
* What are your goals? Do you want to completely disconnect from municipal water? Do you just want to use rainwater to flush your toilet?

Here are some storage options:

* Plastic storage tanks are common and easy to find in a variety of sizes and colors. It is important to choose dark colors to keep out light and prevent algae growth.
* Metal cisterns are often cylindrical and more expensive but last longer and can be more attractive than plastics.
* Ferrocement, which is a mixture of Portland cement plastered over woven metal mesh, is a great material for larger systems but can be more labor intensive to build.
* Slim cisterns are becoming more popular in urban areas where space is limited. They can be much more expensive than agricultural storage units but can double as a privacy screen or wall in many cases.

In addition to these constructed storage options, it is also possible to store the water you've harvested directly in the landscape through the use of infiltration strategies and mulching.

Rain barrel systems

Depending on your site's water budget, it's always a good idea to collect as much water as possible for resilience. Before designing a complex and costly water system, though, it's a good idea to start small. For anyone getting started, we recommend a simple rain barrel system as a first step upon which you can build.

In the rain barrel system illustrated here, the downspout from a rain gutter empties into a clean, new garbage can with a coarse filter on top to catch debris. This coarse filter, which should be easy to clean, is often window screen pulled tightly over the opening. Two garbage cans are plumbed together for additional storage. It is important to have that connection watertight and as low as the containers allow so the water fills both tanks at once. You can chain together as many cans as you'd like in order to expand your capacity. The spigot can be on one or both tanks and should be low on the tanks. Without a pump and additional filtration,

(top left) This painted plastic tank at a school in Portland, Oregon, supplies water for the school garden and serves as a demonstration of rainwater harvesting.

(top middle) If you're short on space, consider connecting and stacking smaller tanks vertically as done at the Planet Repair Institute in Portland, Oregon.

(top right) When you don't have room for a huge water tank or cistern, you can connect several smaller ones together for the same effect. In this case these water tanks fit nicely into a small space between a building and a sidewalk.

(left) This beautiful ferrocement tank at Inanltah, Isla de Ometepe, Nicaragua, has an infinity edge. It provides irrigation water for food production systems as well as a great place to cool off on a hot day.

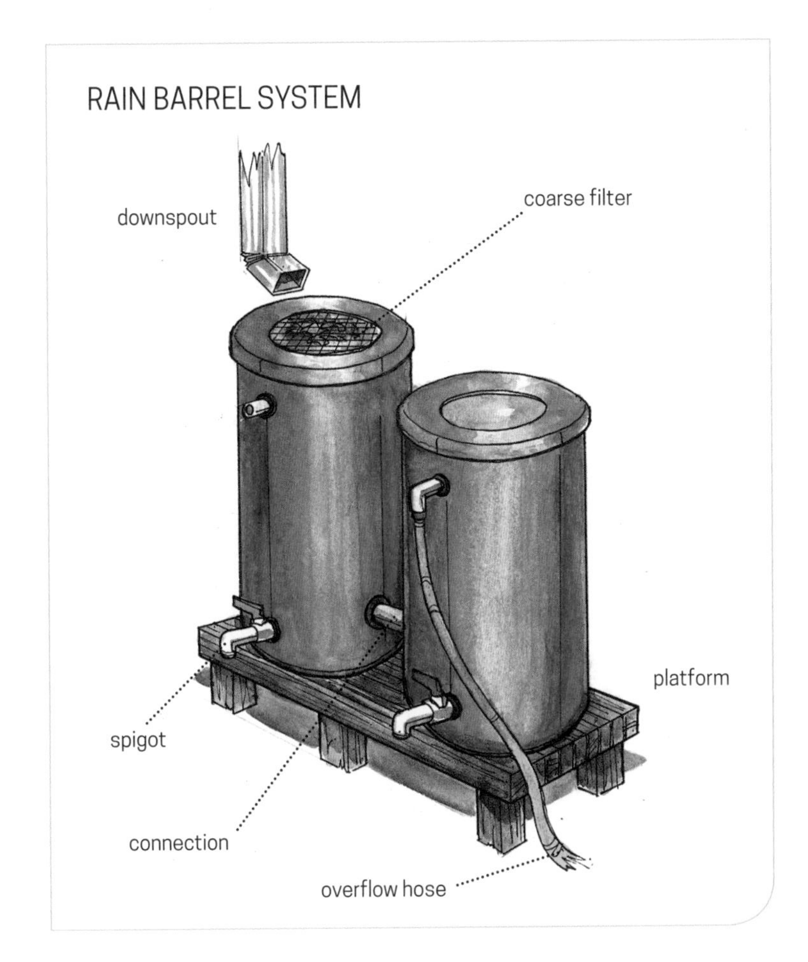

A simple rain barrel is a good starting point for collecting rainwater. Here, a barrel painted to match the house was plumbed to overflow into a salvaged water bed bladder under the deck for more storage.

the water from these tanks will most likely be used for irrigation or other outdoor use. However, the water can be purified in case of an emergency as well.

The garbage cans are placed on a platform so that the additional gravity from being at a slightly higher elevation will add pressure to the outflow. The garbage cans can also be elevated on blocks, pallets, or another man-made structure; just keep in mind the weight of water once these are full. The overflow is located at the top of one of the tanks. It is important to have the destination of the overflow water determined ahead of time so the water has an appropriate place to go. This could be a rain garden or infiltration swale, or the overflow could be directed back into whatever stormwater management system is in place.

Integrated gravity-feed water systems

A step up from the rain barrel system is the fully integrated gravity-feed water system, which is the most efficient and resilient design for a water system. With each of its components at the correct elevation, a system like this uses gravity to move water to where it's used. This kind of system works well for collecting potable water and using the surplus to good effect. The system is elegant in that it can take multiple sources of water and sort it into storage tanks for different uses. If gravity is an issue, not all the components of the system need to be under one roof. You can collect water off of a building higher in the landscape and use that water to supply a building down below.

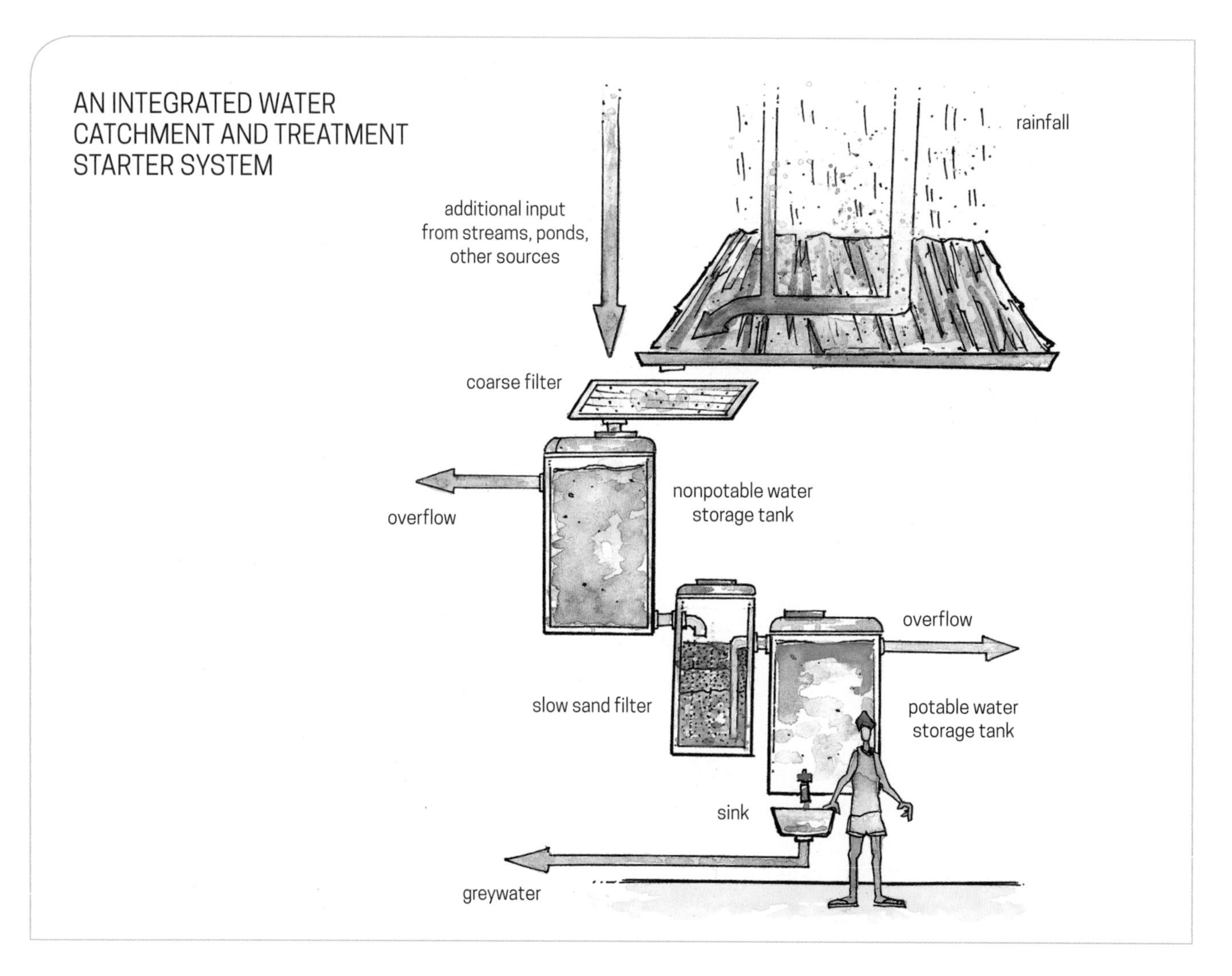

In a starter system of this kind, rain falls into a coarse filter, then another kind of first flush system (not shown in the illustration). Then it is stored in the first tank, which holds nonpotable water. Water in this tank can be used for irrigation, infiltration, and such. From here the water is filtered through a slow sand filter (sized for your needs), which cleans the water for potable use. The discharge from a hand sink fed by the potable water storage tank can then be used in the landscape as greywater. Any storage component needs to have an overflow planned that is lower in elevation than the inflow of the next component.

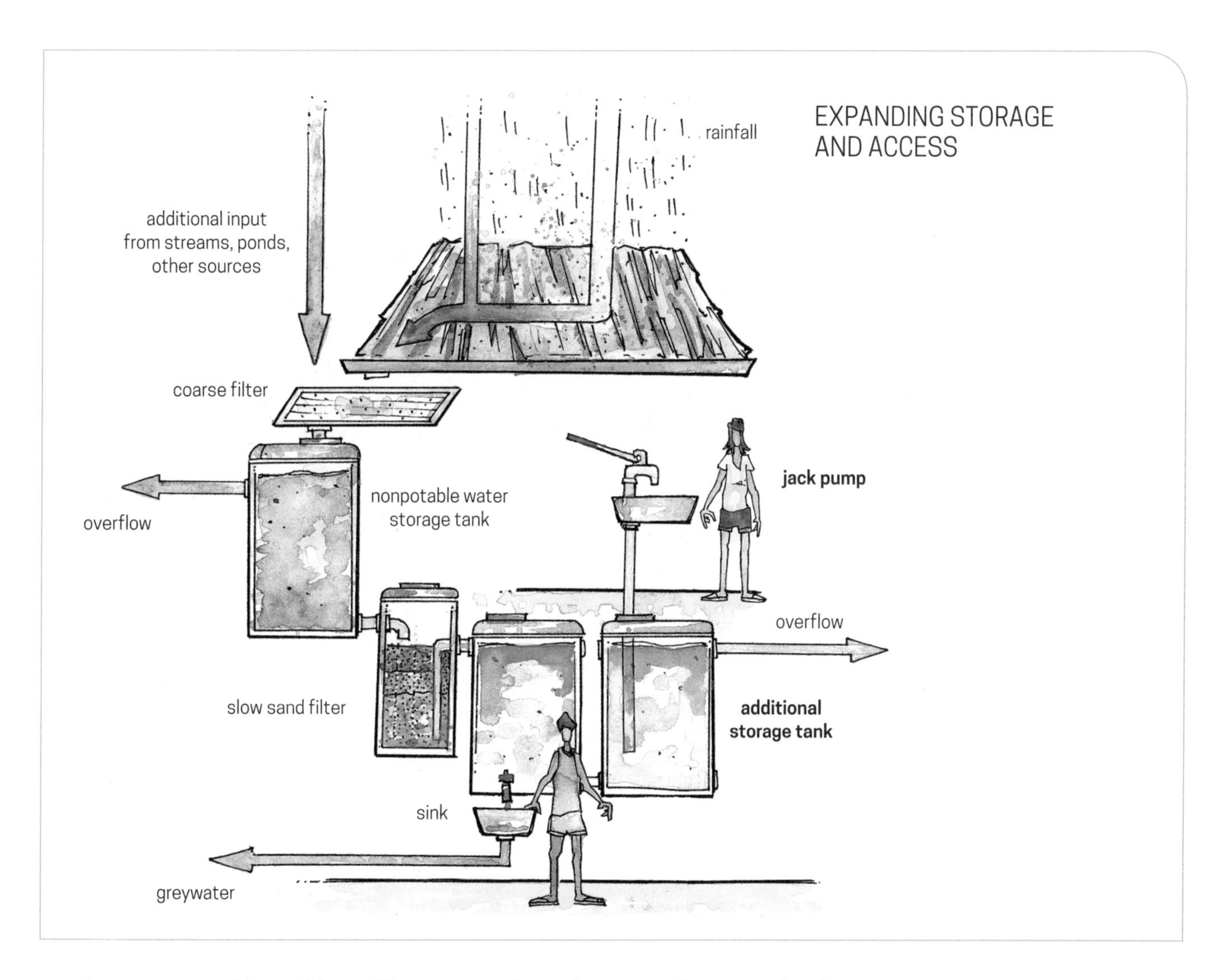

As a next step, it's useful to add storage tanks and connect them together for maximum storage capacity. If they are connected at the top of the tank, the second essentially acts as overflow storage while the first one may not be emptied regularly. If they are connected at the bottom, the tanks' water levels will remain the same, essentially acting as one tank. If the point of use is higher in elevation than the storage containers, a simple jack pump is a good solution for accessing that water in a low-tech way that doesn't require the use of electricity.

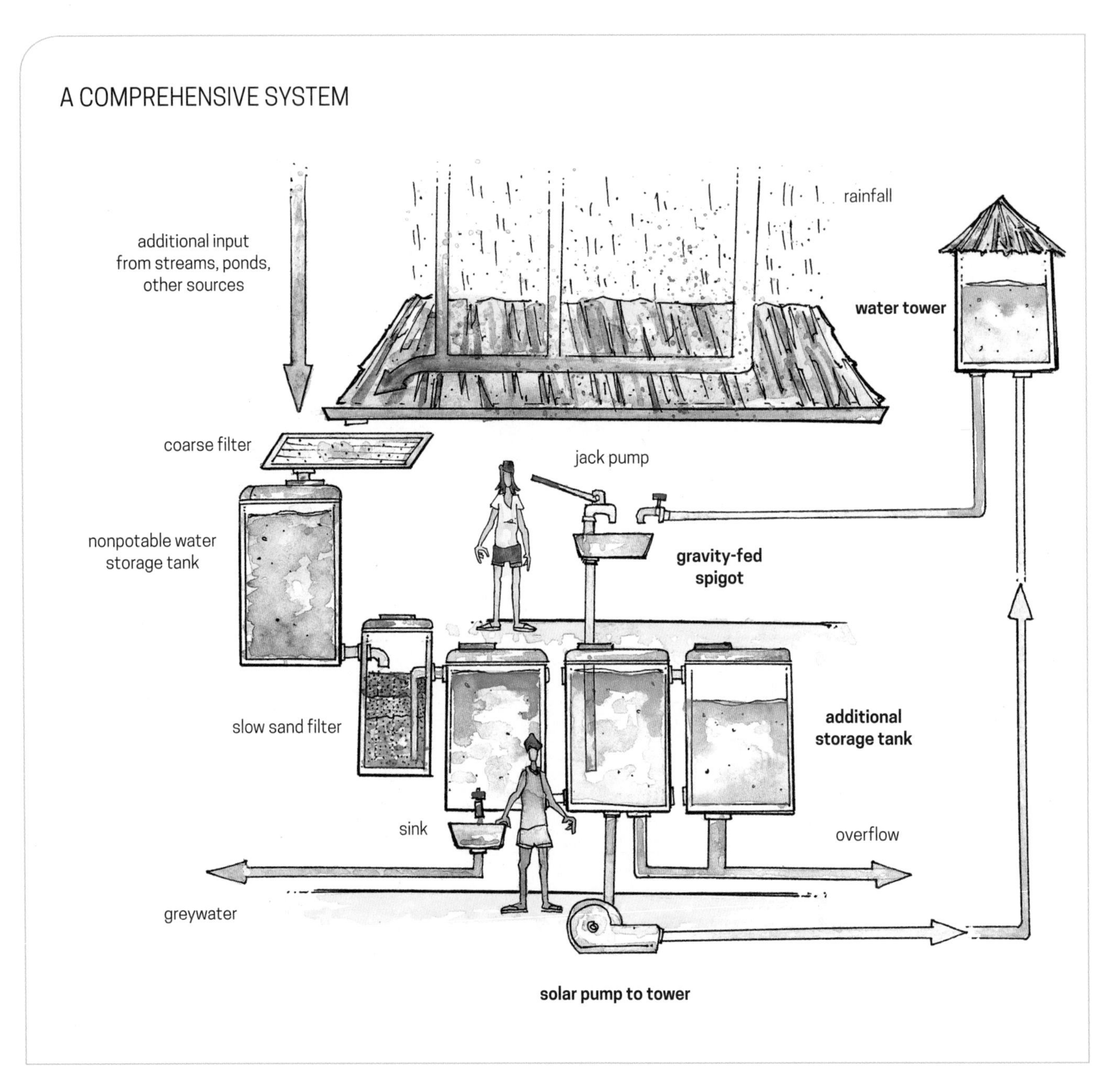

In a more comprehensive water system, water is pumped into a water tower with a low-tech pump (preferably run on renewable energy such as solar). A water tower can provide enough pressure for a gravity-feed system so there is no need for the jack pump (although it can still be a good backup). The main reasons to have a tower are if your point of use is higher than your storage or if you want to increase water pressure. Also, if your power system goes down and your water is already stored in a tower, that volume will gravity feed to your point of use. Another tank can be added to the system to store more water for secondary use.

WATER CONSERVATION

Conservation and protection of fresh water is becoming more and more important in our world. According to the World Health Organization, more than one billion people on earth are affected by water scarcity. The destabilization of climate could result in greater frequency of drought in some areas. The water supply systems in many urban areas are overtaxed. Some areas are just arid all or part of the year. Whatever the context, conserving water in the landscape is always a good idea as a buffer against uncertainty. The table suggests ways to conserve water, from the easiest (changing behavior) to the most challenging (building new systems).

WAYS TO CONSERVE WATER

	Changing behavior	Upgrading what you have	Building new systems
Bathroom	Change your flushing habits: if it's yellow, let it mellow; if it's brown, flush it down.	Get a dual-flush toilet.	Compost all human waste (no water needed).
	Take shorter showers.	Use high-efficiency shower-heads.	Add a pedal sink.
	Don't leave the water running while you brush your teeth.		
Washing dishes and laundry	Hand wash using as little water as possible.	Use high-efficiency machines.	Add a pedal-powered washing machine.
Irrigation	Use plants with water needs that match your climate.	Use drip irrigation.	Install a greywater system along with earthworks to capture runoff.

Waste

Plugging Leaks in the System

This boxed bed growing food at the Picardo P-Patch community garden in Seattle, Washington, was made out of salvaged building materials, which now won't end up in a landfill.

In your permaculture design, you want to shoot for a near-zero-waste system. That doesn't have to happen overnight, but it is definitely a primary goal. If the systems you design are wasteful, they will forever be reliant on large quantities of external inputs to keep them running. Most lawns are like this. Without chemical fertilizers, city water, and gasoline to run the mower, they would very quickly cease to look the way they do.

Many of the external resources we rely on every day are nonrenewable (at least in a human timescale). Once we use them, they're gone. Relying heavily on these resources for day-to-day operations means we are more susceptible to market fluctuations and supply chains, and thus less resilient. In an emergency the lawn can just grow and become weedy, but what happens when we rely heavily on external inputs for our food, water, and heat?

This chapter focuses on where leaks often appear in systems and how we can minimize them, thus eliminating waste. The idea is to integrate those surpluses (another name for waste considered from a different perspective) back into our systems in some way. For instance, if we produce compost, apples that go bad can't really go to waste. If we apply the principle of efficient energy planning and the concept of next highest use, we don't really waste energy. Overall, the goal is to manage the inflows and outflows of our systems. We aren't going to create completely closed-loop systems (where nothing enters or leaves), but we want to get a lot closer to that than where we are right now. Ultimately, we want to be very conscious of how the outflows of our systems can be used as inflows. Any outflows we do end up with should not harm the environment nor our neighbors.

Types of waste to address in your design include human waste, greywater, food and yard waste, and heat. We have already explored some ways to turn food and yard waste into compost in the earlier chapter on soil fertility. We'll look more closely at the other topics here.

HUMAN WASTE

The topic of managing human waste, also known as humanure, is pretty much considered taboo in Western culture. You don't talk about it in polite company. However, it is imperative that we begin to take responsibility for the humanure we produce. Unfortunately, the centralized systems upon which many of us rely and conventional home septic systems do not score that well on their ecological report card. In many parts of the world, waste collected by municipal sewer systems is dumped into the ocean or injected into the groundwater. Even the municipal systems that are ecologically kinder often have enormous energy inputs. The amount of fresh, clean water wasted by these systems is staggering. Consider, for example, that the Colorado River no longer reaches the Gulf of Mexico, thanks partly to all the flush toilets in huge desert metropolises like Las Vegas and Phoenix.

The good news is that there are some excellent ways to take responsibility for your own humanure instead of flushing it away where it both pollutes water and becomes someone else's problem. In fact, with the right system, you can wrap this outflow right back into another part of your permaculture system in a way that's neither gross nor unsanitary. Taking responsibility for your own humanure means managing it in a way that doesn't risk anyone's health or unintentionally cause pollution.

Before you get really excited and start ripping the toilet out of your bathroom, it is critical that you fully educate yourself on this topic. Laws and codes vary from location to location. Whether you choose to obey them or not, you should at least know what they are. Details of timing and temperature can vary from situation to situation. Do plenty of research before you try any of these options at home and start small. And realize that any humanure system requires a core personality trait: detail orientation. To safely operate any of these systems, you must pay close attention to the specifics of temperature, time, and cleanliness. If that doesn't sound like you, everyone may be better off if you just keep flushing it.

Making sure your humanure compost spends enough time above a certain temperature is crucial to making sure it is safe.

In the meantime, there are intermediate options to help minimize negative impacts, such as flushing your toilet with greywater. You can even buy a toilet tank lid with a built-in sink that runs the greywater into the toilet tank.

When you are selecting a humanure management system, two extremely important factors will help to guide your decision: water table and culture.

Whether you are in an area with a high water table or one that is quite deep will determine what options are most appropriate. For instance, an outhouse where you dig a hole in the ground, fill it up with your humanure, then cover it and move on is not appropriate in an area with a high water table, where it may pollute your groundwater (as an aging septic system might also do in this case). Conversely, if your water table is quite deep, this system can work great as it leaves behind big, spongy plugs of composted material that surrounding plants can tap into while allowing you to avoid actually handling the humanure in any way.

The predominant culture in your area is also a factor when choosing a humanure management system. For instance, in a broader, global context there are some places in the world where people use toilet paper and other parts of the world where people use water to clean themselves. This will impact your choice of system. Similarly, the culture of a downtown office building is radically different from a family farm or campground. Hauling around sawdust and wheelbarrows of humanure is appropriate in one setting but not the other. Always think about who the users are and what their level of open-mindedness will be. Also, think about how much education a new user will need to avoid messing the system up (nobody wants to deal with poop in a urine excluder). As with all of the permaculture techniques we recommend, context is crucial. If you choose the right system for the context, you will curtail waste and create a positive experience for both users and managers.

Several options we find particularly good for dealing with humanure are waterless composting toilet systems, engineered wetlands, and biodigesters.

Waterless composting toilet systems

Waterless options are the best for water conservation. In these types of systems, each time you make a deposit you must throw some carbonaceous material such as sawdust or wood chips down the hole to maintain a good carbon-to-nitrogen ratio and keep down smells and flies (just like a cat burying its poop in kitty litter). Throwing a handful of red wiggler worms down there can help, too. The worms will stay away from the hot center of the pile and stay close to the outside, and will help to move material around in the pile so it gets more evenly composted. It is also advisable to use a device called a urine excluder in these systems. That basically routes your urine to another location so it doesn't mix with the humanure and cause bad smells, as the urine is what really gets stinky.

Several models of prefabricated composting toilets are available, with brand names like Clivus Multrum and Sun-Mar. In many places throughout the United States, the building code makes allowances for such systems.

If you want to make your own composting toilet, a two-chambered model is an excellent design. The throne is elevated with a concrete, ferrocement, or stone

vault below. The vault is divided in two to accommodate two deposit holes in the top, and each side has a generous access port at ground level. You use one side to do your business until it's full and then switch to the other side. By the time the second side is full, the first side should be mostly composted. You can remove the material, turning it in the process, and put it into a designated area (protected from both animals and rain) to finish composting. For a bigger group of people, there's no problem with creating a three- or four-chambered system.

A ventilation pipe with a chimney can take the fumes out above the height where people generally spend time. Installing a sheet metal sheath below each throne that extends down lower than the vent pipe will prevent fumes from rising back up through the throne. In fact, if this is done well, the person sitting on the throne will often be able to feel air being pulled down the hole; it's then moved out via convection through the vent pipe, resulting in a stink-free experience.

Another waterless option is a bucket system managed just like the two-chambered system. All you need is a 5-gallon bucket with a toilet seat fitted to the top. You do your business in the bucket, add your carbonaceous material, and every few days empty the bucket into a designated humanure composting area. You let it compost for several months, turning it occasionally. The advantage of this system is that it doesn't require major construction, but it does require more maintenance-intensive management.

So what do you do with the humanure from a composting toilet? It's critical to research temperature and timing for your climate and conditions to make sure you compost it long enough to kill pathogens. It may even be advisable to solarize the humanure at some point under glass. Just in case a pathogen manages to survive,

(left) This outhouse at a community in northern California demonstrates that humanure composting can be colorful and not gross.

(center) At Seattle's Picardo P-Patch community garden, the Picaloo provides gardeners a place to do their business.

(right) Chris Shanks of Project Bona Fide on Isla Ometepe, Nicaragua, designed and built this beautiful two-chambered composting toilet from materials on-site.

you should avoid using the humanure on crops where it may be in direct contact with your food (for example, root vegetables, lettuces, asparagus). Humanure is great for mulching tree crops, ornamentals, and other nonedible crops.

Flush options

Sometimes a flush-compatible system is more appropriate than a waterless one. The effluent that comes from a toilet is commonly referred to as blackwater. Although these systems all have the disadvantage of using water, the wastefulness of this practice can be drastically reduced by finding ways to derive value from the blackwater that is produced.

Engineered wetlands are a great way to manage blackwater by using a biological resource. An engineer will need to help with the design, but in a nutshell, when you flush, your effluent goes into an aboveground series of wetland "cells" that you've created. Much like in a septic system, the blackwater moves through these cells and bacteria breaks down the organic material. Unlike in a septic system, you can fill the cells with nonwoody wetland or prairie plants to absorb those nutrients. The water that comes out the end of a well-designed engineered wetland will generally be cleaner in most respects than what comes out of a municipal wastewater treatment system. This water should be routinely tested, but it can be used in your landscape for irrigation. Obviously, these systems can be somewhat costly as well, but a single system can handle multiple households, so cost can be shared among neighbors.

Another option is a biodigester, or biogas digester, a device that accepts the effluent that comes from your flush toilet and digests it anaerobically in a large tank. During this process, methane is released and captured. This methane can be used as a fuel. The remaining material that comes out of the biodigester, referred to as digestate, can be composted and then serves as an excellent, pathogen-free fertilizer for crops. Anaerobic digestion can actually be used for animal manures and organic waste as well as humanure. The city of Lloydminster, Alberta, Canada, has a municipal system for collecting waste and converting it via biodigestion to electricity, fertilizer, and other uses.

If you choose to use one of these types of systems to manage blackwater, you need to be aware that many household cleaning products contain toxins such as bleach that will actually kill off the microorganisms that are supposed to be processing the waste. The best solution is to find nontoxic alternatives, like hydrogen peroxide, vinegar, and baking soda, for cleaning toilets.

Pee cycling

Urine is an interesting human by-product. Unless a person is sick, it generally leaves the body without pathogens. Mixing it with our humanure or blackwater can cause problems and result in the waste of a very useful substance. Urine is high in nitrogen. The nitrogen in urine happens to come in a form that is quite bioavailable to plants. That means your urine is one of the best nitrogen fertilizers available for use in your garden! According to a study from the University of Kuopio in Finland, one person's urine can supply the nutrients to grow more than 150 cabbages per year. From a design perspective, consider managing urine separately from humanure.

Pathogenic risks of using urine in the garden are minimal. If you are sick or live in the ultratropics, you may want to do more research. Otherwise, for most of us salmonella is the one thing to think about, but it generally dies shortly after excretion. To avoid any danger of salmonella poisoning, you may choose not to harvest urine-fertilized crops for a few days after your last application. You can also ferment your urine for six months before using it. This will cause it to stink, so there is a trade-off. This may be something to consider if you're collecting urine from other people.

When your urine is ready to use, there are several ways you can go:

- You can dilute it and apply it directly to your garden in areas with good drainage. Use cold water and dilute eight parts water to one part urine. If applying your fresh urine daily, spread it around so that no single area gets overloaded with salts. This will also help to avoid odors.
- You can apply your urine to a compost pile and use it to help speed the composting process. If you do this, make sure you make a carbon-rich compost mix to accommodate the extra nitrogen you will be adding.
- You can integrate your urine into a greywater irrigation system.

GREYWATER

Greywater refers to the water that comes from sources not containing human waste. This includes kitchen sinks (in most places), bathroom sinks, dishwashers, clothes washing machines, and showers. In places where water conservation is a high priority, this greywater can be used to help irrigate a productive landscape, minimizing the use of clean tap water for that purpose. Think of using greywater in the landscape as the next highest use for it after it has been used to scrub vegetables or wash clothes.

For any greywater system, two cardinal rules should never be broken:

No surface greywater. Although greywater can be applied to the soil surface, it should not stay there. If the system you've created involves pools, puddles, aqueducts, or ponds of greywater, it's time to rethink the design. All of these can go septic, and animals of all types can become vectors after they interact with the greywater, then interact with you.

No greywater storage. Storing your greywater in a big tank until you're ready to use it will result in it going septic. This means it will stink and possibly contain harmful pathogens. Avoid systems where greywater is held in storage for more than twenty-four hours. Beyond that, greywater should be treated as blackwater.

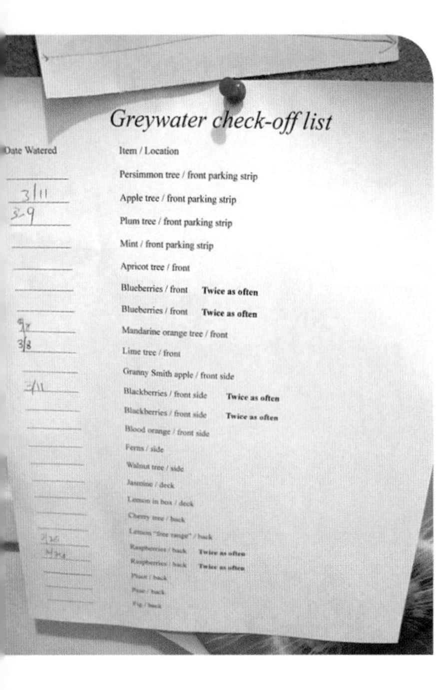

Greywater check-off list

Date Watered	Item / Location	
	Persimmon tree / front parking strip	
3/11	Apple tree / front parking strip	
3-9	Plum tree / front parking strip	
	Mint / front parking strip	
	Apricot tree / front	
	Blueberries / front	Twice as often
	Blueberries / front	Twice as often
	Mandarine orange tree / front	
3/8	Lime tree / front	
	Granny Smith apple / front side	
3/11	Blackberries / front side	Twice as often
	Blackberries / front side	Twice as often
	Blood orange / front side	
	Ferns / side	
	Walnut tree / side	
	Jasmine / deck	
	Lemon in box / deck	
	Cherry tree / back	
	Lemon "free range" / back	
	Raspberries / back	Twice as often
	Raspberries / back	Twice as often

Members of a community housing cooperative in Berkeley, California, water specific plants directly with greywater. This sheet is how they keep track of which plants have been watered.

BRANCHED GREYWATER SYSTEM

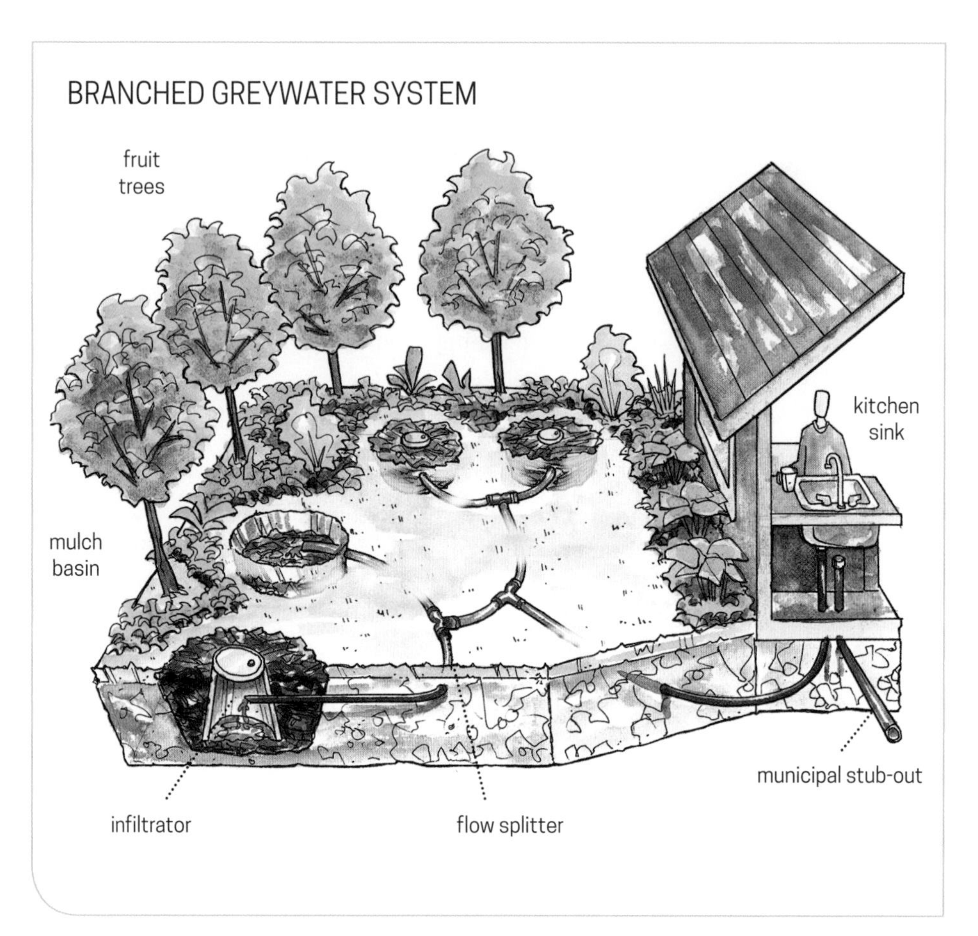

In addition, if you are going to reuse greywater you need to avoid using household cleaning products that contain toxins that will come out in the greywater.

A couple of greywater systems that work really well are dishpan to yard, and branched drain system to mulch basins.

Dishpan to yard is just as simple as it sounds. You can wash your dishes in a dishpan and then dump the water at the base of a fruit tree. You can also disconnect the p-trap in your bathroom sink and collect the greywater in a plastic bucket, emptying it every day. (If you do this, make sure to plug the pipe going to the sewer so gases don't come back into your home.) Assuming your soil percolates, the water you dump on the surface should soak in in just a few minutes. A huge benefit of this system is that it costs less than five dollars to implement.

A branched drain system to mulch basins is the Cadillac of greywater systems. Art Ludwig (a.k.a. the Greywater Guru) of Oasis Design produces a plumbing part called a flow splitter that looks much like a T-fitting. If placed dead level in the landscape, the flow splitter takes the water entering it and divides it in two. With three flow splitters, you end up with four pipes each carrying a quarter of the total load.

Each of these pipes should end at a basin you've dug into the ground in a spot with good drainage, ample distance from the water table, and perennials nearby that you want to irrigate. The basin should be about 16 inches deep (actual size

will vary depending on your percolation rate and volume of greywater produced). Each of these basins should be filled with mulch. Choose something like coarse wood chips that won't readily float. When the greywater flows through the system, one quarter of the total volume should end up flowing into each mulch basin. It will quickly sink below the mulch so we don't violate our first cardinal rule. The greywater is cleaned in this system by bacterial action in the mulch basin. The nutrients released will benefit nearby plants along with the water itself.

If code requires it, you can use some sort of infiltration technology to transition between the pipes and the mulch basins, although it is easier to just have the greywater daylight for a brief moment before it sinks into the basin. Occasionally, it will be necessary to remove the mulch material and replace it as it begins to break down. The material removed is an excellent product to use for mulching other nearby trees.

For places with very cold climates, special considerations may be needed for winter months. As a first step, it is possible to install a valve that provides a choice to either send greywater to the branched drain system or to the conventional sewer or septic system. That way you can always reroute the water in the winter or in case you need to work on the system.

(left) This small reed bed is being used to clean greywater in Portland, Oregon. Just like in an engineered wetland, these plants and their media help to filter and clean the water before it is used to irrigate the landscape.

(right) These "flow-forms" at the Solar Living Institute in Hopland, California, aerate water and draw attention to its beauty as it moves through a children's play area.

FOOD AND YARD WASTE

The earlier chapter on soil fertility has already explored many ways to cycle food and yard waste back into your system. Here we will mention just one more great way to use woody debris from pruning, plant removals, and other general gardening tasks. Hügelkultur, roughly meaning mound culture in German, refers to a method of mounding coarse organic material in layers and covering it with soil to create a bed for growing crops. This method can be applied at many scales.

Creating a hügelkultur bed is a simple process of layering organic material of different coarseness, starting with larger woody material like logs and twigs, and then layering finer material on top. The finer layer could be manure, compost, or other organic matter, then finally topsoil. You can plant directly into the soil and mulch with wood chips or straw. As the plants grow, they draw moisture and nutrients from the layers of organic material slowly breaking down inside the mound.

(top left) This is a hügelkultur bed in progress at Lost Valley Education and Event Center in Dexter, Oregon.

(top right) On the north side of Jessi's greenhouse, compost piles contain coils of pipe that bring the heat from the composting process into the greenhouse, where it helps plants thrive in the winter.

(bottom) At her house in Monroe, Washington, Angela Davis made good use of old windows by turning them into a greenhouse.

HEAT

Heat is an often overlooked waste product. Most biological processes generate heat. Any process involving burning involves heat. The sun produces heat. If heat is in short supply all or part of the year where you are, it's important to look at ways to capture heat that's trying to escape your systems.

Heat generated by animals and compost piles can be captured by greenhouses, hot beds, or even by housing arrangements that place human living quarters above animal stalls (remember that heat rises). Unless you regularly take cold showers, the greywater that goes down your shower drain will take heat with it. Drainline heat exchangers can capture some of that heat and put it back into your hot water tank. The take-home message here is to think creatively and recognize heat at a resource.

SALVAGE

Many opportunities exist for us to make use of the leaks in other people's systems. In many places, especially in the developed world, it is easy to tap into the waste stream to gain access to things you can use to boost fertility and functionality on your property. In urban areas this waste stream is especially huge. Before you buy new things, make sure you first see if you can salvage something that already exists. This can mean trading a crate of apples for wood chips from your local arborist. It can also mean getting grass clippings from neighbors who don't use chemicals on their lawn to use as mulch in your gardens.

The Akiyama family in Marugame, Kagawa, Japan, employs used chopsticks from udon restaurants as fire starters in their wood-burning cookstove.

We definitely want to use these resources, but at the same time it is important not to become reliant upon them. If permaculture design takes off in a big way (and we think it will), many of these "waste resources" may become hot commodities. There may be competition for that 5-gallon bucket of coffee grounds you pick up from your local coffee shop every week.

Imagine a day when the waste stream is significantly smaller because it is no longer cost effective to move things around the globe with ease. Imagine that the waste products you rely upon dry up because the processes that produced them become unfeasible or unprofitable. And remember the transitional ethic. Use these wastes as a way to set yourself up for greater resilience in the future but don't become reliant upon them for your everyday life. That means not relying on external sources for your soil fertility but moving toward generating the fertility your property needs on-site (or locally) whenever possible. The same goes for energy, water, food, and shelter systems as well.

Energy

Minimizing the Work We Do Ourselves

This solar array generates electricity to run a water pumping system at Bullock's Permaculture Homestead.

Energy refers to the ability to do work. In our permaculture systems, a lot of work needs to be done. If we design based on what we see in nature, we can create systems that require minimal external inputs, create little pollution, and perform with great efficiency. Our overarching goal is to design systems that cycle and recycle energy. Ideally, we will minimize the amount of work we need to do ourselves. Imagine a system where you don't need to fertilize your orchard because sheep do it for you while simultaneously cleaning up fallen fruit.

Today, when we hear the word *energy* most of us immediately translate that word in our mind to *electricity*, which for many of us really means fossil fuels. In permaculture design, energy means a lot more than just electricity. In fact, electricity is just one form energy can take; it can also take the form of heat, light, chemicals, or mechanical action. While a large part of this chapter will deal with electrical systems, we will start with other important ways to produce energy as they often lead to more elegant and lower-tech designs.

Just make sure you think about your energy systems at the master plan level. If you begin by assessing your situation, making small changes, and having success, you will find that you can quickly decrease negative impacts and learn more about your overall energy picture at the same time. As you learn more, you can do more.

NONELECTRIC ENERGY

Many of the most efficient energy systems don't use electricity at all. Any form of electricity, even generated by renewable energy systems like photovoltaic panels, comes with a cost in efficiency. The second law of thermodynamics says that every time energy changes form, some is lost as heat or light. For instance, if you turn sunlight into electricity via photovoltaics and then turn that electricity into hot water, the process is likely less than 15 percent efficient because of the two conversions. In other words, 85 percent of the solar energy hitting the photovoltaic panel is lost before it actually results in hot water. Using the source to do a job directly (for example, heating water directly with the sun's rays) is often the most efficient way to get it done. If you use a flat plate collector to directly heat the water, the energy conversion can approach 75 percent efficiency, meaning only 25 percent of the solar energy is lost during the process.

Given that using electricity isn't the most efficient way to get things done and that it takes a fair amount of technological savvy to produce your own electricity, it makes sense to look at nonelectric ways to get work done first. Let's revisit the concept of appropriate technology again from a new perspective. What makes technology appropriate? The answer to that question will be different for different technologies and different people. Here are some key questions:

- Can you understand how it works?
- Can you repair it with local materials if it breaks?
- Can someone in your community do these things?

If the answer is no, the technology in question may not ultimately be that appropriate for you or your situation.

OPERATING WITHIN YOUR CURRENT SOLAR INCOME

The current solar income of your property refers to the amount of energy hitting it in the form of sunlight. When designing energy systems, we want to try to operate within current solar income as much as possible. That means moving away from using fossil fuels (past solar income), grid energy (usually reliant at least in part on fossil fuels), and energy resources like firewood that are brought in from elsewhere (someone else's current solar income). While you may not achieve this in your lifetime, it is still worth striving for.

Sunlight can be transformed into forms useful to us through photovoltaic or solar electric panels (for electricity), greenhouses (for heat), or biological organisms (via photosynthesis). Using biological organisms to harness solar energy is the most regenerative way to go. Biological systems have no inherently nonrenewable components, such as copper wire, plastic sheets, or steel support frames (in this case we mean nonrenewable in the sense that they don't regenerate in the ground after we extract them). When designing your energy systems, pay close attention to how your biological systems tie in so you can move toward operating within your current solar income.

Many nonelectric ways exist to harness energy either passively (without moving parts) or actively (with moving parts). In fact, before the Industrial Revolution, everything from grinding grain to lighting homes had to be done without electricity. Thus, history combines with modern ingenuity to give us a host of good passive and active approaches to energy. Anywhere we find movement in nature, there are creative ways to convert that movement into getting jobs done.

Active systems

Active systems are essentially ways of converting one type of movement in nature into another. Examples include waterwheels, windmills, pedal-powered devices, water-powered pumps, and compressed air technology.

Waterwheels. Converting the force of moving water to a spinning shaft via a waterwheel allows a lot of work to be done. With proper gearing, it's possible to use that spinning shaft to run many shop tools, such as reciprocating saws and lathes. Waterwheels have also traditionally been used to grind grain for whole communities and pump water uphill.

Windmills. In the same fashion as a waterwheel, a windmill can harness the power of moving air in windy areas. Traditionally used for grinding grain, these systems can also be used to pump water up from wells, churn butter, and accomplish many other tasks.

Pedal-powered devices. The bicycle has been around for a long time and is one of the most efficient machines ever invented. It is an excellent way to use biological energy to get jobs done at a small-to-medium scale. Designs and off-the-shelf products exist for pedal-powered blenders, grain mills, oil presses, water pumps, and nut shellers.

Water-powered pumps. If you have flowing water on your property and need to move some of it uphill, you have a choice of mechanical pumping technologies to get the job done. Water-powered pumps use the energy of the water flowing in a stream to move a small amount of water uphill in a pipe. Although the amount of water that gets moved with each pump of the piston is small, these hydraulic pumps operate continuously and can move quite a bit of water in a very durable, electricity-free way.

Compressed air technology. While not a source of energy in and of itself, compressed air (or pneumatic) technology is one way to store energy that doesn't require electricity. Think of the sound you hear at a garage when the mechanic removes or replaces the lug nuts on your car's wheels; this is usually a pneumatic wrench. Dentistry tools are also pneumatic, showing that such tools can be used for fine detail work as well. Any nonelectric mechanical technologies, as well as many electrical systems, can be used to compress air. Compressed air can run many different tools in a shop setting and even vehicles for transportation.

This old-style windmill sits on the island of Shodoshima in Japan. Devices like this were used to do a wide variety of jobs for hundreds of years.

Passive systems

Passive systems have no moving parts, meaning no friction and less wear. These systems are incredibly resilient and unlikely to fail due to mechanical problems. They include solar dehydrators and solar hot water heaters.

Solar dehydrators. For thousands of years, people have preserved food using methods such as fermentation, cold storage, salting, and more recently, canning and freezing. Dehydrating fruits, vegetables, and herbs is another way to make your harvest last a lot longer. In sunny climates, this is something you can accomplish within your current solar budget.

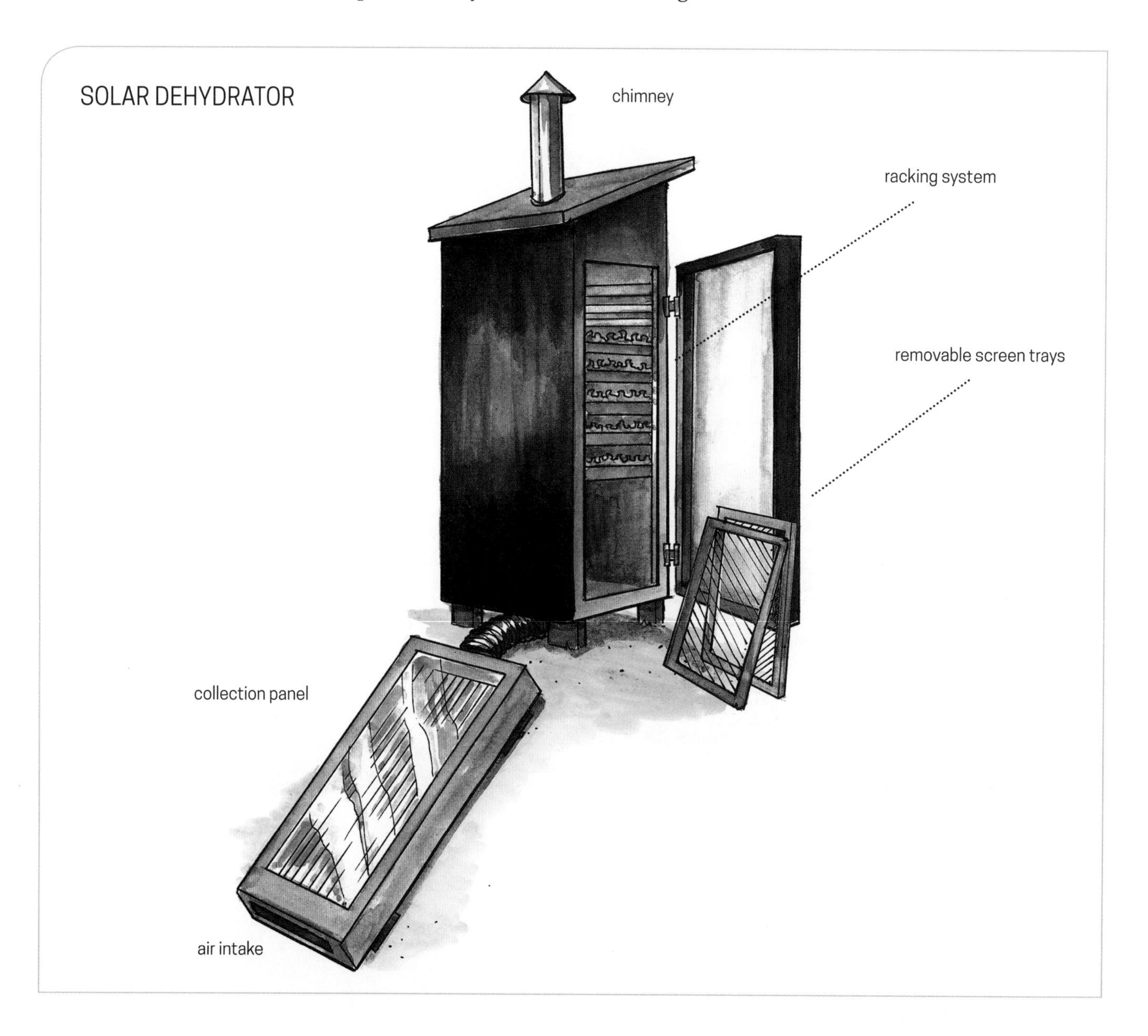

Here are the steps to create your own solar dehydrator:

1. Start by finding a location on a slope with good solar orientation. You will place the dehydrator on the slope so that the solar panel is below the dehydrator unit.
2. Gut an old refrigerator so all you have left is the hollow shell. If you don't know how to remove the refrigerant safely, get help from someone who does. Refrigerants such as freon are greenhouse gases, and it is illegal to release them into the atmosphere in most places.
3. Cut a hole in the bottom large enough to attach a metal tube leading from the solar collection panel.
4. Cut a hole in the top large enough to attach a piece of stovepipe (6 to 8 feet long). Attach the stovepipe and make the union waterproof.
5. Paint the outside of the refrigerator and the stovepipe black and put a cap on top to prevent rain from coming down the chimney.
6. Create a racking system for screen trays inside the refrigerator.
7. Make the collection panel with a piece of roof metal you've painted black and mounted inside a box with a pane of glass on top (the longer, the better). This box should be relatively airtight except for holes at the bottom and the top; the holes at the bottom let air into the panel to be warmed and the holes at the top let the heated air pass from the panel into the refrigerator. Use wire mesh and/or screen to keep animals and insects away from the holes at the bottom of the collection panel.
8. Connect the top of the collection panel to the hole in the bottom of the refrigerator in an airtight way using a length of flexible metal tubing.

When the sun hits the collection panel, the air inside will heat up, causing it to rise. As it rises, air will escape through the stovepipe and new air will be drawn in through the holes in the bottom. This means as long as the sun shines, air will be continuously drawn up through the screens where your produce is drying. Depending on what you're drying and how sunny it is, your produce should be ready in one to four days.

Solar hot water. Solar water heating technology has been around for years and is gaining in popularity. In terms of renewable energy infrastructure, solar hot water technology is among those technologies that offer the fastest payback on your up-front investment. Although you can make your own, most people use prefabricated evacuated tube or flat panel solar collectors. These collectors use the sun's energy to passively heat water and put it into your existing hot water tank. Even if it isn't incredibly sunny where you are, solar hot water systems can at least heat the water part way, which means less electricity, propane, or natural gas will be needed to get it up to household hot water temperatures. These systems can be used in cold climates by putting antifreeze in the panel and a heat exchanger in your water tank.

For the ultimate low-energy system, placing the collection panel lower than the storage tank will allow the water to cycle through the system without need for pumps via a thermosiphon.

A thermosiphon is a passive way to circulate fluids using natural convection. It's based on the fact that heat rises. In a solar hot water system, putting the storage tank above the heat collection panel allows for a thermosiphon because the water heating up in the panel naturally wants to rise. The only place for it to go is into the storage tank above. Conversely, the water cooling down in the storage tank sinks, ultimately ending up back at the bottom of the heat collection panel if you plumb things correctly. Systems using thermosiphons are inherently more resilient as there are no mechanical pumps, which can fail.

This outdoor solar shower is a great way to clean up after a hard day's work at Lost Valley Education and Event Center, Dexter, Oregon.

Biological systems

We can take advantage of biological systems to harness energy in many ways. Pedal-powered devices harness human biological energy. Some other sources we can tap into are animals and plants (for biofuels).

Animals. Often when we think about incorporating animals into our permaculture landscapes, we think of their physical products such as food, fiber, and fertilizer. Many animals also offer us energy that we can use to get work done. Traction in fields can be accomplished with oxen, water buffalo, and horses. For thousands of years, people have used animals such as horses, camels, and oxen to transport themselves and their goods. All warm-blooded animals also produce heat, which can be used to help keep spaces warm. Even animals like domestic dogs and cats have a long history of doing work. In fact, most dog breeds exist to do a specific job, such as retrieving game or even finding truffles. Make sure that you think about how your pets can do work for you.

Biofuels. Small-scale biofuels have a place in most permaculture designs. For instance, the wood you use in your woodstove is a biofuel. This is probably better than using fossil fuels or nuclear power to heat your home, but we must also acknowledge that burning releases carbon dioxide into the atmosphere, contributing to climate change. Therefore, we should try to choose stoves that are efficient and low-pollution as a viable way to cook food and heat living spaces. Biodigesters are another way to use biological products by converting manures and organic material into methane that can be used for cooking.

PLANTS FOR BIOFUEL

Acer saccharum
sugar maple
zones 4–8
excellent firewood

Alnus spp.
alder
zones vary
firewood

Betula alleghaniensis
yellow birch
zones 4–7
good firewood

Calliandra calothyrsus
red calliandra
zones 10–11
excellent small-diameter wood for cook fires

Coccoloba uvifera
sea grape
zones 10–11
firewood

Elaeis guineensis
African oil palm
zone 13
oil for culinary use and biofuel

Fraxinus americana
white ash
zones 6–9
excellent firewood

Gliricidia sepium
madero negro
zones 10–11
firewood

Helianthus tuberosus
Jerusalem artichoke
zones 7–9
sugar for alcohol biofuel

Jatropha curcas
physic nut
zone 9 and warmer
oil for biofuel

Larix laricina
tamarack
zones 2–8
good firewood

Populus spp.
hybrid poplar
zones vary
coppiced firewood
for masonry stoves

Ricinus communis
castor oil plant
zones 11–13
oil for biofuel

Robinia pseudoacacia
black locust
zones 4–9
firewood

Saccharum spp.
sugarcane
zones 10–13
sugar for alcohol biofuel

Salix spp.
willow
zones vary
coppiced firewood for masonry stoves

USING A WOODSTOVE FOR MULTIPLE FUNCTIONS

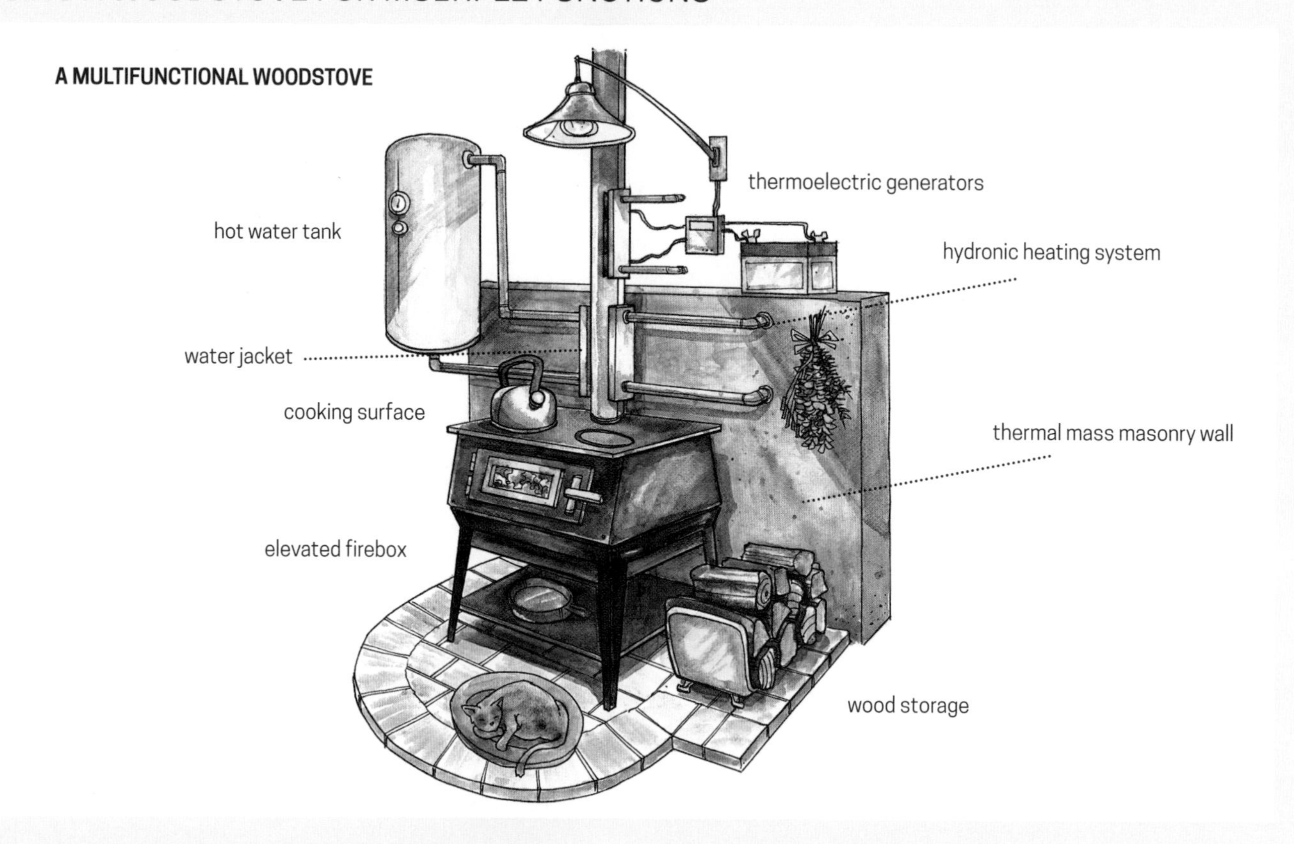

Most people with a woodstove use it to heat their home while it's going, but your woodstove can do many other jobs for you as well. It's a great example of an element that can provide many functions and really earn its keep.

Besides heating the air in a home to make it comfortable during cold weather, a woodstove can be used to heat food and water. A moveable trivet on top can provide a spot to keep food or tea water warm. The glass front of the stove allows burning wood to cast a low, warm light, providing a few extra lumens and a cozy feeling in the room.

A small water jacket on one side of the stovepipe can heat water for washing dishes and taking showers. It can operate via a thermosiphon, with the water tank being higher than the stove. The water jacket should be located on the chimney instead of the combustion chamber to avoid reducing combustion efficiency by causing a drop in the temperature of the combustion chamber and thus leading to greater pollution. When the water jacket captures waste heat from the chimney, this is not an issue. This issue can also be addressed through better stove design. If you want to attach a water jacket to your stove, do further research to really sort out the pros and cons of its location.

Another small water jacket on the other side of the chimney can tie into a loop of polyethylene tubing that runs through the masonry wall behind it. With this setup, when the fire is lit the water heats up and stores its heat in the thermal mass of the wall to be radiated back into the room later once the fire has gone out.

A series of thermoelectric panels on the back of the stove can use temperature differential to produce electricity. The greater the difference in temperature between one side of the panel and the other, the more electricity is produced. One side of each panel can be placed in contact with the 500-degree-plus woodstove, and the other side in contact with a water jacket with 33-to-40-degree water connected to an old car radiator hanging outside in the cold air or in a cold upstairs bedroom that you want to heat. When the stove is going full blast, the little panels can produce 50 watts or more of electricity each. They can be tied into the house's greater renewable energy system.

In addition, consider adding perennial oil and sugar crops to your landscape to produce at least a portion of the culinary oils and sugars you use. For larger landscapes and homesteads, many oil crops such as African oil palm, walnut, and coconut produce edible oils that can also be made into biodiesel for use in diesel engines or used directly in diesel engines that have undergone the proper conversion. This means you can run generators, vehicles, and some tools on the oils you produce. Sugar crops such as sugarcane, sugar beets, and sorghum produce edible products as well. They can also be fermented into alcohol, which can in turn be used as a beverage, a preservative for herbal medicines, or a fuel for modified gasoline engines.

ELECTRIC ENERGY

Much of the modern world is dependent on electricity. In fact, writing this book turned out to be very dependent on electricity. In many cases, using electricity represents using nonrenewable (fossil fuel) resources or creating radioactive waste (in the case of nuclear power plants). If we don't conserve this precious resource and think about our impacts, we will most likely be supporting systems that harm other people and ecosystems. The basic permaculture ethics of redistributing the surplus and regulating consumption make it essential that we take responsibility for our own electricity production and consumption.

In permaculture design, our aim is to produce and use electricity wisely. Of all the forms of energy we've discussed, electricity is both the most difficult to understand and the most technologically intensive. The trade-off is that it is also the highest-grade form of energy to which we are likely to have access, with unparalleled versatility in its uses. In other words, although making electricity is complex, it is incredibly easy to convert that electricity into many different forms of work.

The concept of next highest use suggests that if a job can be done directly with sun, wind, or moving water, you should think about ways to make that happen without the extra step of making electricity first. You should save the electricity for things that can't run without it (such as computers, stereo equipment, and televisions). Heat is one of the lowest-grade forms of energy. It's limited in application but really easy to make (you're making some right now). As a rule of thumb, if your design involves using electricity to create heat, you could probably find a less wasteful, more efficient way of getting the job done.

Auditing and minimizing your energy usage

The first step in designing a renewable energy system is to measure your usage and work to minimize it. By doing this first, you may find that you can actually get away with a much smaller renewable energy system than you would otherwise have needed. That translates into real savings in cost and nonrenewable materials. For existing buildings, an energy audit is a great place to start. You can do your own simple energy audit or have a professional consultant do a more thorough, technical job. Some cities, states, and utilities offer incentives for doing an energy audit on your home. Ultimately, an energy audit will tell you where inefficiencies show up and suggest ways to fix them. It makes sense to start with the simplest and cheapest; changing to more efficient light bulbs and installing some weather stripping should come before replacing all your windows.

An energy-monitoring device known as a Kill A Watt can be used to measure how much electricity your various appliances and devices use. This can be an eye-opener. Try plugging your stereo into the Kill A Watt and playing with the volume to see how the power demand fluctuates when you change the volume. You may find that some of your devices even continue to draw electricity after you turn them off. Such demands are called phantom loads. Identifying and eliminating them by unplugging devices when not in use or plugging them into power strips that can be switched off can be a good first step toward conservation.

WAYS TO CONSERVE ENERGY

	Changing behavior	Upgrading what you have	Building new systems
Heating and cooling	Set the thermostat a couple degrees lower and wear a sweater. Take a swim when it's hot.	Improve weather stripping and insulation. Install a programmable thermostat. Build trellis or shade house to keep cool.	Install a home power system to supply electrical needs.
Lighting	Turn off lights when not in use.	Install solar tubes for passive lighting. Use low-wattage task lights when possible.	
Electronics and appliances	Turn devices off when not in use. Use clothesline instead of dryer when weather is nice.	Find and eliminate phantom loads so energy is not used when devices are off.	

Considerations in designing a home power system

Using available resources and making the most of the energy you harness are two essential aspects of designing a home power system.

You should have noted available energy resources in your site A + A. Does your site have good, consistent wind? Is the site oriented for good solar access? Is there a year-round stream on the property that could be used for microhydro? Does availability of any of these resources change at different times of year (do streams dry up, are winds seasonal, are winters overcast)? As most places do have variation by season, it is important to note that you can create hybrid systems to increase resilience and meet power needs year-round. For instance, solar and wind power make good pairings in some places because during the winter when there is little

(top left) This wind generator at New Forest Farm in Viola, Wisconsin, provides for some of the site's energy needs.

(top right) At the Fertile Ground Guesthouse in Olympia, Washington, linens are dried the old fashioned way, without using any electricity at all.

(bottom left) This solar stereo cart at Bullock's Permaculture Homestead is a great example of a small, simple photovoltaic project. You can load your tools in the cart, head down to the garden, and rock out while you plant your crops!

(bottom right) This air compressor at Bullock's Permaculture Homestead will be used to run pneumatic tools in the shop. Electricity from the photovoltaic system is used to compress air after the lead-acid batteries are full.

sun the winds pick up, and in the summer when the winds are relatively calm it is much sunnier.

In addition to having multiple inputs into your power system, you should design the system to maximize your use of the energy you do harness. Off-grid solar-powered systems in sunny climates may have filled their batteries by as early as 10 a.m. The rest of the energy generated by those solar panels for the rest of the day just gets wasted unless you build a system that uses all the power harnessed in order of priority. For instance, the first priority is topping off the batteries. Once that's done, extra power could be used to fill the compressed air tank in the shop. If both the batteries and the compressed air tank are full, excess power could be used to pump water to a tank or pond high in the landscape. If there is still extra power, perhaps it could be used to run water features, fans, aerators, or some other nonessential, non–time sensitive but helpful technology. Other, more advanced ways you can research to sock away some of that surplus energy include hydrogen fuel cells, flywheels, and spring compression, although most of these require some sort of expertise to work with. Think about all the jobs that need to be done at your home and try to find ways for extra power to help make them happen.

Start small when putting in power systems but plan for growth. Many people get a very minimal system at first and use it to learn before sizing up. If they ruin their batteries somehow, at least they ruin only one or two instead of eight or twelve. If you have an idea of what your ultimate, dream power system would look like, you can design something that can grow toward that goal.

Some components are worth oversizing from the beginning. For instance, modestly oversizing the charge controller and inverter at the beginning will allow you to add more solar panels to your system and grow it over time without having to get all new equipment. You can also consider an upgrade path where old hardware is traded in or used elsewhere. Just make sure you think about it before you start and call a professional when you need help.

Grid-tied and off-grid systems

One core decision most people need to make when designing a renewable energy system is whether to tie into existing grid infrastructure. The main advantage of tying into the grid is that you require no batteries in your system because the grid serves as your "battery." Grid-tied systems don't run out of juice; you just end up paying for the electricity you use beyond what you produce. In some places, it is even possible to sell surplus power back to the grid at a wholesale rate. Conversely, when you are tied into the grid you lose power when the grid goes down (during power outages), which may be when you need it most. That means you lose out on much of the resilience benefit of having your own power system, unless you have battery backup anyway or a generator with fuel.

The primary advantage and disadvantage of an off-grid system is that you have batteries. A battery is basically a way to store electricity, and it can take many forms, each with its own costs and benefits. Deep-cycle lead-acid batteries are most common. If you treat your good-quality lead-acid batteries really well, you can expect them to last ten to fifteen years or more. However, most people occasionally draw down their batteries too far, which shortens their lifespan significantly. While recyclable, lead-acid batteries are the weak point in the sustainability of these systems. Perhaps a better (but significantly more expensive) option is nickel-iron batteries, which can last three to four times longer and do not leave an extremely toxic heavy metal behind at the end of their useful life. Off-grid systems also require you to do more maintenance and be more responsible with your use of power.

Basic components of a home power system

Most renewable energy systems have a few basic components:

Collector. Most commonly the collector will be an array of photovoltaic panels, a wind generator, a microhydro system, and/or a generator. Less common and/or experimental technologies such as pyrolysis machines and tidal power also exist and are worth exploring if you are mechanically inclined and amenable to experimentation.

Batteries. This is where you store the electricity that has been produced.

Charge controller. This essential piece of equipment is the brain of the system. It tells the system where to send incoming electricity. It keeps batteries charged and prevents them from overcharging, which can damage them or cause safety concerns.

AC inverter. The electricity produced by your system (whatever kind it is) will typically be produced as direct current, or DC. However, most of our appliances and electronics are wired to accept alternating current, or AC. Therefore, unless you purchase all DC appliances and equipment, your system will need an inverter to provide AC power. Do your research, analyze your needs, and choose an AC system, a DC system, or both based on your findings. Also, make sure you know your appliances and equipment. Some, such as laser printers and sensitive stereo equipment, may have special power needs. Note that many electronics, such as laptop computers, can be run more efficiently on a DC system with a DC adapter than on an AC system with their normal AC adapter.

Fuses, breakers, and shutoff switch. In case something goes wrong with the system, it's important to be able to shut down the power coming in. Fuses and breakers should help this to happen automatically. Shutoff switches allow you to do it manually. These can be located in multiple places in the system for different protective effects.

AC or DC loads. These are the appliances and electronics you are using that consume either AC or DC power.

COMPARISON OF AC AND DC POWER

AC	DC
Best for long wire runs (>100 feet)	More suited to short wire runs (<100 feet)
Generally 120 or 240 volts = higher risk of electrocution	Usually 48 volts or lower = minimal risk of electrocution
Works with almost all common appliances and equipment	Works only with DC appliances and equipment, which can be more difficult to find and/or expensive
Requires an inverter = expense	Does not require an inverter = savings of money and energy
Generally easy to create a system that meets code	Challenging to create a larger system that meets code
Many inefficient or wasteful appliances available	Many high-efficiency appliances available
Possible health risks from long-term exposure	No suspected health risks

HOME POWER SYSTEM

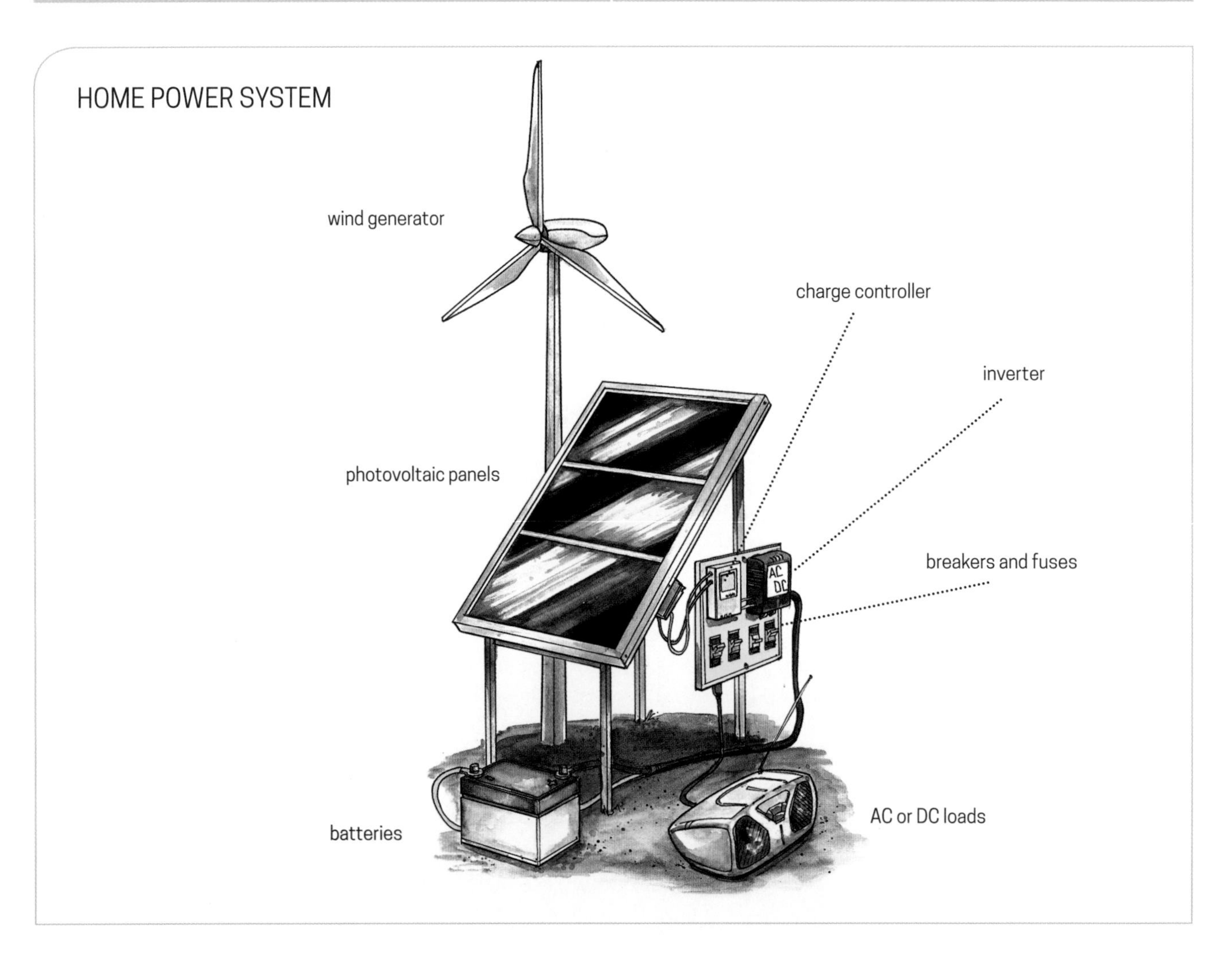

Remember, figuring out your energy systems is a process of designing from patterns to details, just like everything else. Once you've made good basic decisions, the details will likely fall into place fairly easily.

Photovoltaics can serve more purposes than just generating electricity. These panels are providing covered parking and showcasing renewable energy at the First Alternative Natural Foods Co-op in Corvallis, Oregon.

Shelter

Building Functional, Efficient Structures

The design of structures and their integration into our permaculture systems require careful thought. Because structures tend to be very energy and materials intensive, they offer incredible opportunities for energy savings and multifunctionality. The key thing to remember is that your structures should make sense in your context. Looking at how traditional and indigenous people build (or built) their houses is a great starting point. Identifying locally available materials, fuel sources, and local climatic patterns are all essential to designing buildings that perform efficiently and provide healthy, feel-good places to spend time.

THE IMPORTANCE OF CLIMATE APPROPRIATENESS

Before the Industrial Revolution, for a building to be functional it had to be sited and built in a way that was responsive to its climate. Simply put, those who built climatically appropriate buildings were more likely to survive and pass on their craft. This meant people developed houses that stayed cooler in the hot desert, houses that had airflow in the humid tropics, and houses that stayed warm (at least somewhat) in cold climates.

With the Industrial Revolution, much more energy became available at low cost in the form of fossil fuels. That meant inefficient buildings could be made habitable just by throwing massive amounts of resources, such as wood, coal, and oil, at the problem. We still follow this trend today. We tend to build structures that meet our aesthetic desires while paying little attention to climate appropriateness, site conditions like aspect and orientation, or availability of local materials. In many parts of the world, homes are reliant on fossil fuel–powered heating and cooling strategies to remain functional. Without fuel inputs, many of these buildings would become extremely unpleasant for at least part of the year, if not wholly uninhabitable due to excessive moisture, heat, or cold.

We can learn a lot from vernacular architecture, a method of designing buildings that draws heavily upon traditional, local techniques and available resources. The building traditions of the native and traditional peoples in an area come from a time when people didn't have access to as much energy and couldn't waste energy through poor design choices. Their techniques and material choices developed over time as a reflection of their environmental and cultural context, the ultimate in climate responsiveness. For most traditional peoples that meant living within current solar income and working with what was locally abundant. That's exactly what we're trying to do as permaculture designers.

So the question for permaculture designers is: How do we make buildings that are more efficient in terms of up-front and long-term resource consumption while still serving their intended purpose? To answer this question we want to closely examine three aspects of designing structures: siting and basic design, materials, and heating, cooling, and temperature moderation strategies.

Nestled into a hillside, this house takes maximum advantage of solar energy through its siting and orientation.

(top left) This yoga studio at Inanltah, Nicaragua, has an earthen floor, round wood construction, and a thatched roof, all made from local materials. A building made of materials that reflect the landscape gives occupants a grounded feeling and a sense of natural connection to that space.

(top right) Balinese-style architecture consists of small rooms separated from one another and connected by covered walkways. This allows people to move from room to room in the rain without getting wet and provides good airflow.

SITING AND BASIC DESIGN

The key to good building siting and design is a thorough assessment of the natural forces at work on the site. View lines are often important, but you want to find ways to take advantage of spectacular vistas (or block unsightly ones) that don't compromise the performance of the building. That means when siting and designing a building, you should consider factors such as sun, wind, and fire danger first. That great view won't do you much good if the house burns down, gets sweltering hot, or causes you to burn a lot of fuel to stay warm.

It's also important to remember our principle of relative location. How does the placement of your building relate to other buildings on the property? How about utilities, water sources, general topography? How does the placement of your building affect your ability to create gravity-feed water systems? How does your building placement work in terms of access? Buildings are usually quite permanent features. Therefore, test several options before you decide where to put them. Poor building siting can result in type 1 errors that can hinder a project for a long time.

Let's look more closely at how sun, wind, and fire can affect building location.

Sun and passive solar design

The sun is often one of the primary drivers when designing buildings. The sun can keep your home warm when it's cold out or make it so hot that you don't want to be inside it. Consider your climatic context and your site's aspect when figuring out the relationship you want between your building and the sun. In the temperate Northern Hemisphere, south-facing slopes tend to be excellent locations for homes because the sun's rays provide heat and light. The exact opposite is true in the temperate Southern Hemisphere. In equatorial climates, you may want to avoid the sun. Right on the equator, north and south have far less impact than east and west because the sun's path doesn't change much throughout the year. However, the afternoon sun is much hotter than the morning sun.

At permaculturist Jude Hobbs's house in Cottage Grove, Oregon, the indoor living spaces all have beautiful views out into an amazing, productive landscape.

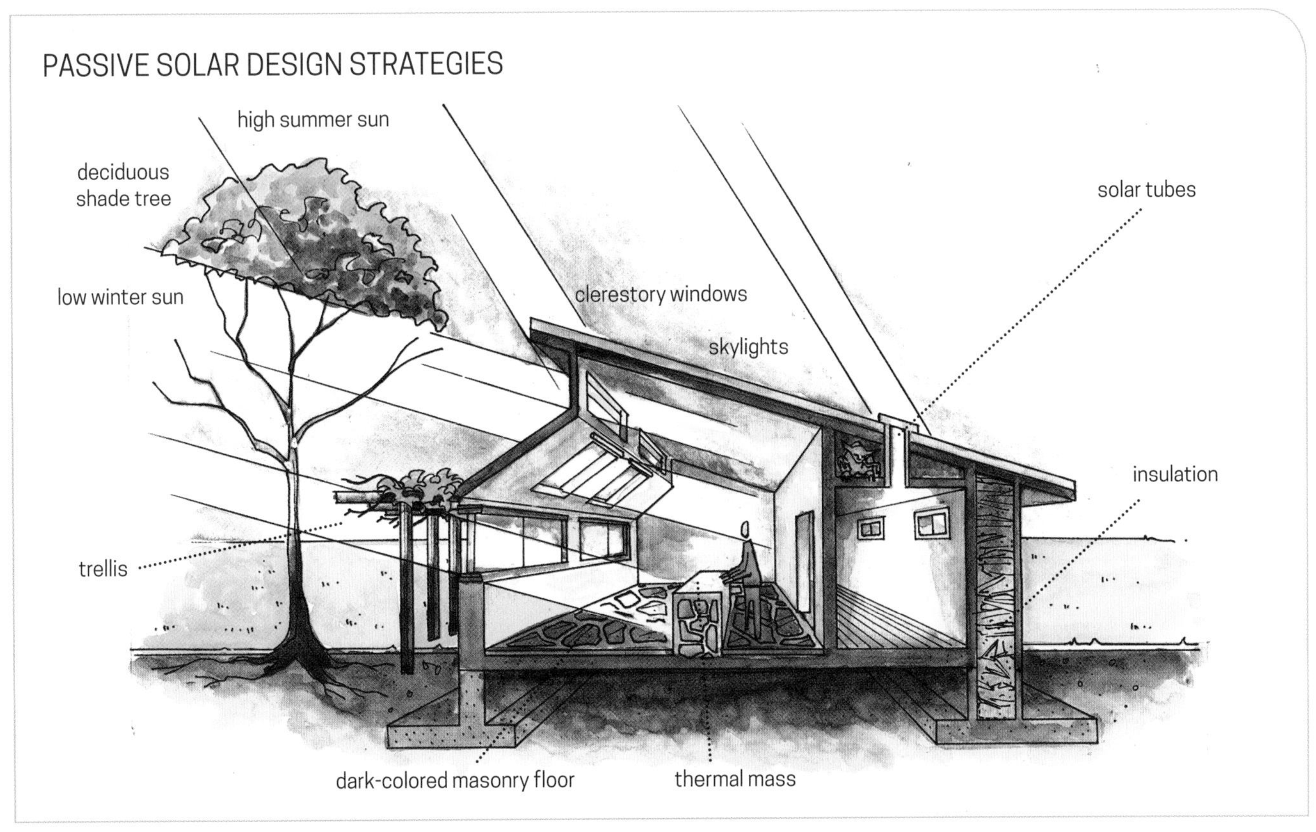

Relationship to the sun should also be based on what kind of building you're designing. A house or office may want extra sun to stay cozy and avoid heating costs, but a fruit cellar will want to be sited in a cool place out of the sun.

Passive solar design refers to including design features in a structure that work as a system with the sun to increase the building's efficiency in a passive way (with no moving parts). As with all good permaculture design, it is important to include a balance of multiple features in a passive solar design. For instance, if you have lots of sun coming into a space without mass to absorb it, it can become uncomfortably warm on a sunny day—even in the winter.

Orientation. The orientation of a passive solar building is the starting point for realizing greater efficiency. Generally you want to lay the building out so that the broad side faces the sun (south in the temperate North, north in the temperate South). You can also place windows on the sunny side for increased solar gain in the winter. Minimizing windows on the side of the house that gets little sun will prevent heat loss, as will increasing insulation in these walls. Clerestory windows allow light to come into a building from on high, which minimizes the need for active lighting systems. If you can open the clerestory windows, they can do a wonderful job of allowing hot air to vent in the summer.

Eaves, trellises, and shade trees. By having properly sized eaves on the sunny side of the house, you can create a situation where the low-angle winter sun comes through the windows and heats the interior space but the high-angle summer sun is blocked from coming in, which helps to prevent overheating. You can also include trellises on the sunny side of the building to help with this effect. By choosing deciduous vines, such as grapes or kiwis, to cover the trellis, you will create shade during the hot summer and early fall but allow sun to filter through in the winter and early spring when you really want that extra heat. Shade trees can have this same effect if you choose the right species—trees that cast light, dappled shade rather than dense, dark pools of shade.

Masonry floors and thermal masses. Dark-colored masonry floors work well in tandem with south-facing windows in the winter. As the sun comes through the windows and strikes the dark-colored masonry, the floor absorbs that heat and then reradiates it after the sun goes down. Thermal mass structures such as trombe walls, cob benches, and masonry half walls inside the windows can also help with this absorption. All of these techniques result in less need for active heating systems.

Deliberate color choices. You can also realize passive solar benefits through the colors you choose. Dark colors absorb heat, and light colors reflect it. Choosing a light color for south-facing walls can help keep buildings cool in summer. Inside the building, choose light wall colors to create brighter spaces where ambient light has more surfaces to reflect off of. Needing to use less electric lighting or fewer candles results in energy savings.

Solar tubes. Minimizing windows on the shady side of the house prevents heat loss but also decreases natural light. This can be addressed for these spaces as well as interior rooms by the addition of solar tubes. Solar tubes in the roof route daylight to these dark spaces and minimize the need for other light sources.

Together all of these techniques, and a host more, can improve the efficiency of buildings to the point where no active heating or cooling systems may be needed in some climates. Even if you do still need those systems in your climate, you will likely be less reliant on them. On top of that, these spaces tend to feel great. The extra light and comfortable conditions have positive impacts on the human psyche.

Wind and windbreaks

Like sunlight, wind can have positive or negative consequences for your buildings, depending on your context. On a windswept plain in a cold temperate climate, wind can have a huge impact on the amount of energy needed to heat a home. Conversely, in the calm, humid, lowland tropics you often want as much access as possible to any air movement at all to keep cooler and minimize mold.

If you must situate your buildings in a windswept area, windbreaks can mitigate many of the wind's negative effects. Windbreaks can not only decrease heating costs for buildings but can also

- make it more pleasant to be outdoors,
- protect crops from wind damage,
- create wildlife habitat,
- block pesticide drift and blowing dust,
- prevent plants from desiccating, and
- prevent weight loss in livestock.

Another benefit of windbreaks for temperate climates is that snow drops out on the lee side, causing larger drifts there; in the spring, those drifts melt and provide water that infiltrates into your land. This infiltration can either supply your tree crops with ample moisture for the growing season or recharge your local aquifer.

Designing an effective windbreak. To offer these benefits, your windbreak must be well designed. A single line of trees serving as a windbreak may work for a long time, but it is a risky move. What if one of those trees dies? That will leave a hole in the windbreak. Due to the Venturi effect, wind blowing through that hole will actually speed up. This will eventually damage the adjacent trees and whatever is on the other side. The added stress on the adjacent trees can cause them to die as well. You can see how a single row of trees could turn into a costly mistake.

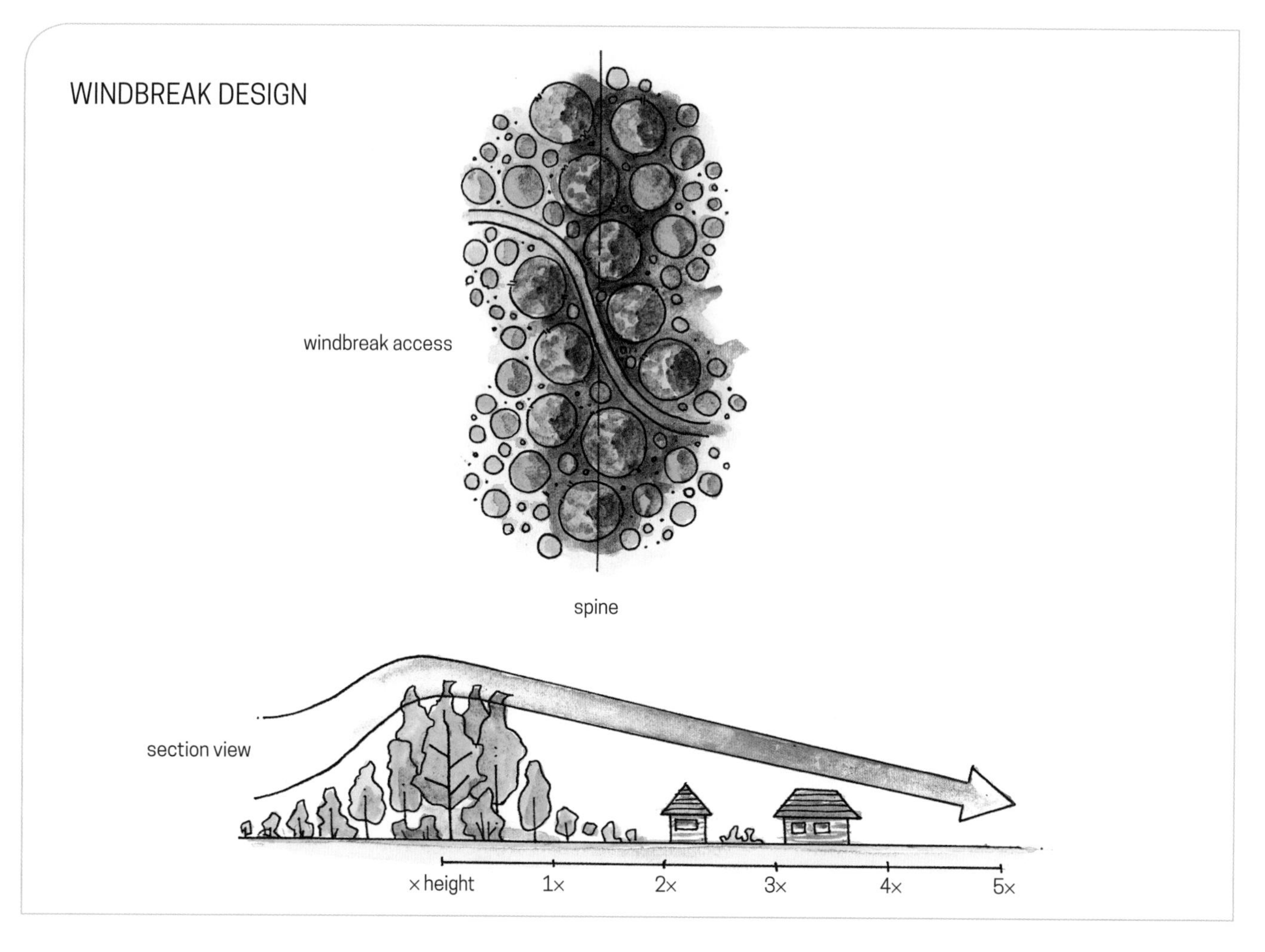

Using nature as a model, think about how windy some shorelines can get and notice that in general, the vegetation ramps up in size instead of presenting an immediate tall, vertical barrier. This helps to lift the wind over the windbreak gently and prevents damaging turbulence on the other side.

The amount of area a windbreak protects depends upon its tallest elements. In general, a windbreak protects a distance two to five times the height of the tallest point, with minimal protection up to ten times that distance. If your windbreak has a row of 60-foot-tall fir trees running down the center, it will protect the area 120 to 300 feet beyond it once it is mature.

If you need to protect a larger space than your windbreak allows, you can plant a series of windbreaks so that each one will push the wind back up before it touches down again. Systems like this are often called shelterbelts. Much of the midwestern United States was covered with shelterbelts at one point. Many of these have been lost over the years with agricultural policies that incentivize planting commodity crops "fence line to fence line." This is a real shame because the shelterbelt forests (which were quite sizeable) provided timber, wildlife habitat, and ecological services, and protected row crops in between, such as corn and wheat. If you're dealing with a large, open, relatively flat landscape, do more research on shelterbelts and consider incorporating them into your design.

For a windbreak to function well, it is also important that you don't try to block all the wind. A windbreak should actually be in the ballpark of 40 to 50 percent permeable. If it is not, damaging turbulence will result on the other side. It's advisable to mix wind-tolerant evergreen and deciduous trees and shrubs when designing your windbreak. If more permeability is needed later, pruning can help to open some spaces; if you prune, avoid removing lower limbs from the trees, which may look tidy but defeats the purpose of the windbreak and allows the wind to tear through underneath.

PLANTS FOR WINDBREAKS

First dimension given is height and second is width.

Akebia quinata
akebia
zones 5–9
vine
to 40 ft

Artocarpus heterophyllus
jackfruit
zones 10–13
deciduous tree
25–75 ft × 50–60 ft

Azadirachta indica
neem
zone 9 and warmer
evergreen or briefly deciduous tree
30–45 ft × 20–30 ft

Bambusa lako
Timor black bamboo
zone 9b and warmer
giant bamboo
50 ft × indef.

Bambusa oldhamii
Oldham bamboo
zone 9 and warmer
giant bamboo
55 ft × indef.

Caragana arborescens
Siberian pea shrub
zones 2–8
deciduous shrub
20 ft × 12 ft

Casuarina equisetifolia
ironwood
zones 11–13
evergreen tree
15–80 ft × 10–25 ft

Celtis occidentalis
hackberry
zones 3–7
deciduous tree
65 ft × 65 ft

Cercis canadensis
eastern redbud
zones 6–9
deciduous tree
40 ft × 30 ft

Chilopsis linearis
desert willow
zones 8–9
deciduous tree
6–30 ft × 6–30 ft

PLANTS FOR WINDBREAKS *(CONTINUED)*

Cocos nucifera
coconut
zone 11 and warmer
evergreen tree
30–90 ft × 15–20 ft

Cornus sericea
red twig dogwood
zones 3–8
deciduous shrub
5–6 ft × 5–12 ft

Erythrina variegata **'Tropic Coral'**
coral tree
zones 11–12
deciduous tree
60–80 ft × 25–50 ft

Fargesia robusta
zone 7a and warmer
clumping bamboo
16 ft × 8–12+ ft

Gliricidia sepium
madero negro
zones 10–13
deciduous tree
20–32 ft × 15–25 ft

Hippophae rhamnoides
sea buckthorn
zones 3–7
deciduous shrub
16–20 ft × 8–20 ft

Juniperus scopulorum
Rocky Mountain juniper
zones 3–7
evergreen tree
30 ft × 12 ft

Lablab purpureus **'Lignosa'**
lablab bean
zones 9–13
vine
to 6 ft

Maclura pomifera
Osage orange
zones 5–9
deciduous tree
50 ft × 40 ft

Myrciaria cauliflora
jaboticaba
zones 9–11
deciduous tree
15–30 ft × 12–18 ft

Nastus elatus
sweet shoot bamboo
zone 9b and warmer
giant bamboo
60 ft × indef.

Quercus macrocarpa
bur oak
zones 2–8
deciduous tree
75–100 ft high and wide

Passiflora edulis
passion fruit
zones 9–11
vine
to 30 ft

Pinus nigra
Austrian pine
zones 5–8
evergreen tree
80–100 ft × 25 ft

Pinus pinea
Italian stone pine
zones 8–11
evergreen tree
30–70 ft × 20–40 ft

Potentilla fruticosa
bush cinquefoil
zones 3–7
deciduous shrub
3–4 ft × 4–5 ft

Prunus tomentosa
Nanking cherry
zones 2–7
shrub
5 ft × 6 ft

Pterocarpus indicus
narra
zone 10 and warmer
briefly deciduous tree
82–115 ft × 82–115 ft

Often it is necessary to provide driveway access through a windbreak, but leaving a gap will cause the same problem as a missing tree in a single-row windbreak. Instead provide access through windbreaks by using a zigzag pattern that takes you into the windbreak in one spot and out in another.

When we create windbreaks, we are actually creating microclimates. One great example of this idea in action is the suntrap. A south-facing slope at Bullock's Permaculture Homestead was covered in young fir and madrone trees when the property was purchased. One of the first things the permaculturists did was to cut down some of the young trees to make a series of scallops into the forest, like a series of fish scales. In one of these scallops they put a home and garden; in another a community gathering space; in yet another they planted fruit trees. The remaining trees around each of these scallops protected the cleared areas from cold winds from the north, east, and west. The surrounding madrones also have glossy leaves, which reflect sunlight into the open space. In the cleared areas, much more dark-colored rock is now exposed to the rays of the sun. All of these things add up, and now the areas in these suntraps have microclimates that heat up faster in the spring to aid production.

At Aprovecho Sustainability Education Center in Cottage Grove, Oregon, the main gardens and residential structures are nestled into a cozy suntrap.

Building a temporary windbreak. How do you go about establishing a windbreak when the wind is so strong it blows the young trees out of the ground? The answer is temporary windbreaks. If you've chosen species for your windbreak that are well anchored and wind tolerant, they should be able to succeed if you can just give them a chance to set their roots before they get blasted.

For starters, think of anything you could use as a temporary windbreak. Old hay bales work great, as do piles of brush. You could even use old junk that prior residents left on the property. (Pull that old Buick out of the woods and use it to establish that oak tree before you send it away for scrap.) Buildings can also serve as windbreaks for anything downwind. If drainage is good, you could even dig pits to plant into. The dirt you remove from the pit could be used to form a mound on the windward side.

Another kind of temporary windbreak is a structure called a Filipino fence. This structure is particularly effective on sites where a constant driving wind makes it difficult to work and to establish plants (or anything else). To make a Filipino fence, start by sinking some posts into the ground in a line almost perpendicular to the wind. Once the posts are set, attach concrete reinforcement wire (6-inch squares) to the posts and weave some sort of vegetation through the wire

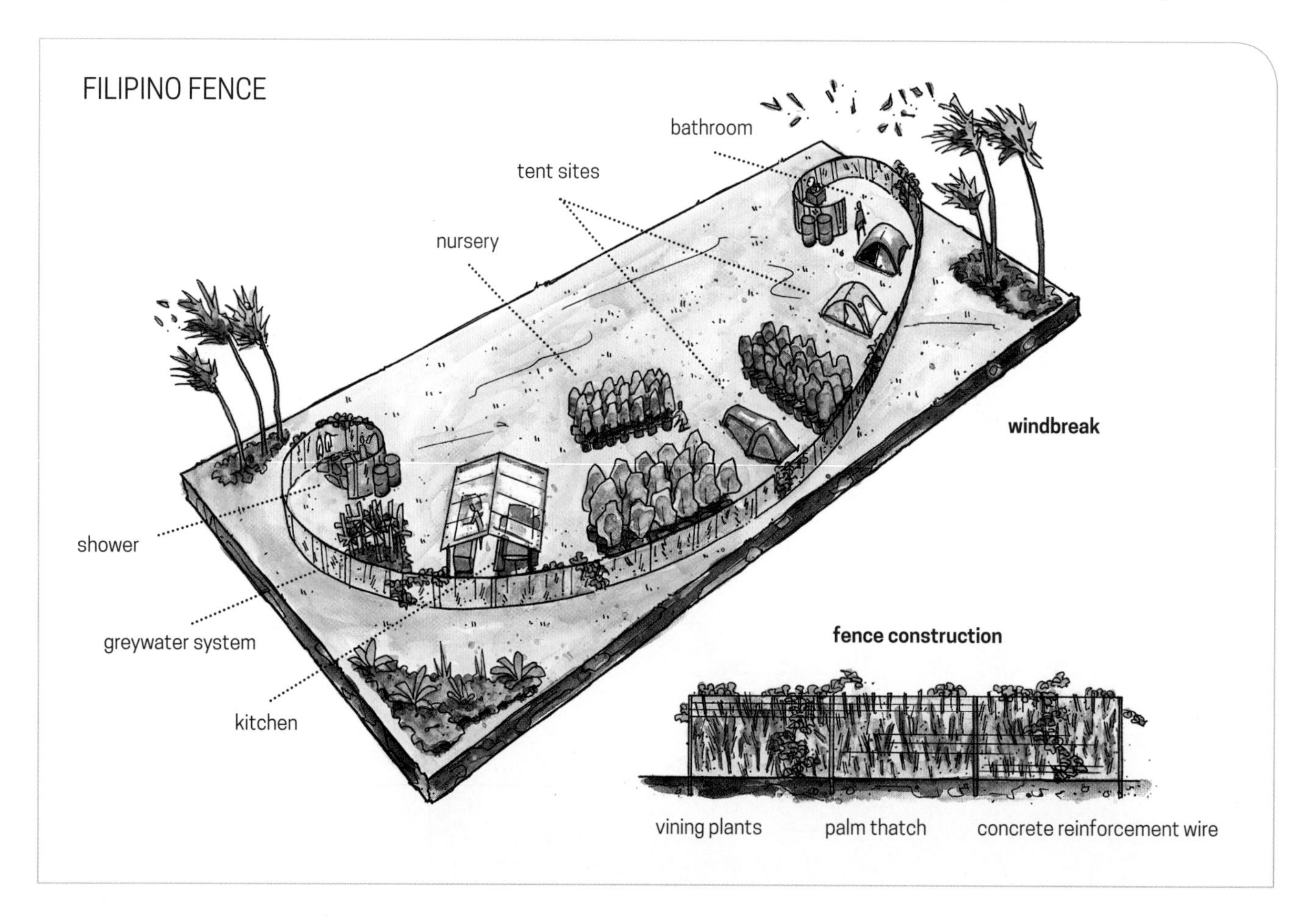

(or lash it on). We like to use palm fronds, small-diameter bamboo, reeds, and grasses. Now you've got a windbreak that will last for a few weeks to a year. At this point we like to plant a vigorous, fast-growing vine such as passion fruit (*Passiflora* spp.) at the base of each post. As the vegetative screen becomes tattered and starts to break down, the vine fills the space. Eventually, the entire fence is covered with the vine, and the vegetation has fallen to the base of the fence to serve as mulch.

You can create an arc shape with your Filipino fence to encourage the wind to go around the edges as well as over the top. On the lee side, you can keep your nursery stock close to the fence to protect it from the wind. You can also set up a tent camp, outdoor kitchen, composting toilet, and shower to make a fully livable establishment camp. Plant the first layers of your eventual permanent windbreak just on the leeward side of the Filipino fence.

Fire

If you live in a place where the natural ecology was once largely governed by fire, you'll want to consider carefully where you put your buildings and what you put between your structure and a known fire sector. Places with dry summers are often more prone to fire. Building your home at the top of a hill in these climates will increase fire danger as fires can move quickly uphill.

MATERIALS

The primary function of most buildings is to provide shelter for us or for something else. However, there are some other prerequisites that seem like such common sense that we don't even think of them. That's often where we run into trouble with modern building materials.

First, we don't want our buildings to be toxic to us. It may sound alarming, but many common building materials contain substances harmful to human health. Issues with indoor air quality are common due to off-gassing from newly made materials such as carpeting and finishes. In fact, there is even a phenomenon known as sick building syndrome where people become ill from spending time in a building made with toxic substances.

Second, we want to avoid using building materials that cause ecological damage through their manufacture, transport, or disposal. For instance, the creation of polyvinyl chloride (PVC)—a material commonly used for pipes, window frames, flooring, and siding—has serious environmental impacts, as does its disposal. Materials transported from around the world have high embodied energy due to shipping. Both of these issues steer us toward local materials that cause minimal environmental damage throughout their life span.

We don't have to look far into the past for a time when building materials were both nontoxic and local. Our goal now is to synthesize the best of what modern building materials have to offer with the best local, natural materials that have been used for thousands of years.

This green roof in Berkeley, California, uses plants and soil to insulate, minimize runoff, and be aesthetically pleasing, but it does use a synthetic membrane to separate roof from soil. This is a good balance of natural and sensible.

Balancing local and natural with sensible

One thing to keep in mind when thinking about your building materials is the tension between local, natural, and sensible. If you can choose building materials that meet all three of these criteria, that's excellent. However, you might run into situations where one of these becomes elusive, extremely costly, or impossible.

For instance, you may have local access to both good clay and straw, making cob an excellent building material for you. However, it may not meet code where you live. That can mean many years of expensive dueling with your local permitting office. Or perhaps you find that you need to use nonlocal materials to provide earthquake resistance. Realistically, very few of us have local sources for cement and steel, but we may need to use some amount of them in our buildings.

The key is to do your best and become very conscious of all the materials you use in your structure. Learn about each material before you use it.

Thinking about thermal mass and insulation

When planning your buildings, make sure you select materials based on your needs. Using an insulative material when you really want a thermal mass or vice versa just won't create the performance you want.

Thermal mass is used in homes to mitigate temperature fluctuations. Imagine a big black rock sitting in your garden during a sunny day. As the sun beats down all day, that rock absorbs quite a bit of heat. After the sun goes down, the rock radiates the stored heat back out, keeping the area nearby a bit warmer. Natural materials such as cob, adobe, and stone are good choices for thermal mass.

A good insulation material prevents the passage of heat (or cold). A down jacket insulates because the fluffy down has lots of air space in it, which traps your body heat and prevents it from escaping so easily. Conversely, a styrofoam cooler insulates because the air space in the material keeps the cold in and prevents your beverages from getting warm. Natural building materials such as straw, wood chips, wool, feathers, or even old blue jeans tend to act as insulators. In construction, R-value (R for resistance) is a measure of how well a particular material resists the passage of heat. The higher the number, the better insulator the material is.

Here are the R-values of some common natural building materials:

straw bales: 1.45 per inch

straw-clay: 0.9–1.8 per inch

wood chip-clay: 1.0–2.0 per inch

granite: 0.05 per inch

brick: 0.2 per inch

blown cellulose: 3.8 per inch

wood: 1.0 per inch

wool: 3.5–3.8 per inch

This ceiling made from caña brava (*Arundo donax*) insulates a metal roof in the tropics. Such roofs are notorious for resulting in unbearably hot spaces on a sunny day and unbearably noisy spaces during rainstorms. This insulating layer helps to mitigate both and looks sharp to boot.

Choosing natural materials

Most natural building uses some combination of biomass and earth or stone as its base materials. Biomass includes wood, straw, sod, and bamboo; earth or stone includes earth (adobe and cob), clay, brick, ceramic, stone, lime, and sand. These resources, independently or in combination, give you a wide choice of natural building techniques including, but certainly not limited to, cob, straw bale, wood, stone, wattle and daub, straw clay, and wood-chip clay.

Cob. Cob creates monolithic walls composed of a conglomeration of sand, clay, and straw (not corncobs). You start by creating individual cobs, which are balls of the ingredients mixed together uniformly. These are stacked in layers and smoothed together. About 1 vertical foot of cob can be laid up per day to give it time to dry and strengthen. Your soils may be appropriate for cob and earth plasters if they are rich in clay. (Do the shaker jar test described earlier in the book to determine your soil type.) A great coblike starter project is to make an earth oven in your yard that you can use to fire your own pizzas. (Earth ovens don't actually mix the cob ingredients; they use clay-sand on the interior, then light clay straw for insulation, then chop straw and earth plaster for the finish.)

Straw bale. Straw bale construction involves building walls out of bales of straw, which are highly insulating. The walls are then coated with plaster inside and out. Natural earth plaster is often used, with a topcoat of lime plaster for moisture resistance. Straw bales can be used as a wall infill material or they can be load bearing. If you want to start small with a straw bale project, they make great walls for workshops and studios. You can practice on these types of structures before you build your house with these materials. Practice where it matters less.

Wood. Techniques such as pole construction, timber framing, and post and beam all involve working with wood, which serves as an excellent local, renewable building material anywhere trees grow. Building structural elements from full branches and small trunks, or "round wood," is a great way to use wood without milling it first. This saves energy. Wood in the round also has more inherent strength than dimensional wood of the same cross-section. If you want to learn woodworking, you can start small with outbuildings, outdoor furniture, and landscape structures such as trellises. The principles you learn from building these simple structures can be applied to larger buildings.

Stone. If you live in an area where you are blessed with good stone, masonry techniques can yield some of the most long-lasting buildings on earth. Stone has no insulation value, though, and is best used where the need for strength or mass is the primary concern. Think about foundations, unheated outbuildings, retaining walls, and other outdoor infrastructure.

Wattle and daub. The technique known as wattle and daub involves using thin-diameter trees (usually coppiced) or bamboo to weave a wall, much like a basket.

(top left) This earthen pizza oven lives at the home of permaculturist Erik Ohlsen of Permaculture Artisans. It even includes a great cob bench to sit on while you wait for your pie.

(top right) This building at the Aprovecho Sustainability Education Center near Cottage Grove, Oregon, was built using small-diameter round wood. It is insulated with recycled paper fiber and plastered with local clay.

(bottom left) At this site in New Hampshire, rocks are abundant. When simply moved to the right place, they become retaining walls that hold soil and define roads.

(bottom right) This summer cottage made of stone in Tihany, Hungary, takes advantage of local resources and provides a cozy place to sleep on a chilly summer night in a cool Mediterranean climate.

(left) Test blocks allow natural builders to see how different mixtures of ingredients perform before they "go big" and use the materials on an entire wall.

(above right) This straw-clay wall on a building in Kasilof, Alaska, adds needed insulation. When the form boards come off, an earth plaster can be applied as a finish.

(right) This house in Devon, United Kingdom, is getting a new thatch roof. Thatching has been widely used throughout the world and is a fading skill set that we would be wise to preserve.

These wattles are then covered in an earthen plaster, the daub. This method is used for non-load-bearing walls. To start small with wattle and daub, you could put up a very simple room divider in your house.

Straw clay and wood-chip clay. These techniques use a matrix of straw or wood chips and clay to fill in the space between posts to create an insulative wall. You start by putting some boards up on the posts to create a trough in between (just like forms for pouring concrete). If you're using straw, you can coarsely chop it. Then you lay out the straw or wood chips and pour a loose clay slip over it, just like dressing a salad. Mix it together and pack it firmly between the form boards with a stick. When you remove the forms, the straw should stay where you put it, and you can move the forms up to start on the next layer. This technique goes very fast. Once you've got all the straw or wood chips in place, you can cover it with an earth plaster to keep it where you put it. This can be a good retrofit for houses with poorly insulated walls. You can just tack a frame onto the outside of the poorly insulated wall, form it up, and add your insulation.

PLANTS THAT PROVIDE STRUCTURAL MATERIALS

Borassus flabellifer
toddy or Palmyra palm
zones 10–11
one of many excellent thatch palms for the tropics

Castanea dentata
American chestnut
zones 4–8
excellent timber

Cedrela odorata
Spanish cedar
zones 9–11
timber and shade tree

Corylus avellana
hazel or filbert
zones 3–9
excellent for wattles
can make 180-degree bends

Eucalyptus deglupta
rainbow eucalyptus
zones 9–11
fast-growing timber

Guadua angustifolia
guadua
zones 10–11
look for the less thorny variety

Lolium perenne
ryegrass
zones 3–6
long history of use for roof thatch throughout Europe

Phyllostachys spp.
zones vary
temperate timber bamboos

Prunus serotina
black cherry
zones 4–8
great for furniture, cabinetry, and finish work

Quercus spp.
oak
zones vary
top-notch lumber for a wide variety of construction uses

Salix spp.
willow
zones vary
excellent coppice source for wattles

Swietenia macrophylla
Honduran mahogany
zone 11
excellent timber

Tabebuia impetiginosa
ipe or pau d'arco
zones 9–11
good for timber

HEATING, COOLING, AND TEMPERATURE MODERATION

Keeping buildings comfortable during hot and cold weather is a very important part of planning for energy efficiency. How can we design buildings that require only minimal heating and cooling? When we do need additional heating or cooling, how can we meet that need with minimal energy inputs and pollution?

Heating

Ideally, your house will be heated through passive solar design and require no additional heat. However, this may be a tall order for some situations. One of the simplest ways to warm a space is a woodstove. Even if this is only a backup for you, having one can be of critical importance in remote, cold-winter areas where service disruptions are likely. In many parts of the world, wood is a widely available, renewable fuel source. Make sure you research the stove before buying, however, as you want one that maximizes efficiency and minimizes pollution. And remember that even though wood is a renewable resource, burning it throws carbon dioxide into the atmosphere. That means contributing to climate change, so if you have options that don't require burning anything, use them first.

Masonry stoves and rocket mass heaters combine a wood-burning firebox with a chimney that runs through a large thermal mass. That way you capture and store a lot more of the heat generated by your fuel. Burning a fire once a day can result in an entire day's warmth being given off by the large thermal mass.

Distributing heat to where it's needed throughout the building is an important aspect of heating technology. Old-fashioned boiler systems piped steam throughout buildings and radiated heat via radiators. Today people often accomplish this through the use of hydronics. Polyethylene tubing carrying water heated by a stove or other strategy can be run through floors and walls to distribute heat throughout a building. If this technology is combined with thermal mass (such as an earthen floor), lots of heat can be stored for slow release throughout the day and night. In combination with an earthen floor with hydronics, it would make sense to insulate below the earthen floor so the heat isn't lost into the earth but rather reradiated into the living space.

An attached greenhouse or solarium can be used to capture heat in places with sunny winters. As the sun passes through windows, it heats the space inside (a perfect illustration of the greenhouse effect). You can tap into that heat resource by opening doors and vents to bring warm air in during the day and closing those doors at night when the greenhouse cools down.

Thinking about smart ways to create and reuse heat can lead you to design a single structure with multiple functions, which also saves on materials and creates beneficial relationships. In the illustrated garden structure, the chickens keep the greenhouse warm in the winter when they use their indoor run. The chicken coop creates a place for the chickens to lay eggs and get out of the sun. The toolshed provides a dark-colored, insulative backdrop for the greenhouse, minimizing heat loss. The toolshed gets extra heat from the greenhouse in the winter, making it a decent place to work on cold, rainy days.

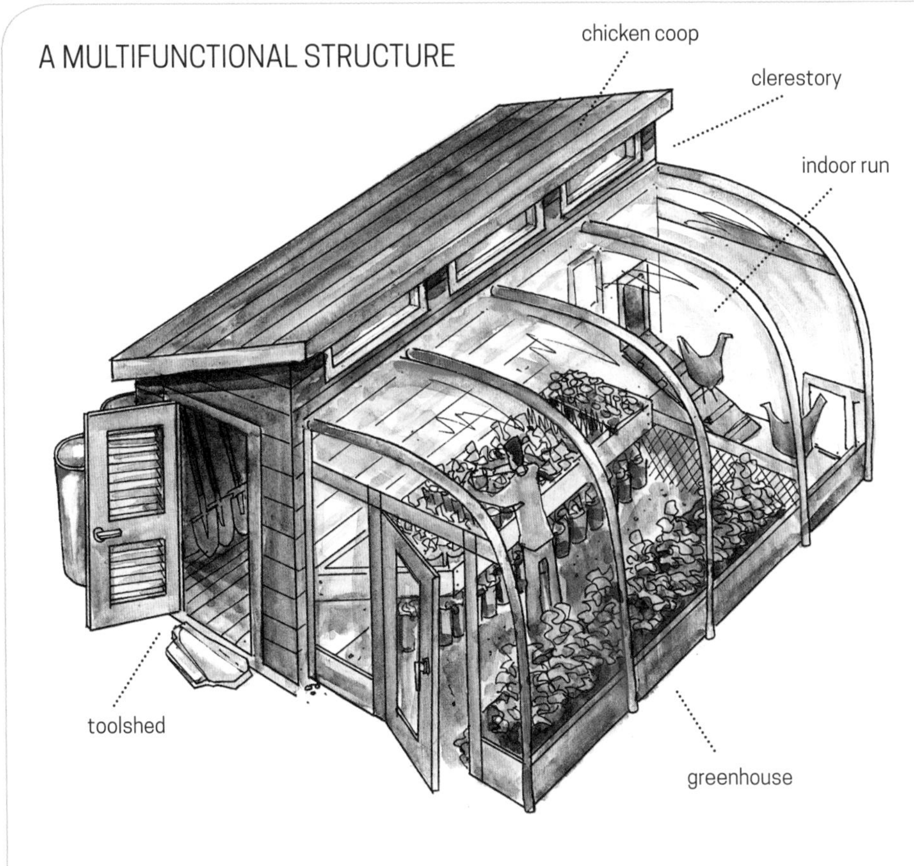

(left) This rocket mass heater has a chimney that runs through the earthen bench attached to it. Firing the stove heats up the bench, which operates as a thermal mass to radiate heat long after the fire goes out.

(right) This greenhouse attached to the south side of an older house at Bullock's Permaculture Homestead picks up a lot of heat, making it possible for crops to produce throughout the winter or be started earlier in the spring.

Cooling

In dry climates, evaporative coolers (or swamp coolers) are a simple, energy-efficient way to keep buildings cool. Swamp coolers basically run a small electric fan and a pump to move water. The water flows over a membrane and the fan blows on it. As the water evaporates, it cools the air around it (just like wearing a wet t-shirt on a windy day). Evaporative coolers use far less electricity than air conditioners. Note, however, that in humid climates these don't work well due to the ambient humidity in the air.

Masonry floors can help to keep a space cool as long as the sun doesn't shine directly on the floors. Anyone who has ever spent time in a basement in a hot climate can attest to this. Shady courtyards off a house can provide a cool place to spend time during hot weather; in a dry climate, running misting sprinklers in a shady area full of plants can enhance the evaporative cooling in the area.

Shade trees that are carefully selected and properly placed can also be a big help in keeping a building cool. Think about the mature size of the species, the best placement based on that size, the type of shade cast (dappled versus deep), and the compatibility of the selected species with the rest of the landscape. An evergreen gives shade year-round, whereas a deciduous tree selected specifically for its leaf-out and leaf-drop dates can provide shade when you need it and sun when you don't. Trees that cast light, dappled shade are great choices as they let enough light through that you can still have a well-lit interior and a productive landscape underneath them. For example, in temperate climates thornless honeylocust (*Gleditsia triacanthos*) and silk tree (*Albizia julibrissin*) are good shade trees. In the tropics, consider monkeypod (*Albizia saman*) or narra (*Pterocarpus indicus*). In addition to casting dappled shade, all of these trees are nitrogen fixers as well.

Temperature moderation

Some strategies can simultaneously keep your building warmer when it's cold and cooler when it's hot. These include earth tubes, cisterns, and ample insulation.

Earth tubes (or ground-coupled heat exchangers) are tubes installed underground near the building in which they are used. Most places on earth have a constant subterranean temperature around 55°F (with some variance for climate). That means the temperatures in those tubes will be similar. If the air from those tubes is pumped into your home directly on a hot summer day, it will feel nice and cool (55°F sounds pretty good when it's 98°F outside). If that same air is pumped into a heating system on a cold winter day, less energy will need to be used to bring that air up to a comfortable temperature (from 55°F to, say, 68°F) as compared with drawing in outside air (which may need to be heated from 5°F to 68°F).

Cisterns located below buildings can moderate temperatures throughout the year because water serves as an excellent thermal mass. Having a large tank of water below your home can make a notable difference in the energy you use to heat and cool it. In fact, if it's well designed you can even use a "thermal cistern" to store heat from a heat source so it radiates back over a long period.

Ample insulation can also help moderate indoor temperatures. Insulating a floor can help to prevent heat loss in the winter and help keep the space nice and cool in the summer. Consider also insulating ceilings, walls, windows, and doors. If you feel cold air at the edges of windows and doors, use weather stripping to keep drafts out and heat in.

RETROFITTING BUILDINGS FOR HEALTH AND ENERGY EFFICIENCY

Ideally, you will plan for a healthy and energy-efficient shelter from the beginning. But what if you aren't starting from scratch and most of your structures are already in place? Tearing down your house and rebuilding it to be more efficient would waste tons of energy, thus defeating the purpose of doing so in the first place (not to mention that it would break the bank for most of us). In making decisions about shelter, you need to balance long-term vision with practicality. If your house was built in the last ten years and isn't going anywhere soon, you can start thinking about retrofits to increase efficiency and decrease toxicity.

WAYS TO INCREASE EFFICIENCY AND DECREASE TOXICITY

	Making simple changes	**Upgrading what you have**	**Adding new features**
Toxicity and indoor air quality	Get houseplants (which can help to remove toxins from the air).	Repaint with low- or no-VOC paints or natural finishes. Remove carpet. Seal off openings to unfinished areas (such as crawl spaces).	Add good exhaust fans to rooms with moisture issues.
Thermal performance	Add weather stripping. Paint outside walls a light color to reflect intense sun.	Increase wall insulation. Incorporate a thermal mass in a sunny spot.	Add or remove windows. Add earth tubes. Redo heating system. Build masonry heater.
Siting for fire	Cut firebreak.	Plant fire-resistant vegetation in fire sector.	Install pond in fire sector.
Siting for wind	Add weather stripping.	Add insulation to windward side.	Plant a windbreak to windward.
Siting for sun	Put blinds outside windows that cause overheating.	Build trellises on the sunny side. Modify eaves on the sunny side.	Lift and turn the building (can be feasible for small buildings). Build an attached greenhouse on the sunny side.

Food and Plant Systems

Providing for Our Own Needs

(left) A productive landscape can also be beautiful.

(right) Growing food in your landscape presents opportunities for diversity you seldom see in the supermarket. Clockwise from bottom: 'Blanka' white currant, golden raspberries, 'Illinois Everbearing' mulberry, 'Red George' gooseberry, blackcurrant, and 'Santa Rosa' plum.

Food production is a main focus of most permaculture designs. Issues with our global industrial food system are so widespread that the best way to ensure food quality and security is to grow at least some of it ourselves. As it is, if the long supply chain is disrupted, we may be left without food. What if semi loads of food stopped coming to your town? Would there be enough food produced locally to go around? Producing at least part of our food supply is one of the most powerful steps we can take toward meeting more of our needs at home and building resilience. Knowing where our food comes from and how it's grown is another very important reason to grow food in our backyards—or front yards, side yards, rooftops, or wherever we find space. If we aren't sure where our food comes from, we don't know if it's being grown with chemicals or how fresh it really is.

This chapter focuses on incorporating food production and other plant systems into our permaculture designs. We explore the place of perennial and annual food crops in the landscape as well as techniques for urban agriculture. We cover various management and maintenance considerations involved in getting our landscapes to reach their full potential. Food processing and storage methods to handle the abundance that results from a successful food production system are also covered here.

PERENNIAL VERSUS ANNUAL CROPS

Much of the world's food supply comes from annual crops that are produced in a way that wastes resources and causes environmental problems. Tillage of the soil on a regular basis is destructive, and the use of chemicals and freshwater irrigation is often rampant in these systems. An annual plant grows for one season and then must be replanted. Biennials take two years to complete their life cycle. Perennials, by contrast, live and yield for many years before they die. Foods from perennial plants can come in the form of tree fruit and nuts, berries, canes, shoots, tubers, and more. Permaculturists advocate obtaining at least part of our food supply from perennial crops because they require fewer resources and inputs than annuals, and they allow the successional process to move past square one (which is what annuals maintain in perpetuity).

Let food be thy medicine and let medicine be thy food.

—HIPPOCRATES

(top left) This bed is designed with perennial vegetables (asparagus) in the background and trellised annuals in front. Surplus slate roofing is used to keep watermelons warm in the foreground.

(top right) Dahlias are beloved for their large and colorful flowers, and many dahlias also have edible petals and tubers.

(left) Many gardeners do not realize that several species of hostas are perennial vegetables, and the new shoots can be eaten.

DESIGN CONSIDERATIONS FOR PLANTING THE PERMACULTURE LANDSCAPE

Design considerations for the permaculture landscape, whether planted in annuals or perennials, include planting polycultures rather than monocultures, companion planting, designing with guilds, selecting species and cultivars or varieties to match site conditions, and using an appropriate mix of natives and nonnatives.

Companion planting

Polyculture is the practice of growing multiple species in the same space and at the same time, as opposed to monoculture, where a single crop is grown over a large area. Companion planting is a form of applied polyculture that uses plant interactions to increase productivity. A traditional example of annual companion planting is the "three sisters": corn, beans, and squash. The corn needs nitrogen, and beans fix nitrogen and thus can give the corn what it needs if planted at the base of the corn. The corn also acts as a trellis for the beans to grow on. Squash covers the ground, providing shade for the roots of the other plants.

In the right combination, plants can provide a number of ecological services for one another. Such services include the following:

Pest control. All pests have a predator, and some plants attract specific pest predator insects.

Resource sharing and availability. Some plants make resources such as nitrogen available to others that need it. Some plants also provide the shade that others need or complementary root-growth habits.

Pest deterrence. Certain plants have been found to deter specific pests through compounds and exudates. If these pest deterrents are planted near plants prone to pest issues, they can help to prevent damage. Some aromatic plants such as onions and marigolds are also suspected to confuse pests, keeping them away from other plants or garden areas.

Trap cropping. Certain plants can be used to actively attract pests and prevent pest damage to desired crops. For instance, where wireworms are a problem, if radishes are planted along with turnips, the radishes will mature first and the wireworms will attack them. The radishes can then be removed, along with the wireworms, to minimize damage to the turnips.

Not all plant interactions are beneficial. Plants have the ability to inhibit the success and survival of other species by releasing chemical growth inhibitors through their tissues (bark, roots, leaves, stems, and such). This allelopathy, as it is termed, is the plant's natural defense to protect itself from competition. Trees in the walnut family (Juglandaceae) and some running grasses are examples of allelopathic plants. Every part of a black walnut tree, for instance, releases a substance known as juglone that suppresses plant growth under the tree. Allelopathic plants are nature's weed killers. If you want to include these species in your designs, you can buffer their effects by locating them away from susceptible crops. Alternatively, you can use them to your advantage by planting them between your crops and an area with noxious weeds. One allelopath, ryegrass, is often part of overwintering cover crop mixes used in annual gardens; it suppresses germination of weed seeds in the early spring.

Designing with guilds

In permaculture, a group of mutually beneficial plants assembled into an interactive community is called a guild. Dave Jacke and Eric Toensmeier, co-authors of *Edible Forest Gardens*, identify three different types of guilds to make the point that permaculturists define guilds somewhat differently from ecologists:

Community function guild. This type of guild groups all of the species that do a specific job or fill a specific niche (for example, plants that host pest predators, nitrogen-fixing ground covers). This is how ecologists generally use the word. In the human context, all of the bakers or all of the masons in a community belong to this type of guild. In the design context, all of the elements that provide a specific, necessary function (for example, generate electricity, provide food, extend seasons) can be identified to give us palettes of options to draw from to solve specific problems.

Mutual support guild. This type of guild groups several species with complementary functions that work together and support each other. This is how permaculturists generally use the word. A mutual support guild in the plant community could be made up of one nitrogen fixer, one plant that hosts pest predator insects, and one that attracts pollinators. In design, we might make lists of community function guilds and draw species from them to create functional mutual support guilds.

Resource partitioning guild. This type of guild groups several species that can share essential resources through niche differentiation. Examples are planting a bulb that needs sunlight early in the season under a late-leafing tree so that they can share the sunlight resource, or using a fibrous-rooted plant next to a tap-rooted plant so that they can draw from different levels to access soil nutrients. This is an added layer to think about with our designs. The idea is to maximize cooperation and minimize competition so that all plants in our polycultures flourish.

Each of these types of guilds can guide you as you select plant combinations for your permaculture landscape. You can make lists of plants grouped into community function guilds, all the species that serve a particular function, so that you have a palette of options to consider. You might have lists of weed barriers, mulch makers, nitrogen fixers, dynamic accumulators, windbreak plants, wildlife food or habitat, pest repellants, fire retardants, insectaries, fragrant plants, and plants for basketry materials, wood products, fibers, medicines, animal forage, dyes, oils, food, and cut flowers. These are the lists from which you can select plants when you assemble mutual support guilds. For instance, one particular guild might have a food plant, a nitrogen fixer, a couple of dynamic accumulators, and a weed barrier. Once you have selected some species for your mutual support guild, you can check for competition using the concept of the resource-partitioning guild. Do the species you have selected do a good job of sharing resources or do they compete with one another? If there is competition, can you manage it through cropping the more vigorous species or would you be better off swapping one species for another? In this way you can use the guild concept to assemble functional communities of plants in your landscape.

In addition to these concepts, we also frequently differentiate between establishment guilds and mature guilds. An establishment guild is a guild you use to get a landscape established. In that case, you generally have one or more target species in your guild that you are trying to get established along with pioneer species to help the target species succeed. Many, if not all, of these helper species will eventually be shaded out or removed as conditions change over time. A mature guild refers to the compatible species you add to the landscape to replace those helpers that have been removed. For instance, if the tree layer in your landscape closes its canopy, you may decide to plant an understory of shade-tolerant medicinal herbs, such as ginseng, black cohosh, and goldenseal. You'll read more about establishment and mature guilds later in the chapter.

Selecting species and cultivars or varieties

When selecting species and cultivars or varieties, you'll want to consider a number of important features of crop plants that may impact their appropriateness for your site and/or situation.

Nutrient density. Some crop plants have a much greater nutrient density than others. With more nutrients and fewer empty calories, those plants do a better job of nourishing the body and supporting long-term health. Some particularly nutrient-dense annuals include kale, spinach, and Swiss chard. Nutrient-dense perennials include strawberries, blackberries, plums, and artichokes. Making sure our landscapes are full of nutrient-dense foods it is like having a full pantry from a nutrition perspective.

Self-fertility and cross-pollination. Some plants are self-fertile. This means they are able to produce fruits or seeds by themselves without another plant of the same species around. However, other plants require two individuals of the same species but different genetics in order to produce. For instance, peaches are self-fertile, but most apples require two different cultivars that flower at the same time (such as 'Fuji' and 'Gala') in order to produce fruit. Make sure you know which category your plants fall into.

Male-female plants. Some species of plants are actually male or female. Generally the males do not produce fruit and the females only produce if they receive pollen from a male. It's important to know this ahead of time. If you're planting from

seed, you may want to plant denser than adult spacing so you can select out the excess males at some point. Examples of plants like this include kiwis (*Actinidia deliciosa*), papayas (*Carica papaya*), and ginkgos (*Ginkgo biloba*).

Chill hours. This is particularly important to understand if you live in a subtropical climate where it seldom freezes in the winter. Some plants, such as apples, cherries, and peaches, require cold weather to form resting buds. If this never happens, they don't respond well when spring comes and they often won't flower. If you live in these climates, you should know if your plants require chill and, if so, how much. A good rule of thumb for calculating chill hours is that every hour a tree spends between 32°F and 45°F is considered one chill hour. If you live somewhere where chill hours are an issue and you want to grow fruit trees that require chill, look for "low-chill" varieties. For example, the 'Pink Lady' apple requires only two hundred to four hundred chill hours to produce, making it a good selection for many parts of the southwestern United States.

Disease. When selecting plants for your landscape, it is also important to know disease issues that are present in the area so you can select species and varieties or cultivars that are resistant. For instance, European pears tend to suffer from fire blight. If this is a potential issue in your area, you may want to choose fire blight–resistant varieties such as 'Harrow Delight' and 'Honeysweet'.

Length of growing season. It is important that you choose crop plants that will be able to ripen their produce given the length of the growing season in your area.

(left) In Jessi's garden this pear tree rises above a mixture of shade-tolerant guild plants including variegated comfrey (dynamic accumulator and weed barrier), carpet bugle (ground cover), hostas (perennial vegetable), and lamium (nectar source for pest predators).

(center) At the Regenerative Design Institute in Bolinas, California, this fruit tree guild is aesthetically pleasing, with repeating color and texture plus a clean edge made of wood found on-site.

(right) This kiwi trellis at the Good Shepherd Center in Seattle has both male and female plants rambling together. This allows for pollination and loads of fruit.

For instance, at Bullock's Permaculture Homestead on Orcas Island, Washington, pecans and 'Fuji' apples grow just fine, but the growing season is seldom long enough to ripen the fruit on either, so they are bad choices. Walnuts and 'Gala' apples both do just fine so are better choices.

Using natives and nonnatives as appropriate

If you're looking for a plant to do a job in the landscape and a native species fits the bill, it makes sense to use it as a first choice. After all, native species tend to be really well adapted to their bioregions. They are well suited to environmental conditions and usually require less maintenance than a nonnative plant. Those natives will also be likely to support a whole host of other native species that we want around. For instance, many native insects have co-evolved with the plants they rely on for food. If we want to maintain the building blocks of a functional ecosystem, supporting native species is key.

Nonnatives also have a place in permaculture design. When you're looking for plants that do a specific job and there are no appropriate native species to choose from, you can explore exotic species for inclusion in your design. There are some good reasons to do this. Permaculture puts a high value on providing for your own needs, and if you limit yourself to using only native species in your landscape, that could compromise your ability to provide for your needs at home. In the greater context, if you don't take responsibility for growing at least some of your own food, fiber, medicine, and so forth, you will always be relying on someone else to do it. Dedicating your yard to natives while importing all your food from afar is not ultimately a sustainable scenario. In fact, if we all produced even a small percentage of our own food, we would be able to let much more monoculture cropland return to the primary function of providing the biodiversity that supports wildlife habitat and ecological services.

Also, in many bioregions, ecosystems have been compromised to the extent that we can no longer live on native species. For example, in the maritime Pacific Northwest, the lives of pre-European-contact indigenous people largely revolved around salmon and old-growth cedar. Today, little old-growth cedar is left, salmon runs aren't what they used to be, and a lot more people must be fed. We can't go back to that way of life, yet we still have to eat, dress, and build shelters. That means using exotic crop plants.

Consider these questions as you think about mixing natives and nonnatives in your permaculture landscape:

- How many native species have you eaten today? Many cultures have diets that include nonnative species. For instance, tomatoes are from Latin America and considered a big part of Italian culture and cuisine.
- How does time play into the concept of native species? Camels evolved in North America but no longer exist there.

- How does geography play into the concept of native species? Is there a line on a map that accurately delineates that a species belongs on one side but not the other?
- Does conserving endangered species where they aren't native have value?

AGROFORESTRY SYSTEMS

Agroforestry refers to systems that grow trees to aid in the provision of agricultural products. Good examples include orchards, nut groves, alley cropping (where alleys of trees are interspersed with other agricultural crops), and forest gardens. Agroforestry systems are based on tree crops and other woody perennials.

Many agroforestry schemes involve mixing trees with meat and dairy animals, apiculture (bees), understory medicinal herbs, basketry materials, and other crops. One key thing to think about when designing agroforestry systems is whether you want the system to produce commercial crops. If not, the forest gardening approach may be an excellent way to go. While exciting, the modern forest gardening concept is still in its infancy; we need more successes and failures from which we can learn.

If your livelihood is tied up in the productivity of these systems (whether focusing on commercial or subsistence crops), your agroforestry scheme should allow for the following:

- easy harvest and transportation
- practical access for machinery and maintenance
- large enough volumes of each potential product to be economically viable
- crop timing that works in tandem with other maintenance tasks
- production of crops that will be marketable

In agroforestry systems, nitrogen-fixing trees and shrubs are often interplanted with crop trees. These can be managed through a system called "chop and drop." In this system, the nitrogen fixers are coppiced or pollarded. (*Coppiced* means cut to the ground with the intention of allowing the tree to sprout again; *pollarded* means cut back to some point higher on the trunk to create new sprouts above ground level.) When the lush green growth on top is chopped off, a proportional amount of the root system dies as well, releasing a charge of nitrogen into the soil that the surrounding plants can access. This is best done just before crop plants go into a growth cycle. Chopping up the removed material and dropping it around the base of the trees gives all the benefits of mulch, and the brush doesn't have to be hauled away.

Overarching goals for any agroforestry system, commercial or not, are to increase ecological function, decrease workload, and increase biodiversity. If we can create productive systems that achieve these goals, we'll be a lot closer to sustainability.

Alley cropping

For production-focused systems, alley cropping is a concept that deserves more attention. Incorporating bands of tree crops into fields of annual grains, pulses, and vegetables offers a number of benefits. One example of this can be seen at Project Bona Fide on Isla Ometepe in Nicaragua. The residents have an area with a gentle slope where they have planted rows of trees on contour with lots of space between the rows. These trees are interspersed with vetiver grass (*Chrysopogon zizanioides*), which prevents soil loss when the annual fields are bare. The tree species consist of nitrogen fixers and other species that provide useful products. One example of the latter is a native tree called madroño (*Calycophyllum candidissimum*), which is useful for making tool handles, firewood, and charcoal.

During the beginning of the wet season, annual crops such as rice, sorghum, and beans are planted in the fields. Simultaneously, the tree crops are coppiced to open up maximum sun to the fields and release nitrogen into the soil to support the growth of the annual crop. Throughout the wet season, the annual crops and tree crops grow (or regrow in case of the trees) together. When the dry season hits, the tree crops provide some protection from the sun; the annual crops are harvested and the fields are left fallow. The process is repeated each year as a means of managing fertility and increasing the total diversity of products from the fields.

At Bullock's Permaculture Homestead, an alley-cropping system intersperses rows of young fruit trees and berry bushes with rows of annual vegetables.

Structural forests and woodlots

Structural forests and woodlots are largely focused around providing wood for lumber, firewood, and/or pulp for paper. Depending on your climate, these may be the best option for large areas of the property that would be classified as zone 4 because they require fairly little maintenance and need to be visited only occasionally. Maintenance activities may include thinning, limbing, pest and disease control, and selecting for preferred species. For these areas, having a well-thought-out management plan is important. Knowing what species are growing there, when you plan to harvest them, and how you plan to harvest them is important.

Timing is an important part of the planning as well. For instance, seasonal weather and soil conditions (such as soil saturation, compaction, and freezing) should be considered when planning management activities to minimize issues with compaction and erosion. Harvest and maintenance methods are also important to think about when going into a forest with heavy logging equipment, as it will have a radically different impact than going into a forest with a team of horses and chainsaws to do the same job.

Forest gardening

Forest gardening focuses more than most other agroforestry approaches on emulating nature. In a historical context, the idea of growing the things we need in a diverse, forested context is not new. However, very few recognizable examples of this are still around today, especially in temperate climates. Permaculturists Dave Jacke and Eric Toensmeier have compiled tons of scientific research and personal experience in their two-volume *Edible Forest Gardens*, which proposes a very detailed way to design forest gardens to increase production for humans while maintaining maximum ecological function.

Forest gardening has several goals that make a lot of sense for anyone designing a permaculture landscape. For instance, we want to design what Jacke and Toensmeier refer to as "overyielding polycultures." That means if we compare a monoculture orchard to a forest garden of the same size, we hope to receive more overall value from the polyculture even though we will receive less of any one product. Another goal is for a forest garden to become a largely self-maintaining system. If we do a good enough job of closing loops, we should be able to let the forest do most of the work while our primary job becomes harvesting.

When designing a forest garden, we want to stack our plants so we can get more production. A forest garden can have seven different layers:

The tall tree layer. The upper canopy layer contains the dominant trees that require the most space and have the biggest impact on resources, including sunlight, water, and nutrients. This layer includes some full-sized fruit trees, many nut crops, and many timber trees.

The lower tree layer. This layer is where most fruit trees live. The trees can be grafted onto dwarf rootstock so the mature plant is low enough to harvest easily and accommodates the sun needs of the layers below it.

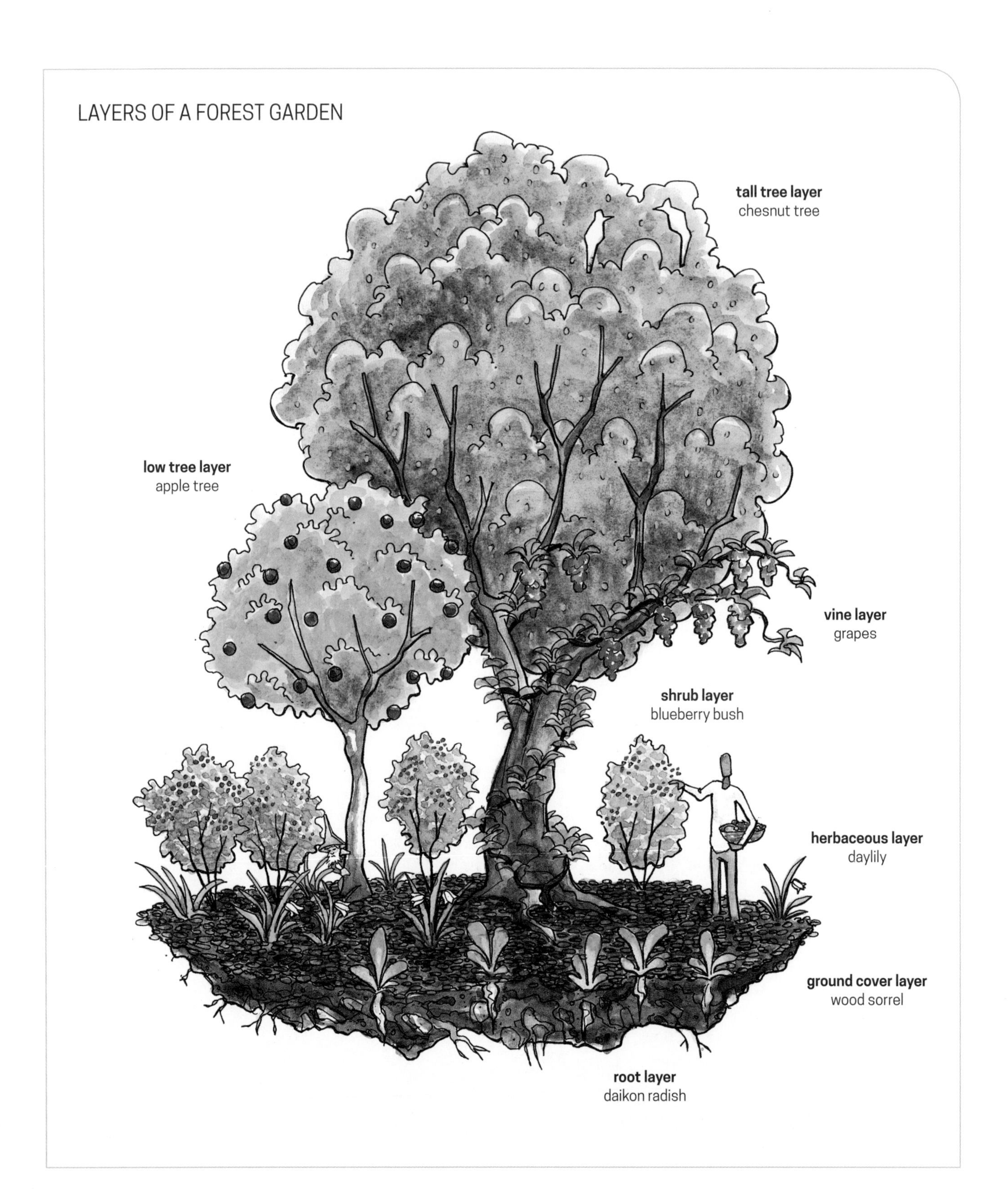
LAYERS OF A FOREST GARDEN
tall tree layer
chesnut tree
low tree layer
apple tree
vine layer
grapes
shrub layer
blueberry bush
herbaceous layer
daylily
ground cover layer
wood sorrel
root layer
daikon radish

The shrub layer. This perennial layer is full of woody plants in a range of sizes that produce food and fibers and provide wildlife habitat. This can include berry shrubs and canes, and even bamboos. Most fruiting plants are more productive with more sunlight, so that should be a consideration in placement of these plants.

The herbaceous layer. This layer contains herbaceous perennials, or in other words, nonwoody plants. Options here include vegetables, herbs, cut flowers, and insectary plants.

The ground cover layer. Covering the ground is important for erosion control and mulch production, and ideally our ground covers will also produce more yield for us such as food or wildlife habitat.

The vine layer. We can grow productive vines in our forest garden for food or wildlife. These can be trellised up the trees to take advantage of an underutilized space but need adequate support for the yield they produce.

The root layer. The underground dimension should not be overlooked. We can benefit from a variety of root crops and mushrooms.

Our goal in a forest garden is to find ways to emulate the architecture of natural forests and increase productivity for humans. However, one of the most common errors in forest garden design is cramming the plants together too tightly. Overcrowding can result in fungal disease and competition for sunlight, both of which will decrease productivity and increase maintenance time. Applying the concept of guilds can help you create systems that balance productivity with ecological function while avoiding undesired competition.

In general, forest gardening is a great permaculture approach for people wanting to move their more traditional landscapes toward more perennial systems and maximizing their productive spaces.

(left) This collection of self-seeding herbs makes a great ground cover in Jessi's garden. It includes chives, red-veined sorrel, and oregano.

(right) At Treesearch Farms in Houston, Texas, demonstration landscapes include a wide variety of productive, functional species that occupy many different vertical layers.

Agroforestry system establishment

Many good strategies exist for establishing agroforestry systems. These include using establishment guilds, contour planting, and succession planting.

Using establishment guilds. Using establishment guilds is a great way to help get your target tree species going. When you plant any establishment guild, you know that most of the species will eventually be departing from the system. Always remember that permaculture landscapes are dynamic.

In a temperate climate, a variety of species can be selected to support a young fruit tree in the center of an establishment guild. For example, comfrey (*Symphytum officinale*) is a great weed barrier and dynamic accumulator, lupine (*Lupinus* spp.) is an ideal nitrogen fixer, and daffodil (*Narcissus* spp.) is a good gopher deterrent. Most of these guild associates are sun lovers. That means once the fruit tree is mature, the conditions to support the associates will no longer exist.

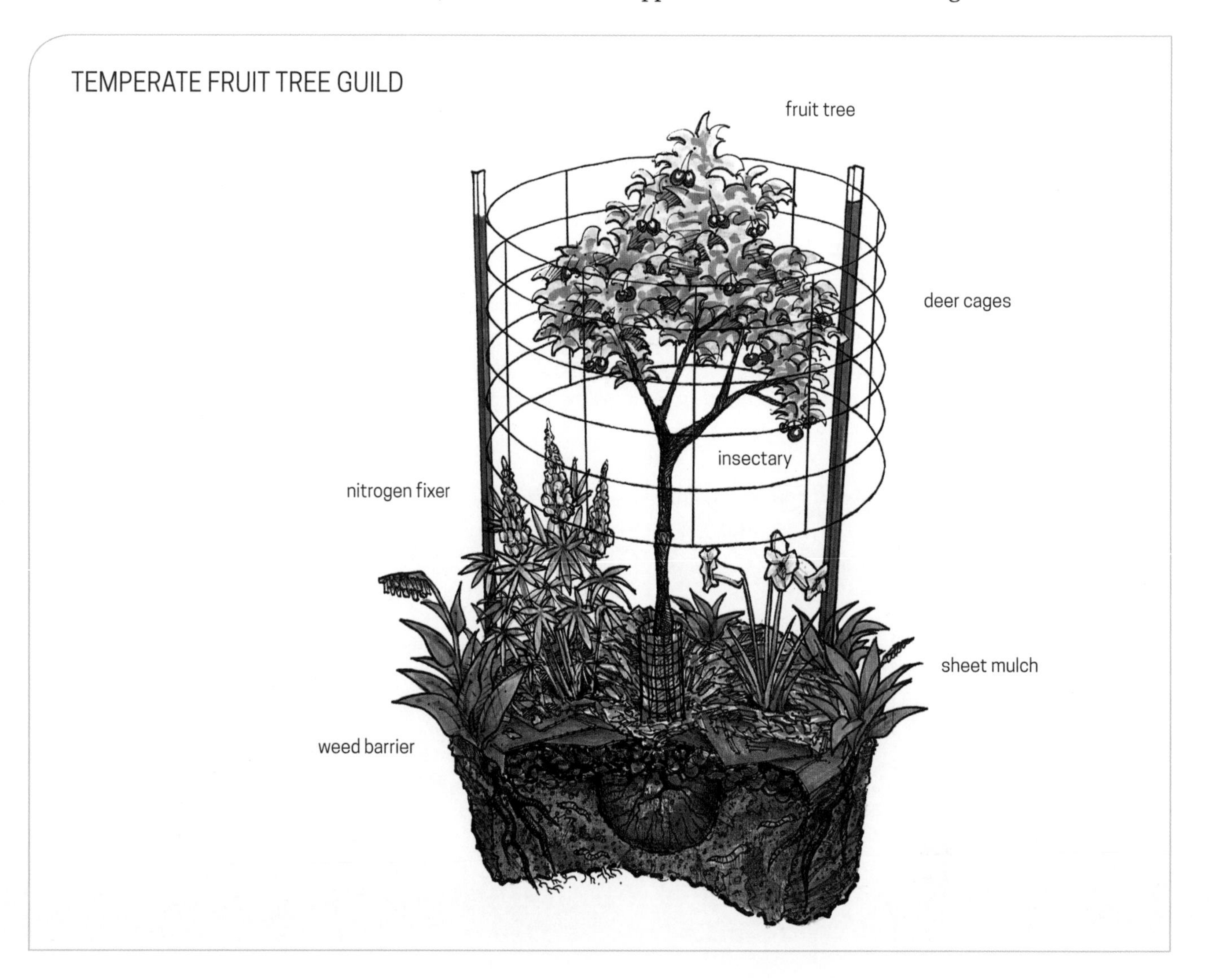

In a tropical banana guild, the banana provides wind protection, sun protection, soil moisture conservation, juicy mulch, and bananas. Meanwhile, all the other plants provide various functions to help the fruit tree (in our illustration, *Artocarpus heterophyllus*, jackfruit) get established. Madero negro (*Gliricidia sepium*) can serve as a chop-and-drop nitrogen fixer, sweet potato (*Ipomoea batatas*) makes a great ground cover, lemongrass (*Cymbopogon flexuosus*) is a good weed barrier, and chile pepper (*Capsicum* spp.) can serve as a human attractant. Attracting people to your establishment guilds with tasty snacks is a great way to bring their attention to the area periodically so they can notice if anything is amiss.

When creating establishment guilds, make sure you don't make them too complex. Too many species or individuals can lead to a management nightmare, so when in doubt, err on the side of simplicity.

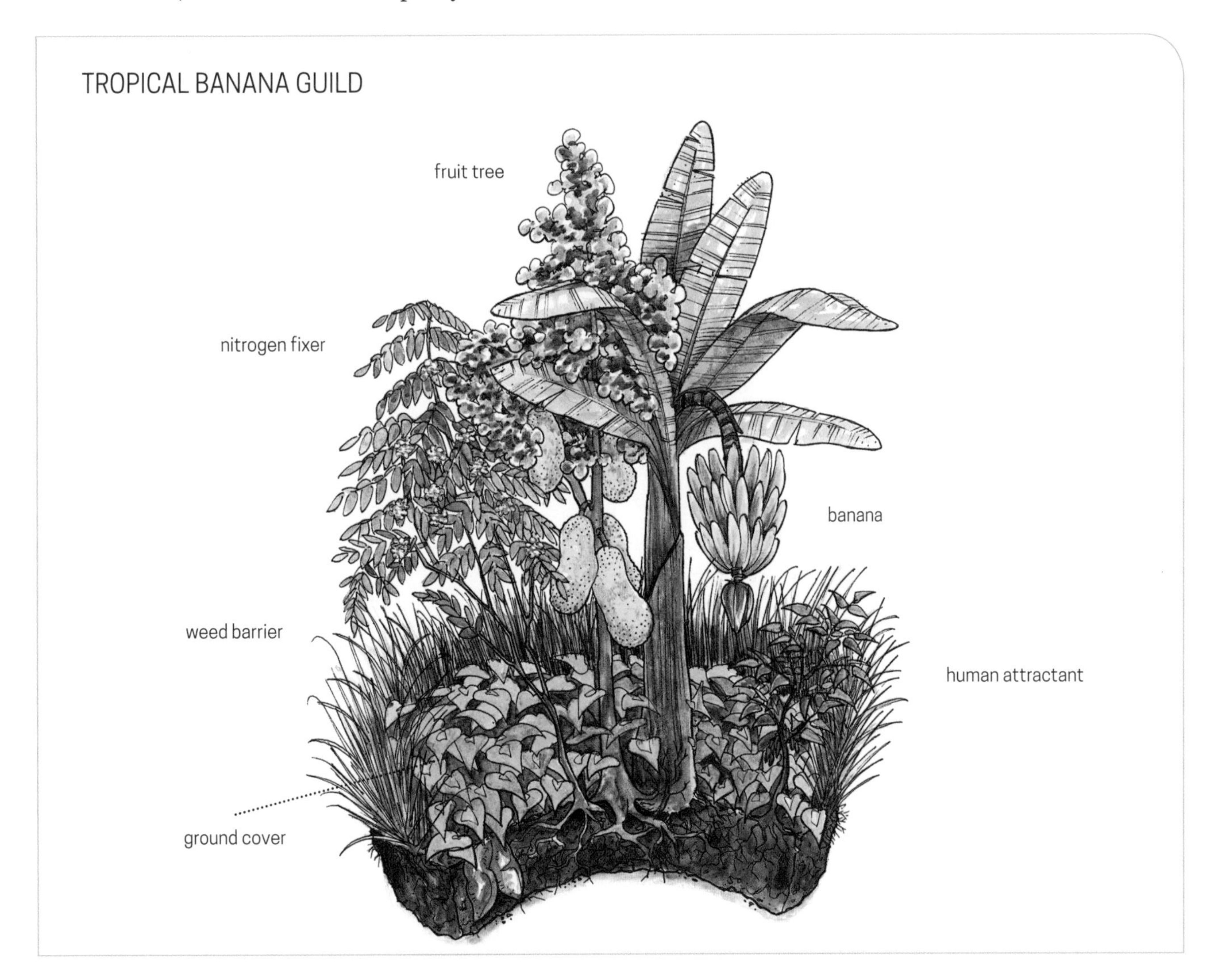

Contour planting. Once you've got the idea of establishment guilds down, you can start to combine them together in ways that lead to a whole landscape. For instance, contour planting is a strategy for establishing tree crops on slopes. Individual establishment guilds are laid out around trees planted on contour (meaning each row of trees is dead level). Contour planting can help to minimize erosion as mulch and materials are laid out along the contours between the trees. This is also compatible with infiltration swales that will help sink water into the landscape. As the system matures, guild associates slowly fade out. Maintenance tasks include "chop and drop" management of nitrogen fixers and establishing emergent palms between fruit trees (appropriate for the tropics).

In the tropics, it makes sense to lay out the whole system with alternating crown and stem bearers, referring to the way trees bear their fruit. Crown bearers tend to have rounded crowns and bear their fruit on the outside of the canopy where it is exposed to the sun. Examples include mango (*Mangifera indica*), litchi (*Litchi sinensis*), longan (*Euphoria longan*), and pili nut (*Canarium* spp.). By contrast, stem bearers tend to have more columnar or pyramidal shapes and bear their fruit along the trunk and major branches away from the sun. Jackfruit (*Artocarpus heterophyllus*), starfruit (*Averrhoa carambola*), cacao (*Theobroma cacao*), and jaboticaba (*Myrciaria cauliflora*) are examples of stem bearers. Trees tend to develop in these forms based on whether their fruit needs sun exposure to ripen well. Cramming crown bearers together tightly means production will be lost on a good part of each tree; conversely, stem bearers will bear good fruit whether the sun hits the fruit or not. In agroforestry systems, you can maximize your trees per acre by interspersing crown and stem bearers.

Succession planting. Succession planting is another great strategy for establishing agroforestry systems. Succession planting is a way to accelerate the successional process and obtain earlier yields by planting plants from many different stages of succession all at once. For instance, you may plant some young fruit and nut trees at their adult spacing. However, that leaves lots of space in between. So between your rows of fruit and nut trees, you might include a row of berries. Between the berries and the tree crops, perhaps you grow some rows of annuals. This fills in all the space with productive crops.

The annuals will come into production the first year. In the second or third year, the berries should come into production. Every year you can continue to grow annuals during the growing season until the tree crops and berries start to create conditions too shady for vegetable production. Little is lost, as the fruits and nuts start to come into production around this time. By the time the tree crops close canopy and shade out the berry rows, their production will be in full swing. At this point you've harvested a lot more than you would have if you had only planted the tree crops at their adult spacing and waited for them to produce. At this point you can also consider shade-tolerant crops for the understory.

Transition to a mature guild. The guild concept that we've demonstrated to establish your agroforestry systems can be augmented by using the same concept to transition to a mature guild. In a mature guild, many of the early functions aren't essential anymore. For instance, nutrients tend to cycle very efficiently in

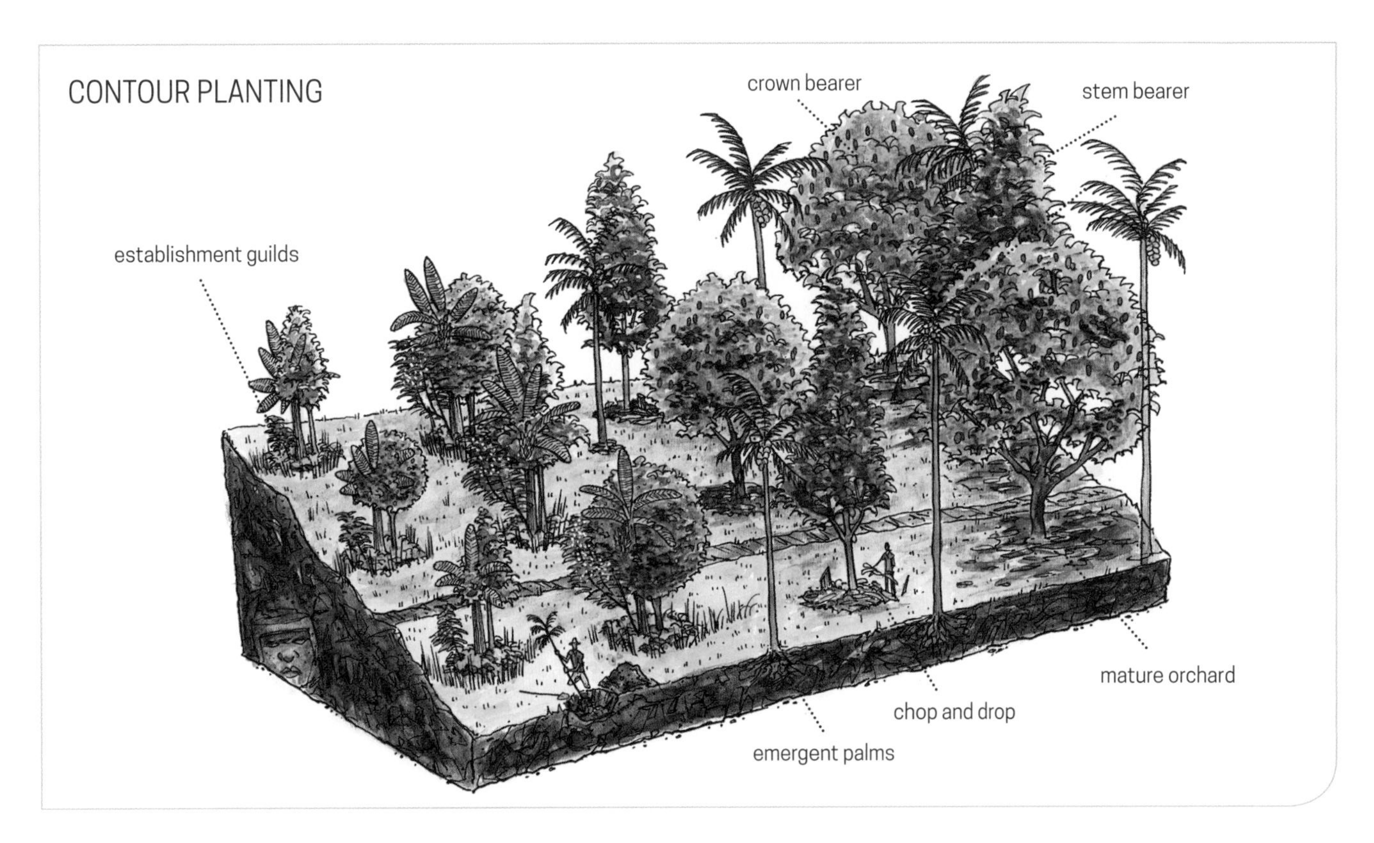
CONTOUR PLANTING
crown bearer
stem bearer
establishment guilds
mature orchard
chop and drop
emergent palms

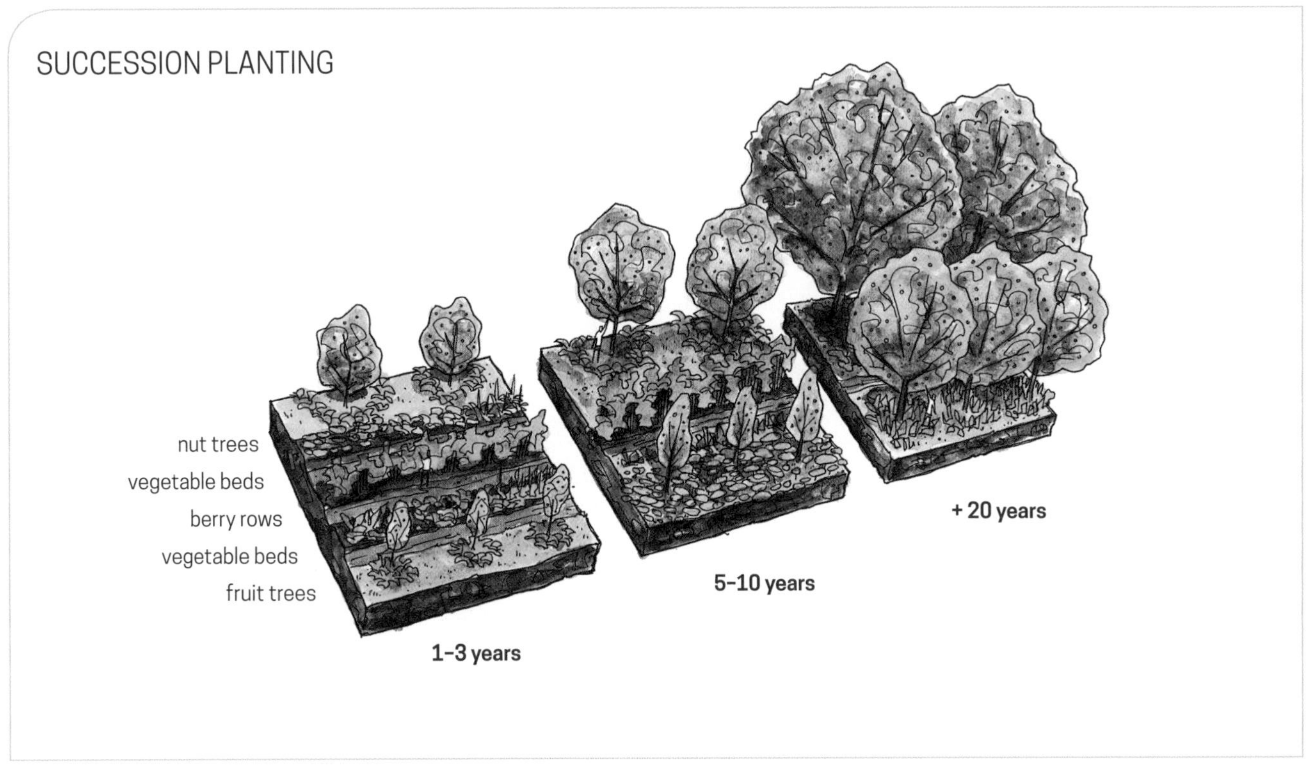
SUCCESSION PLANTING
nut trees
vegetable beds
berry rows
vegetable beds
fruit trees
1–3 years
5–10 years
+ 20 years

established systems so nitrogen fixers are not as important (which is good, since there are very, very few shade-tolerant nitrogen fixers). This is actually a very interesting time for a designer and a great example of how the design process doesn't end when you create your initial plan.

Ideally, by this point you've been able to learn a great deal about what does and doesn't work well for your property and lifestyle. That will help guide you in the process of selecting shade-tolerant guild associates for your plants. Also, note that you can start the shade-tolerant understory plants earlier in the process. They just have to be started in patches close to the tree trunks and they can expand out from there as the pool of shade expands. Some productive understory plants for these shady areas include medicinal herbs and edible and medicinal mushrooms.

ANNUAL CROPS

As much as permaculturists advocate the use of perennial plant systems, growing annual crops will likely always have a place, too. These plants start and complete their life cycle in one season. Overall, the annual garden requires much more work and planning over time than a perennial food system, but if we keep our annual systems small and intensive, the yields can be very rewarding. Designing an annual system starts with choices about soil preparation and bed layout and extends to selecting plants and creating a fertility management plan.

Bed layouts

In permaculture, the keyhole bed and mandala-shaped garden layouts are popular templates for space efficiency. However, if you have a long and narrow space, neither of these may be appropriate. Also, you need to consider access to and through the space. If you need to move wheelbarrows in and out, it's important to design the layout for accessibility.

As mentioned in the chapter on water, in an arid climate you can make the most of limited water by sinking garden beds into the ground. By contrast, raised beds are an opportunity to get roots up and out of saturated soils if you have poorly drained soils. In some climates with a lot of moisture, this is imperative for keeping plants healthy. Such beds also offer an opportunity to provide fertile soil where poor soil may have limited your growing options. Raised beds bring weeding and harvesting closer to the height of the person who does those tasks.

Raised beds don't have to be boxed in with wood or stone or any other material; they can consist simply of piled-up soil. However, boxing in the soil does offer some benefits. The thermal mass from some materials such as urbanite (broken chunks of concrete) or stone can actually help raise and regulate soil temperatures. And the edge of a raised bed can be a great sitting area for people who prefer to have a place to rest while they garden or have physical needs that this helps. On the other side of the coin, using rocks, urbanite, or timbers to raise up a bed can create nooks and crannies that harbor weeds like quackgrass and pests like slugs and snails.

(top left) Rebecca Newburn has laid out her garden beds in the San Francisco Bay Area in an easy-to-access way that is also attractive.

(top right) These boxed beds at Picardo P-Patch Community Garden in Seattle are made with rot-resistant juniper. The vegetables grown here are destined for a local food bank.

(left) Jessi's garden has a variety of beds, some raised and boxed with salvaged granite, local stone, or wood, others consisting of just mounded soil.

This Swiss chard has been mulched with straw to minimize weeding and prevent drying out.

No-till gardening

The idea that we need to rototill the garden every year goes against what nature has shown us to work. True, tilling breaks apart the soil, making it easier to work with. It also hyperaerates the soil, which causes microbe populations to explode and munch through the soil organic material much faster, releasing the nutrients for plants to take up. But many of those nutrients are also lost to leaching, volatilization, and runoff, and over the long haul, tilling actually damages the biological activity in the soil. Better for the soil are methods of gardening that work specifically without tillage.

In the early 1900s, Ruth Stout, an organic gardening pioneer, learned and taught that we need to follow nature's example of using mulch to have healthy annual gardens without tilling. Mulch protects and feeds the soil, which in turn feeds the plants. It also minimizes the work needed compared to conventional gardening: no tilling, no weeding, no fertilizing, and minimal watering.

Masanobu Fukuoka was a Japanese agricultural scientist who was also highly influential in permaculture and natural farming methods. His "do-nothing" approach uses very little tillage and emphasizes letting natural biological systems do the work that humans would otherwise need to do. His internationally known best seller *The One Straw Revolution* not only encompasses natural farming but also connects that to health, diet, and cultural values. While the book contains practical advice, it is ultimately an inspiring philosophical work about human connection to the natural world.

Timing your crops and extending the growing season

For annuals, proper timing of sowing and harvest is essential. Your growing season in temperate climates is the time between the last frost in the spring and the first frost in the fall. In warmer climates, you may be able to grow throughout the year as long as you can irrigate during dry periods and keep plants cool enough during hot spells. This varies from region to region. Some plants can be sowed directly into the ground, while others are best started indoors or with protection before the last frost date or in anticipation of a rainy season. Even though decades of weather data is available, first and last frost dates vary from year to year. Watching weather patterns and forecasts will help. Optimal harvest times can vary from plant to plant in the same garden. Some plants, like peas, lettuce, and basil, can be sown multiple times each year to ensure a harvest throughout the growing season.

In the temperate world, many methods can be used to extend your season in either direction, spring or fall. Cloches and cold frames, which can be made from a variety of inexpensive or free materials, retain heat and protect plants from frost and cold temperatures. You can also grow species adapted to some frost and colder temperatures if you want to extend your season.

(left) The garden at Dave's and Yuko's apartment in Seattle has boxed beds outfitted with moveable cloches to enable year-round production in the city.

(right) Growing watermelons under glass is another way to use season extension infrastructure to help a crop along that doesn't otherwise get enough warmth to ripen.

In the wet-dry tropics, extending your harvest into the dry season can benefit your crops greatly since there can be less pest and disease pressure during the dry season. Season extension techniques in this case include water conservation strategies, efficient irrigation, appropriate shading to minimize heat damage, and proper crop selection. Notice that the concept is the same as in the temperate world, but the actual techniques differ.

Crop rotation

Crop rotation is an age-old practice used by gardeners and farmers to systematically plan for moving annual crops from area to area. Growing the same crop—or crops in the same family—in the same area for years on end can deplete the soil, allow a disease and pest population to build up in that soil, and result in poor yields. It is important to plan ahead to rotate families of crops to different areas. Here are the most common families of annual crops:

Amaranthaceae: beets, Swiss chard, amaranth, quinoa, spinach, sugar beets, orach

Amaryllidaceae: bulbing onions, bunching onions, garlic, leeks, shallots

Apiaceae: dill, parsley, fennel, carrots, parsnip, cumin, caraway

Asteraceae: sunflowers, lettuce, endive, chicory, chamomile, salsify, shungiku

Brassicaceae: broccoli, kale, collards, radishes, cabbage, rutabaga,mustard, arugula, bok choy, brussels sprouts

Cucurbitaceae: squashes (winter and summer, pumpkins), cucumbers, melons, zucchini, luffas

Fabaceae: peas, beans, soybeans, chickpeas, lentils, fava beans, alfalfa, jicama, peanuts

Solanaceae: tomatoes, tomatillos, potatoes, peppers, chiles, pimentos, eggplant

Some of these "annuals" are actually perennials that are treated as annuals in temperate climates (this is true for most of the nightshade—Solanaceae—family). It's a good idea to have a crop rotation plan in place before you start. For instance, in a particular area you might have legumes follow brassicas, which follow alliums. This will prevent buildup of disease organisms and replenish the soil.

SELF-SEEDING FOOD ANNUALS AND BIENNIALS

Antheum graveolens
dill
annual

Arctium lappa
burdock or gobo
biennial

Borago officinalis
borage
annual or perennial

Brassica napus var. *pabularia*
red Russian kale
annual
overwinters in zone 7 and warmer

Chrysanthemum coronarium
shungiku
annual

Coriandrum sativum
cilantro or coriander
annual

Eryngium foetidum
cilantro
annual or perennial

Hibiscus acetosella
cranberry hibiscus
annual or perennial

Lactuca indica
Indian lettuce
annual or biennial

Petroselinum crispum
parsley
biennial

Tragopogon porrifolius
salsify
biennial

FOOD PRODUCTION IN URBAN AREAS

Some techniques for producing food, whether using annuals or perennials, are especially important for urban dwellers. While many of these techniques make sense for rural locations as well, in urban and suburban areas they can significantly increase productivity.

Espaliers. An espalier is a woody plant that has been pruned to a two-dimensional form so it can run along a fence line or spread out against a building. Fruit trees can be espaliered where space is extremely limited.

Rooftop gardens. In urban areas, plenty of flat rooftops go unused. If the buildings have been engineered to handle the load, growing food up there can be a great idea.

Vertical growing. With newer materials, people are now starting to grow food on vertical surfaces. Urban areas have lots of unused vertical surfaces. The vertical growing system needs to make sense with the rest of the design, must be easily harvestable, and must not be too energy intensive or dependent on outside resources.

Container gardening. Many different types of plants, from vegetables to fruit trees, can be grown in containers. This is great for folks who live in apartments. One of the best things about container gardening is that if you move, you can take your plants with you.

(left) Dave espaliered this peach tree on the south side of the apartment building where he lives. Growing flat against the wall allows it to take advantage of the extra reflected light and heat and keeps its foliage out of the rain so as to avoid the fungal disease peach leaf curl, which is a problem in his area.

(right) In Seattle, garden blogger Stacy Brewer used an old bike wheel to create a trellis to grow her beans vertically.

Street trees. Most urban areas encourage the planting of street trees. While some places may have rules against fruit trees in the parking strip, you might want to get creative and push the boundaries on this. If you've got flowering fruit trees in your parking strip already, you can slowly convert them to more productive fruit trees branch by branch using grafting techniques. You could have a future as a guerilla grafter!

Public lands. From community gardens and orchards to city parks, much urban land could be used for food production. Participating in community gardening programs in your area could be a great way to produce more food in an urban situation. Imagine if city parks were planted with low-maintenance, food-producing tree crops like chestnuts, walnuts, and nut pines. To make that happen, you just need to be in contact with your local representatives and participate in the public process when public lands are being designed.

PERMACULTURE FOOD PLANTS FOR SMALL SPACES

Acca sellowiana
pineapple guava or feijoa
zones 8–11
handsome evergreen shrub or tree with beautiful
edible flowers followed by pineapple-flavored fruit

Butia capitata
jelly palm
zones 10–12
attractive
yard-scale specimen tree producing fruits that make excellent jelly

Castanea pumila
eastern chinquapin
zones 5b–9
large shrub with edible chestnuts

Citrus ichangensis × *C. reticulata*
yuzu citrus
zones 7a–11
small citrus that is a popular flavoring in Japanese cuisine

Citrus japonica
oval kumquat
zones 8–12
small citrus tree that can be hedged
fruits don't require peeling making them a great snack fruit

Cornus mas
cornelian cherry
zones 5–8
attractive small tree with tasty
deep red or gold fruits

Dioon edule
chestnut dioon
zones 11–12
very attractive Mexican cycad
females bear edible nuts that can be ground into flour and eaten

Diospyros kaki 'Izu'
Izu dwarf persimmon
zones 7–10
naturally dwarfing Asian persimmon that requires little pruning

Eugenia uniflora
Surinam cherry
zones 10–12
excellent snack fruit perfect for lining trails

Malus domestica 'Scarlet Sentinel' and 'Golden Sentinel'
columnar apple
zones 4–9
nearly branchless apples meant to grow in containers or small spaces

---

Manilkara zapota
sapodilla
zone 11 and warmer
excellent fruit tree with fruit tasting like brown sugar or maple syrup and with special dwarfing varieties available

---

Myrciaria cauliflora
jaboticaba
zones 9–11
incredibly attractive
ornamental fruit tree that doesn't get too big

Pouteria sapota
mamey sapote
zone 10 and warmer
excellent fruit tree with fruit tasting somewhat like sweet potato and with dwarfing specimens finding their way into markets

---

Punica granatum
pomegranate
zones 7–10
attractive ornamental tree with dwarf varieties available that are perfect for containers

---

Pyrus 'Warren' and 'Magness'
columnar European pear
zones 4–9
varieties with a more columnar growth habit than most

LANDSCAPE MANAGEMENT

When it comes to maintaining your food production and plant systems, you need to learn and schedule many tasks and techniques. Nature will throw some curve balls when you least expect it, and it is best to be as prepared as possible and to have a plan in place for when this happens. All plants are subject to damage, whether from insects, animals, environmental toxins, extreme weather, or disease. Most gardeners are lifelong learners and constantly seeking knowledge, which is a fairly important trait to possess. It will become important to learn as seasons change and plants need different care for different reasons, from dealing with weeds to pruning to managing pests and more.

Dealing with invasive or opportunistic species

A small fraction of exotic and native species are opportunists that take advantage of available niches and spread rapidly via vegetative means and/or seed dispersal. Sometimes these are called invasive species, but we prefer to avoid that term, which is emotionally laden and often complicates discussions. Ultimately, each permaculturist needs to formulate his or her own perspective on opportunistic species.

Most opportunistic species are adapted to disturbed conditions, so their appearance tells us something about the health of the site. If we want to keep opportunistic species from proliferating, we need to address the underlying cause, which is a disturbed and degraded landscape. If we focus on eradicating opportunists without addressing the problem, we aren't likely to solve anything. This would be much like spraying an herbicide on your lawn year after year to eliminate moss without addressing the fact that the area is too shady and acidic to successfully support lawn.

In actuality, opportunistic species often act as nature's repair crew. They show up in disturbed landscapes and cover the earth to prevent soil loss, produce biomass, and improve fertility. They fill open niches, capture nutrients newly made available, and enable succession to move forward. These valuable functions are not often discussed. We must remember that landscapes in nature are not static. They are all in one stage of succession or another, and they are on the move. Trying to freeze them in time means fighting against nature and creating work for us.

Sometimes we do want to stop succession. For instance, an annual garden bed is purposely frozen at an early stage of succession, and we do the extra work involved because we are going to get something productive out of it. However, at other times, halting succession exacerbates a problem we are trying to solve. If we freeze a landscape at an early enough stage of succession, we may always have problems with opportunistic species. In fact, among the best ways to address early successional opportunists is to allow succession to move past them. A hallmark of moving beyond the earliest stages of succession is the appearance of shade, and most opportunists cannot persist in the shade. Therefore, establishing canopy cover can be a chemical-free, low-labor way to eliminate a locally rampant species. Shade-tolerant opportunistic species, such as wild morning glory in western Washington or kudzu in the southeastern United States, present a more difficult challenge. One potential way to deal with these sticklers is to find uses for them. For instance, kudzu root powder is used in place of cornstarch in Japanese cooking.

Often the alternative to this approach is applying massive amounts of time, money, resources, and chemicals. All these things are being used to work against nature. As nature tries to repair a disturbed landscape, we humans fight against it and, simultaneously, continue to disturb more landscapes. This approach makes little sense and is certainly an inadequate excuse for using carcinogenic herbicides in our landscape, supporting the fossil fuel industry through the creation of those herbicides, or spending our very limited tax dollars allocated to land care to fight against nature. In fact, from a practicality perspective, eradicating many of the exotic opportunists that have gotten a foothold in our landscapes would be impossible (especially since many of them actually adapt to become more resistant to herbicides over time).

So how should we approach opportunistic species as permaculture designers? For starters, let's avoid incorporating opportunistic species into our designs that create more work for ourselves and/or problems for our neighbors. In addition, we need to look at landscapes with a perspective that goes beyond human timeframes. With or without humans, plants are always moving around the globe.

As a new plant comes into an ecosystem, it may initially act as an opportunist. However, over time (usually more time than humans think about or plan around) other species adapt and the ecosystem comes back into a state of dynamic balance. What we end up with is a functional hybrid ecology (what permaculture cofounder David Holmgren refers to as ecosynthesis). If we stop redisturbing areas claimed by exotic opportunists with chemicals or other eradication efforts, other species will adapt and balance will again be achieved eventually. The problem is that this can take many years, and few people take the long view on this issue.

Pests and diseases

One approach to managing pests and diseases is integrated pest management (IPM). This strategy seeks the least energy-intensive way to manage pests and diseases. When applying IPM, you explore solutions in order from least energy intensive to most energy intensive.

The first step is to correctly identify the pest or disease. This can be done with the help of trained extension agents who work at local universities and are aware of what is happening in your area with specific crops. Knowing the pest will allow you to research its biology and life cycle to get a better understanding of how you may be able to impact it. With this information you can begin monitoring, which will tell you how much damage is being done to your crops.

At this point you can establish a threshold of acceptable damage. It is entirely possible that the amount of actual damage is negligible, in which case you can just ignore it. Insects' nibbling on leaves here and there is much different from stressing a plant to the point of its needing to be on life support. Sometimes you simply need to take maintenance measures to prevent the pests in the first place. That should be considered a standard practice when you are aware of problems that

(right) In this border along Jessi's annual garden, many of the plants, such as yarrow and aster, support pest predators. The alliums and lavender act as aromatic pest confusers, and the bantam chickens deal with any pests that do manage to stumble into the area.

(left) Jessi has a small nursery for her landscaping company, NW Bloom Ecological Landscapes. She uses poultry power to control pests and weeds in the pots.

could be alleviated that way. When damage goes beyond the acceptable and starts to seriously impact your yield, you can intervene in the system. If you decide to do this, you can start by trying the least ecologically damaging strategies first. This could mean mechanical (hand picking, pruning, trapping), cultural (crop rotation, intercropping, sanitation procedures), biological (releasing pheromones or pest predators), or, as a last resort, chemical controls.

The next step is to evaluate the success of the strategies you've tried. Some difficult pest or disease issues may actually require multiple strategies applied together to achieve success. Don't hesitate to try more than one way of dealing with a pest. For instance, you may need to prune at a certain time of year in combination with trapping at another time of year to get a particular pest under control.

Unfortunately, this sensible approach to pest and disease management is sometimes used as an excuse for using chemical biocides. Chemicals should really be a last resort and avoided entirely, if possible, as they cause damage to other systems (including soil health, water quality, and pest predator populations) you are working hard to keep healthy. In fact, if you can't grow it without chemicals, you probably ought to consider growing something else instead.

Pruning

Pruning is an important part of caring for any garden. Of primary importance is pruning for health—removing dead or diseased branches, letting light through and improving air circulation, making sure the plant is structurally sounds as it grows. Young trees often need pruning and training to make sure they develop a strong scaffolding of limbs. Fruit trees are pruned to encourage fruit production and control the height for easier harvest. Some trees are coppiced or pollarded and the new growth is harvested. Get yourself some high-quality pruning tools, attend workshops, and consult how-to books as you learn this essential art.

Propagation

You can grow your own plants through propagation. In nature, plants can grow from seeds or by vegetative means. Propagation by seed often results in variation, as the process involves mixing genetics from multiple parents, while propagation by vegetative means results in offspring that are exact replicas of the parent plant. Some plants take well to only one form of propagation and have specific requirements. For plants that can be grown either way, your goals also affect your selected propagation method. For instance, if you want to breed a new variety of quinoa for your climate, you will need to propagate the plants from seed so that you can select the progeny that are most successful in your climate. However, if you are trying to create an exact replica of another plant to take advantage of various qualities (such as ripening time, disease resistance, color, and variegation), you should choose a vegetative means of propagation (cuttings, division, grafting). You also need to consider the time of year or season when you choose a propagation technique, as well as what kinds of care these plants will need after propagating.

Seed. Some plants are best propagated through seed. It's important to figure out the natural conditions under which a specific plant germinates. Some seeds require special processing to germinate. For instance, dill seeds require moisture and direct sunlight to germinate. Many perennial seeds have built-in mechanisms that ensure they won't start germinating on a warm day in the fall. Seeds requiring stratification must be subjected to cold temperatures for several weeks to several months, depending on the species, to germinate. You can accomplish this in some climates with fall planting. You can also put such seeds in a plastic bag with a moist

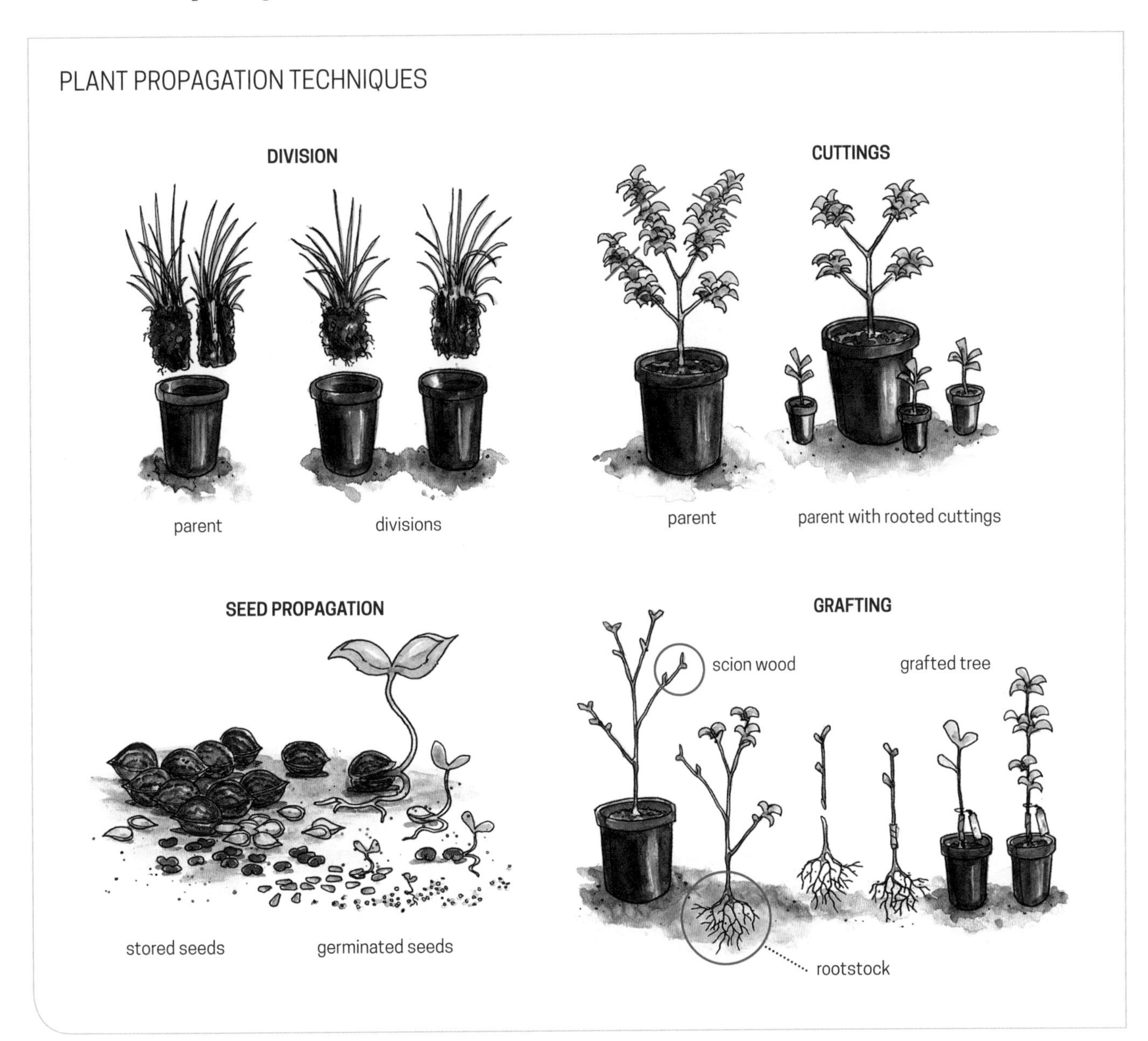

In Portland, Oregon, Marisha Auerbach prepares seed for storage by drying it in a solar seed dryer.

medium and refrigerate them for the required time. Scarification is another treatment that may be needed to physically break through the seed coat in order for it to germinate. By physically damaging the seed coat by mechanical means such as cutting, drilling, or sandpapering, you allow water to seep in. In nature, this process often happens in the stomachs of animals.

Cuttings. Cutting off a branch, branch tip, or root segment and getting it to form roots is an easy way to start new plants from a parent plant. Generally this is done at certain times of the year, and some plants readily grow if you put them in a soil medium. Some plants such as willow (*Salix* spp.) have rooting hormone in their tissues that naturally facilitates root growth from their nodes, while other plants need some help. Rooting hormone can be purchased, or you can make your own in the form of willow tea. Plants like willow with lots of ambient rooting hormone can often be used as an erosion control strategy in a technique called live staking. This involves taking branches of the tree in question and pounding them directly into the soil, where they can root and grow into shrubs or trees.

Division. Propagation through division is a great strategy for many herbaceous perennials that form crowns with multiple stems. Simply take the parent plant and divide it in half with a good blade, and voila—you have two! Our tool of choice is a hori hori knife or serrated garden blade, but a sharp shovel can also work well.

Grafting. Grafting is a near magical plant propagation technique that has been used to reproduce plants with desirable characteristics for more than two thousand years. When you graft, you are actually doing a bit of plant surgery and connecting two plants together to provide benefits for you. You select a rootstock, the plant that will provide the roots, with certain attributes or performance characteristics (for example, size control, drought or water tolerance, disease resistance), and then you select the scionwood, the part that will be attached to the rootstock, to have different attributes such as fruit or nut type, flavor, or ripening time. For instance, you might choose a wild plum seedling for your rootstock because it's tough, and you might graft a scion of a flavorful Santa Rosa plum to that.

Almost every tree fruit you've ever had from the grocery store has been brought to you through grafting. This is also commonly done with ornamentals to produce certain characteristics like leaf color or flowering time. One species can generally be grafted to another member of the same species, and there are even some members of different species (or genera) that can cross, but be sure to do your homework before selecting species for grafting. Native rootstock is a great way to go if you have that option as the plant will be adapted to local conditions such as soil type and climate. Other reasons to graft include repairing older trees, preserving heirloom varieties, speeding production or bearing time (waiting two years for yield instead of five or more), producing seedless fruit, increasing disease resistance, and fitting multiple fruit varieties in a limited space by grafting them onto one tree.

Grafting actually does occur in nature, most often at the root level. In crabapple thickets, for instance, the roots often graft together so the trees can support one another. We have also seen evidence of grafting in Douglas-firs that have been cut down. Usually a Douglas-fir will die when cut, but if the tree has root-grafted to its neighbors, the nearby trees will send healing energy to the stump to heal it over with scar tissue and bark, treating it just as they would if they had a wound in their own tissue. All this is just to say that grafting has been used successfully with no harm done to humans or nature for a very long time.

Seed saving

When planning your gardens, consider seed saving as an important component of your resiliency plan since it helps you avoid the need to rely on seed companies each spring. It's easier to save seed from some species, such as lettuce, beans, and tomatoes, than from others. Focus on these species first and plan to give at least a few individuals time and space to bolt and produce seed you can collect. By saving seed you not only save money but also allow plants to flower, which supports insects.

FOOD PROCESSING AND STORAGE

If your food production system is successful, you will likely have more than you can eat at some times during the year and in some years. Food processing and storage is needed to take the abundance from the lush times and make it available during the lean times. Food preservation options include canning, dehydrating, freezing, cold storage, salting and smoking, and fermentation.

One method that merits special attention when designing your site is cold storage. Fruit and root cellars have been used to preserve the harvest for thousands of years. These cellar spaces need to be located somewhere that will stay cool without freezing. Often this means partially or wholly buried in the earth, where the temperature is relatively cool and constant (around 55°F). In the Northern Hemisphere, cellars are often the perfect use for the north side of buildings and the foot of north-facing slopes.

Similar in function to a root cellar is a cool pantry. The difference is that a root cellar needs to have a relatively high level of humidity (90 to 95 percent), whereas a pantry should be much drier. The pantry is a good place to store bulk dry foods like beans, flours, and grains. It also provides an excellent home for your preserves stored in canning jars. (Putting your canned foods in a humid root cellar can cause the lids to rust over time.) The pantry is also a great place to store your dehydrated foods.

FOOD PRODUCTION STARTING POINTS

	Easiest	More involved	Most involved
Gardening	Make a small vegetable garden.	Expand your vegetable garden and plant some fruit trees. Add insectary plants to your garden.	Create a forest garden. Manage a structural forest.
Propagation	Propagate plants by division.	Propagate plants by seeds and cuttings. Create an in-house nursery.	Propagate plants by grafting. Create a commercial nursery.
Food preservation	Preserve food by dehydration.	Preserve food by canning.	Build a root cellar.

At permaculturist Jude Hobbs's house in Cottage Grove, Oregon, the attached greenhouse is used in the summer for dehydrating produce, thus preserving the harvest for year-round consumption.

Animals and Wildlife

Welcoming Natural Diversity

Animals are part of natural systems. If your permaculture design doesn't utilize livestock or create provisions for wildlife, it isn't mimicking nature as well as it could. Animals are integral to the success of natural ecosystems and have long been used by humans to meet our needs for food, transportation, fibers, and manures, among other things. Animals dramatically speed up soil fertility development by digesting biomass, which is a step in the process of nutrient cycling. Unlike plants, animals are mobile and can deposit manure across the landscape, redistributing minerals, carbohydrates, fiber, and amino acids, important food sources for both plants and the soil food web. With some forethought, we can mimic this natural pattern to our benefit.

Keeping honeybees is a skill worth learning, as bees provide pollination, honey, and wax, along with other by-products.

DESIGNING FOR ANIMALS

Animal husbandry has gone by the wayside for most of society today as animals have become a commodity in the industrialized food system. Before petroleum-fueled vehicles, animals were used not only for food but also for heavy labor. In permaculture design, we want to make use of animals' natural behaviors as well as their abilities to provide other services and products.

All animals have similar basic needs: food, clean drinking water, and protection from the elements and predators. Beyond that, animals have different minimum space requirements, climate tolerances, and fencing needs. Before buying an animal or herd, do some research. Meanwhile, here are some things to consider when planning to incorporate animals into your permaculture landscape:

Function. Are you raising animals for income or simply to be as self-reliant as possible? What products or yields do you want from them?

Land or space. What types of animals are best suited to your environment? A wooded lot may support different species than wide-open pastureland. The way animals access forage in an environment can be carefully managed via harvest and feeding, or animals can access their forage themselves in the field.

Time. How much time do you have to care for animals? Some animals require an intensive amount of labor to care for, and others can be left to their own devices if you occasionally check on them. It's important to match your time available to your chosen management system. As the system you design matures, it is also possible that it will require more or less time to manage, so plan ahead.

Protection. What kind of fencing and shelter do the animals need to keep them safe from the elements and predators? This varies by animal as some are much more vulnerable than others.

Stocking rate. This is similar to carrying capacity in that it refers to how many animals can live in a certain amount of space. However, carrying capacity refers to animals foraging year after year and being supported long-term by the natural ecosystem, whereas stocking rate applies more to animals in a farm setting where they might be rotated in and out of areas seasonally. Any given piece of land can convert only a certain amount of sunlight to forage each growing season. If too many animals live on a piece of land for too long, the soil can become damaged and serious problems can arise, such as land degradation or illness and stress for the animals. The number of animals that can live on a given area of land can vary widely based on the characteristics of the land itself and how it is managed. The type of animal is also an important variable, as some animals, especially those with hooves, can be much harder on the land than others.

Free-range versus confined management

The spectrum of how to manage animals on land is broad, and each choice has pros and cons. The closer to free range an animal system is, the less work is required on

(left) All animals require some sort of shelter. This shelter was built with goat behavior in mind and allows opportunities for climbing.

(right) Jessi integrates her poultry into her free-range garden slowly through the use of "tractors" (small, bottomless, mobile enclosures). This protects the poultry and prevents them from doing damage.

your part. The animals eat from the land and spread their manure as they move across the landscape. However, free-ranging animals can also be more susceptible to losses from predation and unnoticed illnesses. When animals are confined, they are generally safer from predation and get closer health monitoring, but you need to do more work in caring for the space, managing manure, and providing food, among other tasks. Confined-range systems are a middle ground that allows the animals to forage naturally, often in pastures, but in a confined and managed space, as in tractor systems and paddocks.

CONFINEMENT SPECTRUM

free range	**confined range**	**confined**
less work		more work
more predation		less predation
animals get their own feed		feed must be brought to animals
animals spread their own manure		manure must be collected and managed

Forage system design

The more you can feed your animals from your own land, the more sustainable and resilient your husbandry will be. If you broaden your thinking a bit, you can also think of nearby, local options to which you have access and include those in

Mulberries, raspberries, and alpine strawberries from Jessi's forest garden are excellent additions to her poultry forage system.

your planning. This is superior to feeding your animals from distant, commercial sources, in which case you have a higher cost and less control over the quality of that feed. You must also remember that feeding your animals is a dynamic process; you need to carefully monitor the health of your land and animals and adapt as you see changes.

To design a forage system, first look at what the animal naturally eats and is able to digest. Design a system as close to the animal's natural habitat as possible and that provides compatible feed. This will be the most resilient and require the least maintenance. If that isn't possible, you can grow as much of the animal's feed as possible. You can also look at trading with neighbors for feed or even pasture space.

PLANTS FOR ANIMAL FORAGE

Calliandra calothyrsus
red calliandra
zones 10–11
excellent for coppicing

Caragana arborescens
Siberian pea shrub
zones 2–8
seeds are good forage for chickens and other fowl

Ceratonia siliqua
carob
zones 9–11

Erythrina spp.
coral tree
zones vary

Gleditsia triacanthos
honeylocust
zones 3–7
best for sheep and goats which can access both carbs and protein in the seedpods without processing

Leucaena leucocephala
leucaena
zones 9–11

Lolium perenne
ryegrass
zones 3–6

Lycium barbarum
wolfberry or goji
zones 6–9
excellent for chicken forage systems

Medicago sativa
alfalfa
zones 3–8
best feed for dairy animals

Morus alba
white mulberry
zones 4–8
fruit and high-protein leaves

Prosopis glandulosa
honey mesquite
zones 8–11

Robinia pseudoacacia
black locust
zones 4–9
can be coppiced

Rubus idaeus
raspberry
zones vary
excellent for chicken forage systems
most important medicinal plant for dairy animals

Holistic management

With grazing livestock, it's important that we take care of the land they live on. Herd animals can either be destructive or a tool for healing the land. Allan Savory, founder of a system he calls holistic management, was instrumental in discovering that we need to follow nature's example of how herd animals forage to find a symbiotic relationship between the herd, the soil food web, and the grasslands they live on. Holistic management has become an important model for ecologically based agriculture that mimics the processes we see in nature.

For instance, in the wild, herds graze for a certain amount of time and then move on, leaving behind their nutrient-rich manure. Allan Savory found that in certain landscapes where herd animals had been removed due to fear of overgrazing, the ecosystems actually degraded rather than recovered. This may seem counterintuitive, but when you understand how nutrients cycle in landscapes that have a marked dry season, it becomes clear. The seasonal lack of water in these landscapes means a lack of activity in the soil food web, which means the decomposers don't do their job of returning nutrients to the soil. That's where herd animals come in. When they eat grasses and forbs, they process them through their bodies and return them to the soil, partially broken down and loaded with water and microorganisms. In other words, animals are actually an essential part of the nutrient cycling picture in these environments.

In many of Savory's projects, he actually increases stocking rates but manages much more carefully where the herds go and when. Using holistic management, Allan Savory has been able to repair degraded landscapes throughout the world. Perhaps even more important, Savory speculates that animals are actually the most appropriate use for these landscapes rather than vegetable crops that require massive amounts of irrigation. Currently, holistic management is being explored as a means of battling climate change by impounding more carbon in the soils of our grasslands and savannas worldwide.

Cattle are raised on this family farm using a rotational grazing system to manage the pasture resource.

Rotational grazing

Most farmers have a limited amount of land, so it is important to use what they do have efficiently. Rotational grazing is a way to use pastureland efficiently by simulating the natural process of animals moving across the landscape. The idea is to divide the land into cells, whether these are smaller paddocks or larger pastures, and have the animals rotate from one to the next so the land can recover after the animals impact it for a while. In that recovery period, the plant roots deepen, the biomass regenerates, and the organic matter left behind by the animals has an opportunity to break down and feed the soil.

To manage the land with rotational grazing, timing is everything. Factors to consider include soil type, climate, sun exposure, paddock size, and number and type of animals. Other things to be mindful of are exposed soil, weed management, and soil fertility.

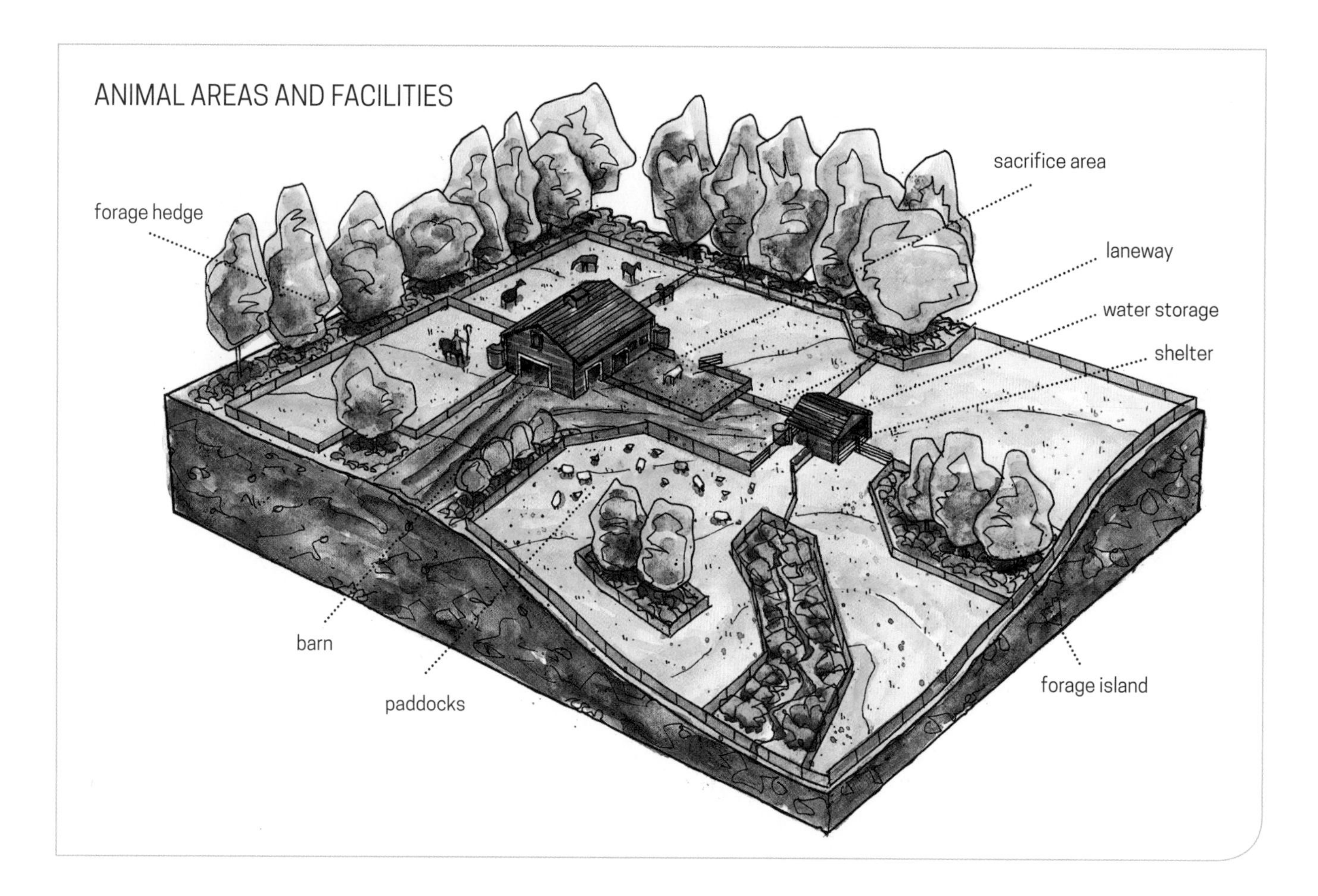

Animal areas and facilities

If you have many different types of animals, you may wonder how to raise them together. Who needs to be separate from whom? How do you separate and divide the land? Here are the main components to include when designing animals into your permaculture system:

- A barn that's easily accessible for people, animals, and vehicles. Hayloft access, trailer parking, and turnarounds may all be considerations. The barn could contain storage for animal-related needs—food, tack, grooming supplies—as well as breeding and brooding quarters and a place for processing products such as dairy, wool, and meat.
- A sacrifice area located adjacent to the barn with easy access for animals stalled indoors. These typically are designed with a rock aggregate, like gravel, to ensure proper drainage. These are set aside from the pasture space and used when the soil is saturated or for animals needing special care.
- A simple shelter in pastures or paddocks for animals to get out of the elements. If water catchment is placed on these structures, piping water from afar may be unnecessary.

* Laneways and gates to make it easy to take animals from one space to the next. Consider having central laneways to access all of the pasture spaces.
* Forage islands and hedges to provide food, shelter, and shade for the animals. These can be created and fenced off in the middle of pastures or along the perimeters.
* Protection zones around sensitive areas to protect natural resources. A hedgerow surrounding the banks of a stream is one example.

Fencing systems

Fencing specific to different animals' needs is required to control roaming and protect them from predators. The size of the animal plus its behaviors determines what type of fencing to use. In general, larger animals need to be enclosed, but sometimes that can be accomplished with a simple visual barrier (for example, a single strand of electric fencing can be enough to contain a herd of cattle). Smaller animals require smaller gaps to prevent them from leaving the space and prevent predators from getting in.

Many different types of fencing material are available, varying in cost and ease of installation. It's important to think about what resources are available and what will be required to maintain the fencing. We recommend that anyone new to animals or an area use temporary fencing before going out and putting in a permanent fence. Temporary fencing is also great for rotational grazing or changing paddock sizes.

Deer fencing. A fence to keep deer out should be at least 6 to 10 feet tall with the top angled outward to prevent them from jumping over, depending on the species you're dealing with and how much running room they have. The materials can be just about any mesh, but beware of materials that they can get tangled in, and remember that the more visible from a distance the fencing is the better. If deer are running at full speed and try to jump, they can get caught in the fencing and get injured or strangle themselves.

Hog fencing. Pigs are heavy and prone to digging, so having a sturdy fence that's partially buried is important. A product called hog panel is common but costly. Alternatives include tubed livestock panels or even well constructed chain-link fencing.

Split rail stock fencing. Split rail is a common fencing style that's great for larger livestock such as cattle, llamas, and horses. The material and construction can vary widely but often consist of weather-resistant wood. Note this doesn't do much to keep out predators, and even normally well-behaved animals can breach it if in a predator panic.

Living fencing. A living fence can be built using hedgerows, pleaching, or pollarding, depending on the animal or livestock and resources available. Hedgerows are

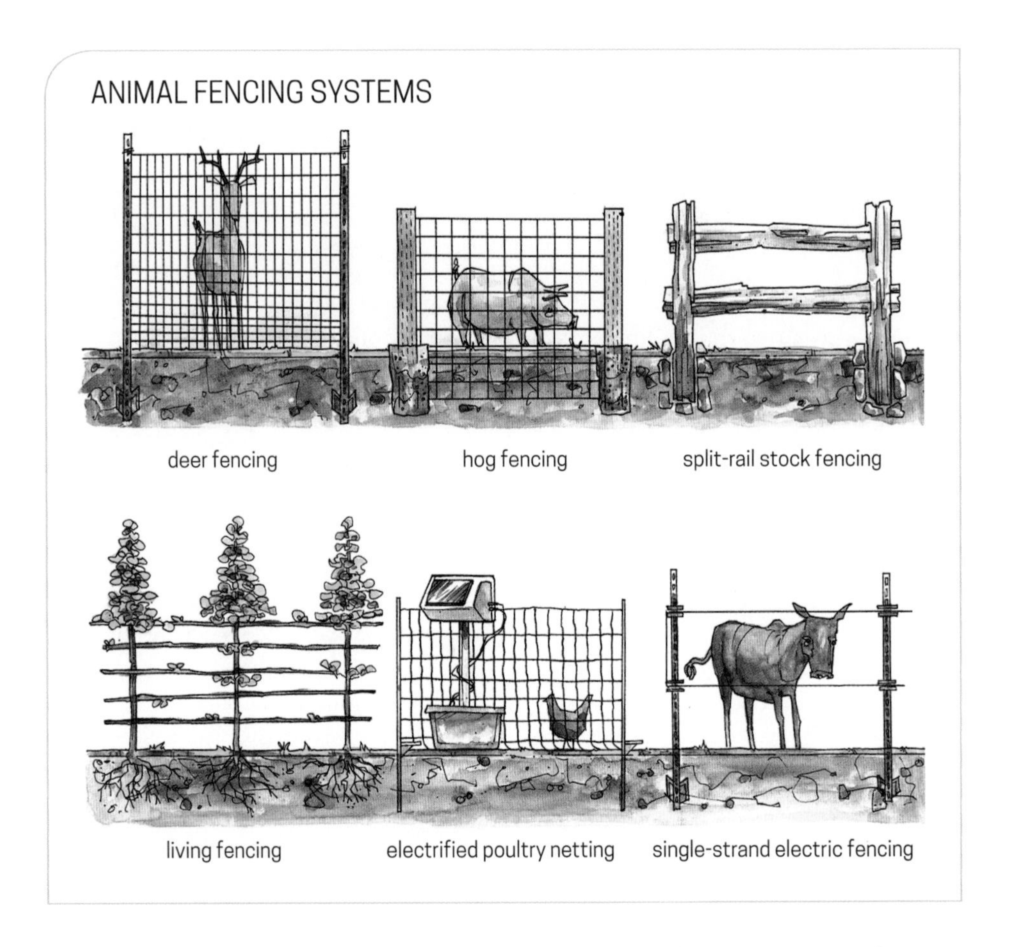

great for property edges and along natural corridors such as waterways. They can contain some types of animals and are excellent habitat for wildlife as well. Plants are placed closely together and often very diverse in species. Plants with thorns are good barriers against larger animals.

Pleaching creates a quickset hedge and was common in the late medieval era and early 1800s. It involves planting trees with fairly close spacing. After they've grown for several years, the trees are cut at the base but not completely disconnected from the root system. Each tree is laid over on top of the previous one. Stakes may be used to keep things orderly. This is a good permaculture practice in some places today as it uses indigenous materials and repurposes waste materials.

Pollarding uses living fence posts of a species that will easily resprout at the desired height after annual cutting back. The material removed can be woven through the posts and serve as a barrier. If the pollarded species is a good match for the livestock, the material removed can also be used as fodder.

Poultry netting. Lightweight electric netting is a good option for small pasture-raised animals. These fences can be charged with a small solar charger and easily moved from one area to another.

Single- (or multiple-) strand electric fencing. Generally used for larger livestock, this fencing is great because it is inexpensive and easy to use for temporary situations. It is easy to roll up, move, and store for reuse. It doesn't work well for small animals and won't necessarily stop a frightened herd or an animal determined to get through.

ANIMAL PROFILES

Historically, animals were bred for maximum self-sufficiency and resilience on family farms. Animals that reproduced easily with mothering instincts, that foraged naturally, and that had disease and pest resistance were highly valued. Nowadays, many animals are bred for commercial and industrial production (which usually just means for the maximum amount of product per pound of feed) and many of the natural characteristics have been bred out. In the last century, many breeds have become rarer or even extinct, which is a great loss in genetic diversity. When selecting breeds of animals to raise, consider heritage breeds adapted to your location and needs.

Also think about the social structure animals need in order to live naturally and to be healthy. Many domesticated farm animals are herd animals. These animals need to be with other animals of their kind, or at the very least, other herd animals. In a space too small for multiple horses but large enough for another small animal, Jessi has a companion goat to keep her horse company. Chickens are more comfortable and happier in groups, even if they do get into tiffs. Jessi has noticed that her hens are more relaxed when there is a rooster in the flock. More animals means more pressure or competition for forage or during feeding time, but it also means there is usually less risk of predation.

Small livestock

If you have a small piece of land, it's important to select animals that have a light impact and make sure you're not overloading your area. Impact is determined not only by the weight of an animal but also by how it's raised.

Rabbits. These cute creatures are a simple addition to most homesteads, including urban environments, which potentially have more restrictions because of space, laws, and proximity of neighbors. Rabbits are quiet, don't take up much room, and can be kept for meat, manure, and fiber. They can also eat surplus scraps and weeds. However, they are very destructive in the garden if they get out. In general they make good breeders, reproducing in prolific numbers, but are not always the best mothers. Their housing can be a simple hutch out of the weather with a wire mesh floor so their manure falls below; this can be placed above a worm composting system. Rabbits can also be put in confined range systems similar to chickens, such as a "tractor" with a fenced bottom, so they can graze in specific areas without burrowing out.

Bees. These hardworking flying insects are a critical part of our ecosystems and are responsible for pollinating many of our food crops. With recent awareness of colony collapse disorder (CCD) in honeybees, it is even more important now to be looking at increasing means of pollination for our food. Bees also provide us with wax and honey as well as other by-products. If you want to keep bees, your hives should ideally be placed in a sheltered location away from wind and harsh weather and facing the earliest morning sun. Beekeeping requires careful study and mentoring, as there is much to learn. Bees have a complex social structure and specific needs at different stages of their life cycle.

Dogs and cats. These domesticated pets have long been valued companions in our lives. Dogs and cats both have a number of useful purposes around a farm. It's important to consider breed traits and historical uses and behaviors to match up with what you need the animal to do. Some dog breeds are wonderful at guarding property and livestock, making the landowner aware of any visitors or dangerous threats. Some breeds are naturals at herding livestock or fetching, or carrying items in a backpack. Some types of dogs make for excellent rodent or pest controllers and great hunters. In cold climates, sled dogs have been used for transportation. Dogs also generate a great amount of heat, making them warm snuggle buddies or space heaters.

Cats are efficient hunters and are great for taking care of voles, rats, and other rodents on a farm. However, their place outside of our homes has become somewhat controversial as researchers are finding that cats contribute to high fatality rates among some songbirds. You can use bells on collars to help prevent this, and you can simultaneously design your garden to encourage healthy bird populations.

Poultry. Birds are a great addition to almost any farm or small homestead. It is possible to allow poultry to forage free-range in woodlots or to tractor them across a pasture. They offer the multiple benefits of eggs, meat, insect and pest control, manure, and more. All need adequate shelter to protect them from predators.

- Chickens are a great starter animal for most people because they are fun and easy to raise with many yields. But because they are scratchers, they can be destructive in a garden setting if it's not designed well. Jessi's book *Free-Range Chicken Gardens* tackles that very issue and goes into great detail about designing chicken habitat using permaculture design principles.
- If you have slugs, ducks may be your poultry solution. They are excellent foragers and don't scratch the soil like chickens do. They do require water and can be messier than other poultry.
- Geese enjoy swimming and can even be good as guardians. They don't make ideal pest control agents as they are vegetarian, but they're great at weed control.

(top left) Garden blogger Stacy Brewer raises both chickens and rabbits in her Seattle backyard, integrating them with food production and lawn spaces.

(top center) It is important to have plenty of time available if you are going to raise baby animals as they require a lot of up-front care. Here Jessi's son, Micah, is getting familiar with some new additions to the farm.

(top right) At Bullock's Permaculture Homestead, permaculturist Doug Bullock enjoys keeping ducks for their friendly nature, ease of maintenance, and love of slug eggs.

(top lower right) Lulu, a pit bull mix, learns at an early age how to interact with other farm animals.

(left) Oregon State University's Sustainable Agriculture Center has a collection of different beehive designs.

Heidi is a baby LaMancha goat, a breed known for excellent dairy production, being raised in an urban environment in Seattle.

- Turkeys produce high-quality meat but are not great egg producers. They are very easy to manage and can forage in a wide variety of landscape systems—woodlots to pastures.

Medium-sized livestock

Medium-sized livestock make it possible for us to produce some of our own meat, fiber, and manure on even a small amount of land. Animals that produce dairy products require special care and management, involving time and dedication. For continuous milk production, you must consistently (normally twice a day) milk the animal, properly care for its teats, and have a well-thought-out breeding program. Be sure to think through what will happen with the offspring and who will milk your animals if you go out of town. A pregnant or milking animal requires specialized care that the average petsitter may not be equipped to handle.

Goats. Goats produce many yields, including fiber, meat, and dairy. They are great browsers and are known for being able to clear brushy land in short order. By converting that biomass to goat manure first, they actually speed the process of composting. Goats are incredibly smart and great climbers, making them escape artists if they are lonely, bored, or hungry. They're also good at pulling carts or small plows and are becoming increasingly popular lower-impact pack animals.

Llamas and alpacas. These animals are especially useful if fiber is what you're after. They make great guardians and can pull a cart for light transportation needs. One interesting aspect of these guys is that they defecate in one spot, making manure harvesting an easier chore than for other animals.

Pigs. Pigs are very versatile and eat a wide array of food, making them great at turning biomass into a useful product: high-quality meat. They are also great at tilling the soil and can be used in preparing new garden spaces. They can be ranged on pasture or in a forest, or even confined in small paddocks. Pigs are extremely smart and can be difficult to handle once they reach a certain size. It's not uncommon for them to be several hundred pounds at maturity, depending on breed.

Sheep. These great grazers produce high-quality fiber, dairy, and meat. Sheep are a good replacement for lawn mowers and can handle cold and wet climates. They are also gentle and easy to work with.

Large livestock

Large hooved animals can be hard on the land, so it's important to make sure you get the correct number of animals for the amount of land you have and the system you have in mind.

Horses, donkeys, and mules. Horses are especially good for pulling carts, logging, and carrying people and goods. They are great grazers and produce high-quality manure. In some cultures, their meat is eaten. Horses can be wonderful companion animals and are highly trainable to do a number of things. Jessi describes her horse as a full-time biomass creator and lawn mower that provides part-time transportation.

Cattle, oxen, and water buffalo. These large animals have been used for many products and services over the course of humans domesticating them. Bovines produce meat, dairy, leather, and manure along with being able to pull vehicles used for transportation of goods and plowing.

Other animals

All over the world, in various regions and cultures, animals are raised and used for multiple functions. The animals we see in North America are not the only ones we can consider in our designs. Here is a short list of other animals we may be able to use:

- Fowl: pigeons, guinea fowl, doves, quail
- Mammals: wild game, kangaroos, yaks, camels, guinea pigs, agouti
- Insects: mealworms, silkworms, grasshoppers, crickets

It's very important to consider the impact of raising animals, both on the land and on the animal's welfare, as we meet human needs. We also need to be responsible about wild populations of animals that we may harvest via hunting, trapping, or fishing. Check with local wildlife and game wardens to learn more about these activities in your area.

AQUACULTURE

Aquaculture is among the most productive methods of growing food in terms of natural and diverse yields from one location. If designed well, an aquaculture system can produce vegetation and meat as well as other highly valuable products such as energy and water storage. It can be great at cycling nutrients. You can create an aquaculture system by working with an existing natural feature on your land or by installing a new system in the ground or even in tanks. Factors to consider when it comes to design include water quality, edges, species selection, and stocking rate. Also consider the different life stages of fish and other aquatic organisms and plan for shelter and ways to keep them safe from predation. Do your research ahead of time to ensure success before you introduce new species into the system you create.

Water quality is the first thing to take a close look at. Different species of plants and animals require specific levels of oxygen and nutrients, a specific pH, and specific temperatures to stay alive and thrive. Oxygen can be added to water with a wide array of aerators, from simple to pump driven and ideally renewable-energy-powered.

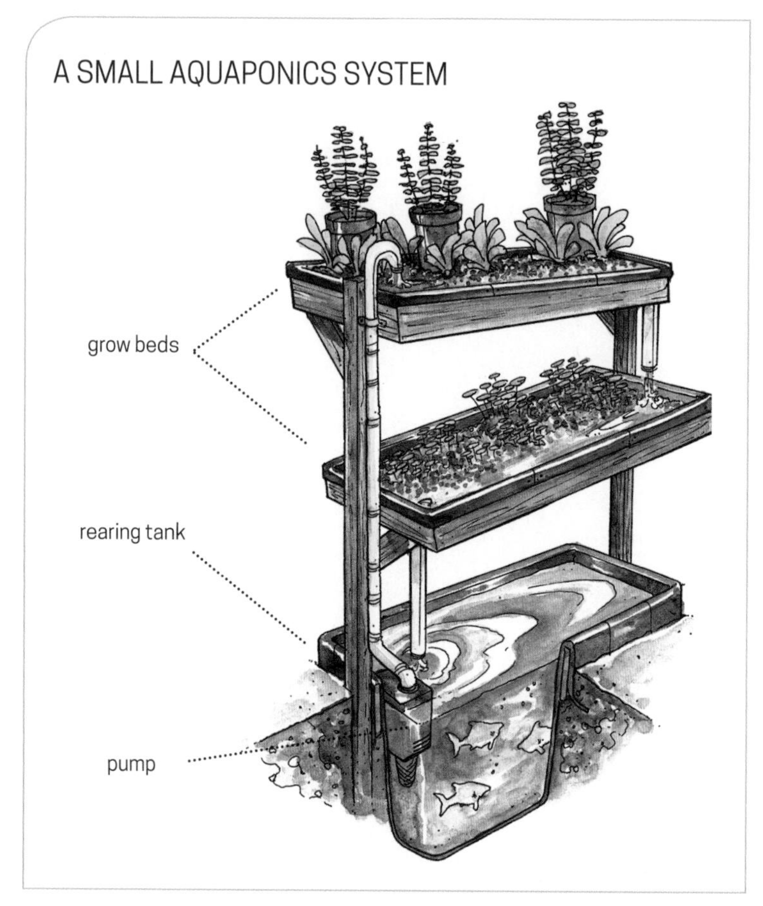

This simple aquaponics system in Jessi's greenhouse grows vegetables and fish. The space underneath the grow beds is used for worm bins.

Amendments can be added to change nutrients and pH. Water temperature can be altered through a variety of means. Water can be warmed up with a solar water siphon by having a smaller dark-colored tank lower in elevation connected to the larger body of water. The thermosiphon convection loop moves the warm water up into the pond while moving the cooler water into the lower tank to be reheated. Shade from trees planted on edges or islands is the best way to cool a body of water. Using refrigeration devices to cool is expensive and unsustainable. Having a mosaic of conditions is ideal, as some species will benefit from one set of conditions and others from different conditions.

To select plants and animals for your aquaculture system, research stocking and growth rates, breeding information, ranges of ideal and acceptable conditions, and food and nutrient requirements. Depending on your cultural conditions, you may consider growing fish, crayfish, shrimp, eels, oysters, mussels, clams, and snails.

Aquaponics is a clever way of combining fish farming and hydroponics to produce high yields of both fish and plants in controlled environments at many scales. A small aquaponics tank system can be set up in a greenhouse. Basically the system consists of a rearing tank that houses the fish and the main body of water, and grow beds that contain media that grows plants. The plants absorb the nutrients from the fish wastewater, and the clean water is then returned to the rearing tanks. Note that electrical power is needed to run a pump to cycle the water through the system, and with aquaponics it's important to monitor water quality on a regular basis.

DESIGNING FOR WILDLIFE HABITAT

Wild animals are part of almost all functional ecologies, so it's important that our permaculture designs support them too. For example, creating rat snake habitat will mean we have fewer problems with rats because the population is being controlled. Also, by encouraging thriving wildlife populations we create opportunities for sustainable harvest down the road. Surplus deer could be used to stock food pantries with venison.

The first step in designing for wildlife is to figure out what you have and what is missing. Start exploring your property and talking to neighbors about what they've seen. Put that excellent set of field guides you invested in to use. Talk to local wildlife managers. Once you know which actors should be playing a role in your habitat, you can figure out what they need and then do your best to provide it. Always pay special attention to endangered and threatened species and do your best to meet their needs for the sake of conserving biodiversity for the future.

Conversely, you might find that you have an overabundance of a particular wildlife element. It is also possible that some of the wildlife in your area is unwelcome (for example, the fox in your henhouse). You can design to keep certain wildlife species out or away from particular areas through use of fencing and guard animals. Wildlife can also be a problem if the crops they like to eat are the crops you like to eat. Birds and cherries are a perfect example. In these situations, it pays to experiment with decoy crops. If you can grow another crop producing at the same time that animals prefer to the crop you want to harvest, you can both have your needs met.

Wildlife needs

To encourage wildlife, the three most important things to provide are food, water, and shelter. Ample food for herbivores is the first step to seeing the return of predators. In fact, you can use the presence or absence of top predators to get a feel for how healthy the surrounding ecosystem really is. Providing food for those herbivores generally means planting the things they like to eat and making sure something is available to eat during all parts of the year when they need to feed. If you want to encourage a seed eater, you need to provide a diversity of plants that produce seeds throughout the year. You can't just meet the needs of wildlife during one part of the year and expect them to stick around the rest of the year.

Having access to open water, from birdbaths to large ponds, is essential for almost all wildlife. A water feature is not only a wonderful aesthetic element but also helps support the wildlife that in turn helps to support you.

Shelter can come in many forms, depending on the kind of wildlife you want to encourage. Most wildlife wants some sort of three-dimensional structure to use as shelter. You can set out human-made structures, such as birdhouses, or seek to create natural spaces where animals can find the shelter they need. Brush piles are

A water source is essential for most wildlife. If you want dragonflies, which eat mosquitoes, you have to have water in the landscape.

great places for birds and small mammals to escape from predators. Rock piles and rock walls in the open provide good places for lizards to sun themselves. Leaving dead flower heads standing through the winter gives arachnids a place to overwinter. Trees in pastures serve as perches for raptors looking for a meal. In areas with cold winters, deer look for stands of evergreens to take cover in for protection from wind and extreme temperatures.

Many types of wildlife need sizeable areas in which to range. It's important to connect habitat blocks using corridors that allow wildlife to pass from one area to another. Hedgerows and wooded streams are often natural spots where wildlife migrate, so aim to enhance these areas by design.

Wildlife functions

One important function that wildlife can provide is pest control, especially controlling herbivorous insects that are intent on damaging your crops. In any garden it is important to provide habitat for pest predator insects. During the larval phase of their life, many of these insects feast on garden pests such as aphids and cutworms. However, during their adult stage many are reliant on plant nectar to survive. If you want the function provided by the little ones, you need to take care of the adults (you don't get more ladybug larvae to eat your aphids if you don't have anything for the ladybug to eat). That means providing ample nectar sources throughout the year so successive generations of pest predators always have something to eat. Especially important are plants in the families Asteraceae (asters), Lamiaceae (mints), and Apiaceae (carrot), which provide copious quantities of nectar.

PLANTS THAT ATTRACT BENEFICIAL INSECTS

Achillea millefolium
yarrow
zones 3–9

Artemisia spp.
wormwood and mugwort
zones vary

Aster spp.
aster
zones vary

Borago officinalis
borage
annual

Eryngium maritimum
sea holly
zones 4b–10

Levisticum officinale
lovage
zones 4–9

Lobelia cardinalis
cardinal flower
zones 4–10

Medicago sativa
alfalfa
zones 3–8

Melissa officinalis
lemon balm
zones 4–9

Monarda spp.
bee balm
zones 4–9

Myrrhis odorata
sweet cicely
zones 3–7

Sambucus canadensis
American elderberry
zones 3–10

Thymus spp.
thyme
zones vary

Tiarella cordifolia
foamflower
zones 3–8

Trifolium repens
white clover
zones 4–8

Pest predators also often take the form of reptiles and amphibians. Reptiles can help to control slugs, snails, and rodents, while amphibians can have a noticeable impact on mosquito populations.

Another function that can be performed by wildlife is pollination. Native and wild pollinators can help to ensure you have good fruit set on your plants. Many types of native bees, wasps, and flies help to pollinate our crops. Find out which ones are in your area and do your best to meet their habitat needs. Again, remember that you can't expect them to pollinate your fruit trees in the spring if they don't have access to flowers for the rest of the year. Make sure to include plenty of flowering perennials that kick in throughout the year.

Aster ×frikartii 'Monch' is one of Jessi's favorite fall bloomers. It also serves as a great nectar source for pest predator insects.

The common garter snake helps to control slug populations. Improving garter snake habitat means fewer slugs to battle in the garden.

(next page) Spanky the Staffordshire bull terrier checks in with new additions to the flock (or in his mind, the pack).

In addition, don't hesitate to come up with novel ways to use wildlife to get jobs done. A strategically placed bat house or dovecote can put phosphorus-rich manure right where you want it. Wild pigs can be lured into fenced areas and used for tilling and manuring. Ideally, you'll come up with many more useful ways to put wildlife to work in a design that supports both you and them.

ANIMALS AND WILDLIFE STARTING POINTS

	Easiest	**More involved**	**Most involved**
Animals to raise	Rabbits, pigeons, mason bees	Poultry, honeybees, home flocks and herds	Cattle (dairy and meat), goats, horses, commercial flocks and herds
Feeding systems	Purchase from local feed sources.	Grow feed (grain crops, fodder plantings).	Set up perennial forage systems.
Management systems	Tractoring	Rotational grazing	Holistic management
Fencing	Temporary electric	Wire fencing	Split rail fences, living fences, hedgerows
Aquaculture	Grow aquatic plants for mulches and fertilizer.	Create integrated outdoor pond systems.	Set up an aquaponics system.
Wildlife	Plant shelter and food plants for wildlife. Plant insectiary species.	Build accessible water features, birdhouses, and feeders.	Do large-scale restoration and create connecting corridors.

50 USEFUL PLANTS FOR PERMACULTURE LANDSCAPES

Food security is a high-profile issue that gets a lot of attention, but food is just the beginning of the supplies we depend on. Relying on industrial systems on the other side of the planet for our clothing, parts for our machines, and fuel is just as precarious in the long run. Our goal as permaculturists is to use our landscapes to simultaneously build resilience for ourselves and support ecological function. Our landscapes can produce much more than just food. Regardless of where your land is, you have the capacity to provide a wide range of the things you need in order to survive and thrive.

Plants are integral to meeting these basic needs. They take in the energy from sunlight, soil, and water, and they make use of that energy in many ways. We can then use that energy for life-supporting activities. Plants have myriad ecological functions from a permaculture point of view. We can use plants to build soil and provide support for fauna (including humans). In the earlier chapter on soil fertility, we considered how plants can help build soil. With regard to fauna support, just about all organisms on the planet require plants in some capacity in order to survive. Plants meet all of these basic needs for humans:

- food—nourishment for our bodies
- medicine—healing for our maladies
- shelter—materials to build places to live
- fuel—products to fuel our homes and even some vehicles
- fiber—materials to produce clothing, ropes, and such
- fodder—food for our animals
- insectaries—food and habitat for beneficial insects

When you select plants for your landscape, look at how they've traditionally been used to learn about how they can be useful to you. An unattractive plant with an awkward growth habit and giant thorns may become much more attractive when you know it produces berries to nourish your body and enhances your soil as well. Many plants have multiple functions, and the more a plant can offer you, the better it is as a selection for your design. With that in mind, we offer the following list of fifty plants we consider to be rock stars of the plant world for permaculture landscapes. We have chosen plants for this list that represent four distinct climate types: temperate, subtropical, tropical, and arid.

This list gives a variety of information for each plant. Besides scientific name, common name(s), family, and USDA hardiness zones, the plant's light and water needs are indicated.

Light needs:

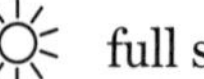 full sun—six hours or more of direct sun

 part sun—three hours of direct sun or six hours of dappled sun

 shade—no direct sun

Water needs:

 low—drought tolerant

 medium—prefers moist soils

 high—prefers wet or seasonally wet soils

The plant's mature height × width is also given. This is useful for figuring out how much space it takes up in the landscape. Keep in mind that these are the maximum dimensions a plant can attain and may vary according to conditions.

TREES

Artocarpus heterophyllus

jackfruit | Moraceae | zone 10 | 25–75 ft × 50–60 ft

Jackfruit is definitely one of our permaculture all-star plants. It is the largest fruit in the world, and each fruit can be eaten multiple ways. It can be cooked and eaten green in a curry or can be eaten ripe. When it's ripe, the seeds are edible after boiling; they are rich in carbohydrates and can also be used for animal feed. Around each seed is a sweet, delicious carpel. This part is the inspiration for Juicy Fruit gum. The tree also makes an excellent windbreak, tolerates drought, and provides high-quality lumber.

Casimiroa edulis

white sapote | Rutaceae | zones 9–11 | 25–60 ft × 25–30 ft

This member of the citrus family bears little resemblance to oranges and lemons. However, it is an excellent tree for subtropical climates. It is relatively drought tolerant and has an aesthetically pleasing form with its glossy green leaves and bright orange fruit. When ripe, the fruit turns from green to yellow or orange on the outside; a sweet custardy treat with a very pleasant flavor fills the inside. Like the avocado, the white sapote comes in varieties that fruit at different times of year, so you can be eating sapote for the better part of October through May (in North America).

Castanea spp.

chestnut | Fagaceae | zones 4–9 | 15–60 ft × 20–40 ft

Each type of chestnut is adapted to a specific temperate climate and successional stage. Almost anywhere in the world that chestnuts occur, they are a staple in the diets of that region's people. Unlike most nuts, chestnuts are actually low in fats and oils but high in carbohydrates. They're like potatoes that grow on trees. They keep for a while and you can make many products from them, so they can add to your resilience big time. The lumber from chestnut trees is also quite good for carpentry and woodworking.

KEY

Light needs:
- full sun
- part sun
- shade

Water needs:
- low
- medium
- high

Ceratonia siliqua

carob | Fabaceae | zones 9–11 | 30–40 ft × 30–40 ft

Carob is a great choice for desert landscapes. This drought-tolerant nitrogen fixer casts dappled shade and enriches the soil at the same time. In addition, the pods of the carob tree contain a sweet substance that can be processed and used for a variety of culinary treats. The seeds themselves are the source of the thickening agent locust bean gum. The seedpods also serve as a good addition to animal feeds in limited quantities.

Citrus japonica

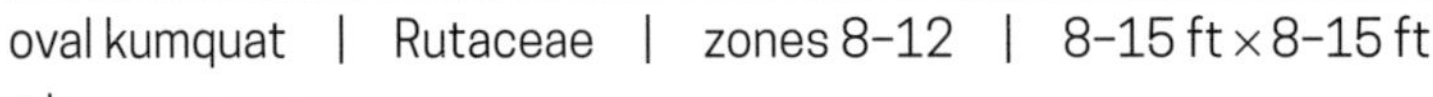

oval kumquat | Rutaceae | zones 8–12 | 8–15 ft × 8–15 ft

Kumquats are great additions to the landscape and are fairly hardy for citrus. The trees tend to stay small, making them a sensible addition to an urban or suburban yard. The small fruits are eaten whole (including the skin). Like other citrus, they are high in vitamin C, and their skin contains cancer-fighting antioxidants. Kumquats can also be planted as a tightly pruned privacy hedge that produces fruit.

Cocos nucifera

coconut | Arecaceae | zone 11 and warmer | 30–90 ft × 15–20 ft

A true staff of life for the tropics, this palm can vary widely in size, from 30-foot "dwarfs" to 90-foot monsters. Some varieties are selected for their nutmeat, while others are selected for their electrolyte-balanced coconut water. The nuts are high in oil that can be used for cooking or fuel. When sprouted, coconuts are one of the finest treats around. Coconuts are extremely wind resistant and salt tolerant, and their fronds are useful for thatch and fiber. The fiber from the husk is called coir and it is used as a horticultural amendment. This may be the most useful palm in the world.

Jubaea chilensis

Chilean wine palm | Arecaceae | zones 8–11 | 40–80 ft × 15–30 ft

The Chilean wine palm is a stunning landscape specimen. Although slow growing, it can tolerate a wide variety of conditions and looks stately throughout its life. It produces nuts with a hard shell and an oily nutmeat not unlike coconut. This is the coconut analog for the subtropical world. Where it's native in Chile, its sap is harvested and fermented to make an alcoholic drink—although one must kill the tree to get at that sap. Fibers on the trunk are also used in textiles, baskets, and paper.

Moringa oleifera

moringa or horseradish tree | Moringaceae | zone 10 and warmer | 20–45 ft × 15–30 ft

Moringa is also known as horseradish tree as many of its edible parts have a hint of that flavor. Roots, leaves, and seedpods are edible; the leaves in particular are highly nutritious. It also makes excellent fodder for animals. The seeds contain a good quality cooking oil. When ground up, the mature seeds can also be used to purify water. This tree is drought tolerant and extremely fast growing. It can be pruned to keep it relatively small. It can also be grown as a dieback perennial in colder climates.

Morus spp.

mulberry | Moraceae | zones 4–10 | 30–50 ft × 30–40 ft

Black, white, and red mulberries (plus their hybrids) are excellent trees for temperate climates on a number of levels. First, their fruit is delicious and prolific, and some species produce fruit for up to two months straight. Mulberries are a great choice for a snack fruit for you and an excellent addition to a chicken forage system or aquaculture pond edge. In addition, the leaves of mulberries (particularly of the white mulberry, *Morus alba*) make excellent, high-protein animal fodder. In addition to all of this, if you ever wanted to raise silkworms, this is just about the only plant they eat, so they are an essential element for your silk farm.

KEY

Light needs:
full sun
part sun
shade

Water needs:
low
medium
high

Myrciaria cauliflora

jaboticaba | Myrtaceae | zones 9–11 | 15–30 ft × 12–18 ft

As with many members of the family Myrtaceae, the jaboticaba makes a stunning landscape specimen. It has mottled bark and glossy leaves. Several times per year it produces a profusion of fragrant, fuzzy white flowers all over its branches and trunk, followed by large grape-sized fruits in either purple or yellow. These fruits have a rich, sweet flavor and are used for winemaking in the tree's native Brazil. Most of these fruits are produced once a year, but significant crops of fruit can be produced throughout the entire year. Although slow growing and slow to begin producing, these trees are definitely worth including in landscapes that can support them.

Olea europaea

olive | Oleaceae | zones 8–10 | 30 ft × 30 ft

The olive is a handsome addition to almost any landscape in a Mediterranean climate. The silver cast to the leaves and handsome fruit make olives an attractive ornamental wherever they grow. We're all familiar with the brined fruit of the olive. The leaves can also be used to make tea that is purported to have calming effects. Perhaps the most important product of the olive tree is the oil pressed from the fruit. Being able to press your own oil, for both culinary and medicinal use, is incredibly valuable. You can also obtain a dye from the fruit.

Pachira aquatica

Malabar chestnut | Malvaceae | zone 10 | 30–60 ft × 30- ft

Found all over the world, this plant grows quickly and can start producing seedpods as soon as one or two years after planting. The seedpods contain clusters of starchy nuts that are eaten raw or cooked. Malabar chestnut is at its best when used as an early production crop in mixed agroforestry systems where the other species won't come into production for several years.

Persea americana

avocado | Lauraceae | zones 9–11 | 30–40 ft × 25–35 ft

Avocados can be incredibly productive. The myriad varieties of avocados have radically different characteristics and range from cocktail avocados that can be eaten whole in one bite to enormous 2-pound monsters. Although the fruit in general is an excellent source of fats, one can also seek less rich varieties (typically noted as "watery" as compared to "buttery"). Enough breeding work has been done that there are avocado varieties that fruit during almost all parts of the year. If you live in a subtropical climate, you can plant a selection of varieties that will allow you to have avocados nearly year-round.

Phoenix dactylifera

date palm | Arecaceae | zones 11–12 | 40–75 ft × 20–35 ft

The classic plant grown at oases in the Middle East, the date palm is one of the most important commercially grown palms in the world. Although adapted to hot, dry climates, it does require regular water and good drainage. Many of us have seen Medjool or Deglet dates in the supermarket, but many more excellent types exist. Plants are male or female, and to get good pollination you need to plant them in about equal quantities. However, if you are willing to get involved with pollinating them by hand, you can have as few as one male per hundred females. Also, be aware that these massively productive desert giants have spines at the base of their fronds.

Pistacia vera

pistachio | Anacardinaceae | zones 9–11 | 30 ft × 30 ft

Pistachios are a great nut tree for arid climates. These drought-tolerant trees require free-draining soils but benefit from regular water. They are either male or female, so you need to plant at least one of each to get production. With their glossy compound foliage and striking red-gold fruits, they make a striking addition to any landscape. The nuts are delicious and help to lower the risk of heart disease. However, be aware that the tree is related to poison ivy and poison oak, containing a similar irritant. The unsalted shells can be used as a long-lasting mulch in the garden but are toxic to dogs so be cautious.

KEY

Light needs:
full sun
part sun
shade

Water needs:
low
medium
high

Salix purpurea

purple osier willow | Salicaceae | zones 5–10 | 15 ft × 15 ft

 | 

Throughout history, willows have been used extensively in temperate landscapes around the world. Willows in general are easy to propagate via cuttings and can help stabilize slopes. They also take well to coppicing. Several varieties of purple osier willow provide some of the best basketry materials around. In addition, their attractive bark color makes them a good addition to the winter landscape. If you intend to do cutting propagation to fill out the rest of your permaculture landscape, willows can also be used to provide a natural rooting hormone to help those plants take.

Tamarindus indica

tamarind | Fabaceae | zone 10 | 40 ft × 60 ft

 | 

Native to tropical Africa, this tree produces podlike fruit that has become a staple ingredient in cuisine around the world. The edible pulp inside of the pods also has medicinal value and can be stored and used in a variety of ways: as a powder, paste, syrup, or concentrate. This robust evergreen tree is also a nitrogen fixer, improving soil health, and its flowers make it a good insectary plant. It can be coppiced, and its wood is useful as firewood and charcoal as well as for making furniture and tool handles, among other things.

Ziziphus jujuba

jujube | Rhamnaceae | zones 6–9 | 15–30 ft × 15–30 ft

 | 

If you get enough heat, this can be an excellent choice for your edible landscape. Jujubes are drought-tolerant heat lovers from India and the dry parts of China. The small fruits can be eaten fresh or dried. When fresh they have the flavor and consistency of dried apples. In fact, since apples are one of the most disease-prone, high-maintenance fruit trees you could plant, this may be a good alternative. The fruit is also used to tonify the liver in traditional Chinese medicine.

SHRUBS AND WOODY PERENNIALS

Aronia melanocarpa

aronia or chokeberry | Rosaceae | zones 3–9 | 3–7 ft × 6–10 ft

 | 

Aronia is one of our favorite ornamental additions to the edible landscape. It is incredibly adaptable, tolerating both seasonal inundation and periods of drought. Due to this adaptability, it can be a good choice for rain gardens. It has beautiful white flowers in the spring followed by dark purple berries favored by birds and spectacular crimson leaf color in the fall. The berries are tasty (when fully ripe) and high in antioxidants.

Cajanus cajan

pigeon pea | Fabaceae | zones 9–13 | 3–12 ft × 5–8 ft

Also known as gandule, this short-lived perennial is a great solution for most sites where producing food and building fertility need to happen simultaneously. It's drought tolerant and easy to manage. It is also a nitrogen fixer, so it will help to build soil over time. It can grow densely, smothering other weeds. In addition, it produces protein-rich beans that are a staple in many parts of the West Indies and Africa. Because it is short lived, it can be used to enrich soil and eliminate weeds before giving way to more permanent perennial tree crops. It is not thorny, as many nitrogen-fixing shrubs are, so it is also easier to work with.

Capparis spinosa

caper bush | Capparaceae | zones 8–10 | 1–3 ft × 1–3 ft

The caper bush is a small, attractive shrub with fleshy leaves. It produces beautiful white or pink flowers with long stamens. The capers you find pickled or brined in stores are the buds of these flowers. However, the flowers are so attractive (and fragrant) that you may not want to pick the buds. After the flowers, fruits called caper berries appear that can also be brined or pickled. This shrub is tough as nails and loves heat. Medicinally, the bark, leaves, shoots, and roots are used to treat a variety of ailments, including rheumatism.

KEY

Light needs:
- full sun
- part sun
- shade

Water needs:
- low
- medium
- high

Caragana arborescens

Siberian pea shrub | Fabaceae | zones 2–8 | 5–20 ft × 3–12 ft

This nitrogen-fixing shrub comes in a variety of sizes. It has attractive white lenticels on its bark and produces small yellow pealike flowers. It is very tough, handling both drought and extreme cold. The small pods produced during summer drop seeds that can be eaten by people but are also an excellent addition to a forage system for livestock, particularly chickens. Even if you don't have livestock, wildlife love the seeds as well. Siberian pea shrub also makes a great addition to windbreaks due to its toughness.

Carissa macrocarpa

Natal plum | Apocynaceae | zones 9–11 | 6–10 ft × 10–15 ft

This evergreen shrub with glossy green leaves makes a great privacy screen or hedgerow in sun or shade. It also grows tough, dense, and thorny, so it lends itself to livestock barriers and defensive perimeters. It is also tolerant of salt spray. It produces waxy white flowers with an almost seductive aroma that intensifies at night. These flowers are followed by edible pink fruits. All parts of the plant exude a waxy latex that is mildly toxic (probably enough to make it unpalatable to animals). The fruit doesn't seem to have these toxic effects.

Cnidoscolus aconitifolius

chaya | Euphorbiaceae | zones 8–11 | 6–20 ft × 6–15 ft

This shrub is native to Mexico, where it can often be seen growing wild. In the colder part of its range, it can be grown as a perennial that dies to the ground each year. Although it can grow quite tall, it can easily be maintained at a lower height. That is desirable because the leaves are a nutritious perennial vegetable, and you want to be able to reach them. The leaves are toxic when raw, but once cooked they are high in protein, iron, and vitamins A and C. As with many plants in the family Euphorbiaceae, chaya exudes a white latex that can be a skin irritant. However, the quality of its edible parts in conjunction with its toughness make it a good addition to the edible landscape.

Elaeagnus multiflora

goumi | Elaeagnaceae | zones 5–8 | 6–8 ft × 6–8 ft

Goumi is native to Japan, where it is sometimes harvested in the wild. The plant itself has an attractive metallic bronze cast to it. Bronzy, cream-colored flowers in the spring are followed by speckled red berries in the early summer. These mucilaginous berries taste good and are very high in lycopene, an essential amino acid that promotes eye health. Goumi is also a nitrogen fixer so it can help build the soil in your forest gardens and orchards.

Eugenia stipitata

araza | Myrtaceae | zone 10 and warmer | 8–10 ft × 8–10 ft

This attractive shrub can flower and fruit throughout the year when given plenty of water and fertile soil. The flowers are fragrant and attractive. These are followed by large yellow-orange fruits, up to 1.5 pounds, that are loaded with juice. They are sour but serve as an excellent base for beverages, sorbets, and jellies. They can grow in full or part shade, so they are a great choice for an agroforestry project where you are stacking multiple species together.

Genista pilosa 'Procumbens'

trailing silky-leaf woodwaxen | Fabaceae | zones 9–11 | 2 in × 3 ft

This attractive ground cover does best in full sun. It is an evergreen nitrogen fixer that creates a dense carpet over bare soil, preventing erosion and shading the ground. Its tiny yellow pealike flowers create a delicate, showy display in spring.

KEY

Light needs:
☼ full sun
☼ part sun
☼ shade

Water needs:
low
medium
high

Opuntia ficus-indica

prickly pear cactus | Cactaceae | zones 9–11 | 15 ft × 15 ft

Prickly pear cactus is a true permaculture all-star due to its many functions. The pads are eaten as a vegetable called nopales. They have a mucilaginous texture and taste somewhat like green beans when cooked. Thornless varieties make the preparation of nopales easier. The prickly pear also produces a high-quality fruit prized for fresh eating, jellies, and liqueurs. Beyond that, prickly pear is a drought-tolerant medicinal plant that can also be grown as a windbreak or living fence for animal containment. Some animals, such as cattle, actually love to eat the pads. Obviously, the thornless varieties are preferred for this, but you can also burn the thorns off the regular varieties to prevent mouth injury to livestock.

Ribes rubrum

red currant | Grossulaceae | zones 3–9 | 5–7 ft × 3–5 ft

Currants are attractive, especially when they have clusters of fruit hanging from their branches, and produce delicious berries. They are at least somewhat shade tolerant, which is unusual for a fruiting plant. However, be aware that they are at their best in full sun. The species includes varieties with both red and white fruit; in our experience, red berries draw a lot of birds, while white-fruited currants are often overlooked by birds. That means you get to eat more fruit without going to extreme measures to protect your crop.

Sauropus androgynus

katuk | Euphorbiaceae | zone 10 and warmer | 4–10 ft × 5 ft

Katuk is an excellent perennial vegetable for tropical regions. The leaves have a sweet, nutty flavor that is perfect for salads. The fresh growing tips can also be harvested and eaten like asparagus. It can grow in both sun and shade, producing well in both. It works well as a hedge that can be slashed to the ground once a year to allow it to revegetate and keep it from getting leggy. Katuk also produces attractive, tasty white fruits with a pink cap. This one is a winner for both ornamental and culinary purposes.

HERBACEOUS PERENNIALS

Achillea millefolium

yarrow | Asteraceae | zones 3–9 | 2–3 ft × indef.

Many gardeners are familiar with yarrow, but not many know why it's such a garden superstar. Yarrow has attractive flowers and can function as a tough, drought-tolerant ground cover. It doesn't stop there, however. Yarrow is also an excellent plant for supporting pest predator insects and is a dynamic accumulator of phosphorous, potassium, and copper. Among other medicinal uses, yarrow is commonly used to treat wounds and stop bleeding.

Alternanthera sissoo

Brazilian spinach | Amaranthaceae | zones 9–11 | 1 ft × indef.

Brazilian spinach, a perennial leaf vegetable, is another excellent ground cover for stacked tropical gardens. Consider planting it underneath tree crops that cast light shade. It can be fairly tough and is probably better when cooked. It forms a dense enough mat that it can help suppress weeds where it is planted.

Apios americana

groundnut | Fabaceae | zones 3–10 | 4–8 ft × indef.

Not to be confused with the peanut, this is an interesting and unusual plant in the legume family. It is a vining nitrogen fixer that produces edible tubers purported to have a flavor somewhat akin to sweet potatoes. Some work has been done at the University of Louisiana to develop cultivars that produce more and larger tubers, so look for named varieties. Bell-shaped, pealike pink or white flowers are produced in clusters.

KEY

Light needs:
- full sun
- part sun
- shade

Water needs:
- low
- medium
- high

Arachis glabrata

perennial peanut | Fabaceae | zones 9–11 | 3–5 in × indef.

Perennial peanut is a great nitrogen-fixing ground cover for tropical climates. Its broad leaves and tiny yellow pea flowers are very attractive. In fact, perennial peanut is a great tropical lawn option. However, unlike grasses, it doesn't require fertile soils to look good. It is also "steppable."

Chrysopogon zizanioides

vetiver grass | Poaceae | zones 9–11 | 3–4 ft × 3–4 ft

Vetiver is a superstar erosion-control plant. It is often planted in contour bands on eroding slopes where it acts as a sieve, slowing down overland flow and causing the water to drop suspended soil particles. It also has a tremendous root system that can run as deep as 8 feet, which helps to stitch together hillsides. It makes an excellent livestock feed when young, and an essential oil used in 90 percent of all Western perfumes can be extracted from its roots. Vetiver is also being used experimentally as a termite repellant.

Crocus sativus

saffron crocus | Iridaceae | zones 5–8 | 6 in × 6 in

We've all heard of saffron, one of the most highly prized spices in the world. Saffron is actually the stigmas of the saffron crocus flower, which blooms in the fall. The bright red stigmas are dried and act as a potent spice. They can also be used as a dye. The flowers of the saffron crocus are a beautiful lavender color, earning a respectable place in any landscape whether the spice is harvested or not. Plus, as fall bloomers they provide a late source of pollen for bees and other pollinators getting ready for winter. Saffron has long been used medicinally and is currently being researched for its anticancer properties.

Curcuma longa

turmeric | Zingiberaceae | zones 7–11 | 3 ft × 3 ft

Turmeric makes a nice addition to the understory of your perennial landscape. It grows and flowers each year and then goes dormant until the following year. At this point you can dig up some of its rhizomes. Many of us know turmeric as the yellow stuff in Indian curry dishes, but it also has medicinal qualities. In Ayurveda, the traditional healing system of India, turmeric has been used for everything from calming upset stomachs to healing wounds. It is thought to have antimicrobial and anti-inflammatory properties. In addition, it is an excellent dye plant.

Eryngium foetidum

culantro | Apiaceae | zones 10–11 | 8 in × 8 in

This biennial can be grown as an annual in hot summer climates outside the tropics. In tough arid places, like beaches, this plant can grow abundantly and cover the ground, self-seeding and returning each year. It has a flavor almost identical to cilantro and has several uses in folk medicine. As a member of the family Apiaceae, it is also a good nectar source for pest predators.

Ferula asafoetida

asafoetida | Apiaceae | zones 8–11 | 6 ft × 6 ft

If you like to cook Indian food, you may be familiar with asafoetida as that pungent (read: stinky) spice you add to your dhal to make it rich and delicious. That yellow powder is actually a dried and pulverized resin extracted from the roots of this plant. Asafoetida is also used medicinally as a digestive aid and flatulence preventer, among other things. It makes the list because it is also a tough plant that makes an impressive display. It produces large feathery fronds, like a giant dill, and then it produces dill-like flower stalks. The striking flowers are excellent nectar sources for pest predator insects, so having asafoetida in your landscape can indirectly help reduce pest pressure.

KEY

Light needs:
☼ full sun
part sun
shade

Water needs:
low
medium
high

Hemerocallis spp.

daylily | Xanthorrhoeaceae | zones 4–9 | 1–6 ft × 2–5 ft

Daylilies are both beautiful and functional additions to the landscape. This popular plant comes in many different colors and forms and is a perennial vegetable. Although many parts of the plant are edible, the flowers and flower buds really shine. Much like squash blossoms, the flowers can be harvested and prepared as fritters (tempura is also great due to their large amount of surface area). Before the flowers open, you can nibble the flower buds—they taste like a combination of onions and peas. Occasionally, some people have an adverse reaction to eating them in this way, so start small. Daylilies also grow very densely, so consider using them for weed suppression.

Musa spp.

banana | Musaceae | zones vary | size varies

Banana is another incredibly productive food plant for the subtropics and tropics. Most of us are familiar with the Cavendish banana we see in the supermarket. However, literally hundreds of different types of bananas can be found worldwide. They vary in flavor, size, color, wind resistance, and more. Naturally, bananas grow in clumps. Each stalk will produce one rack of bananas. Once it finishes its crop, the stalk can be slashed and used to mulch the clump (or other nearby tree crops). Soon the next largest stalk will start to produce a rack, and so on. Wind-tolerant varieties can be used to shelter other young trees on windy sites until they get established.

Myrrhis odorata

sweet cicely | Apiaceae | zones 3–7 | 3 ft × 3 ft

Sweet cicely is a lovely garden herb that has a fragrance and flavor of anise. With its lacy foliage and attractive white flowers, it is an excellent ornamental as well as an excellent support plant for pest predator insects. After the flowers come seedpods. When these are young, they're like the gardeners' candy. Once this herb is established, its abundant foliage can be used for mulch and compost. Cutting it back before the seeds ripen will prevent it from spreading too much.

Petasites japonicus

fuki | Asteraceae | zones 5–9 | 2–6 ft × indef.

Fuki is an interesting addition to the landscape because it is a perennial vegetable that grows in moist, shady locations. The flowers and stalks of fuki are cooked and used as a vegetable and condiment in Japan, where it is native. With its enormous round leaves, fuki is also an incredibly impressive landscape element that looks almost Dr. Seussical. Before you plant this one, however, be aware it is fairly expansive, so have a containment plan before you proceed.

Smallanthus sonchifolius

yacon | Asteraceae | zones 8–10 | 3–6 ft × 2–3 ft

Yacon is an Andean tuber related to sunflowers. It makes the most sense to plant it in a place where you don't mind repeatedly digging the soil to harvest the tubers, which are roughly the size of sweet potatoes. If the plant gets enough heat during the growing season, the tubers are sweet and crunchy. They are at their best raw, although you can make sugar or syrup from them. Their sugars come in the form of inulin, which is great for diabetics and for the beneficial organisms that live in your digestive system, and which increases calcium absorption. When you eat these tubers, think of it as feeding the critters that keep you healthy.

Symphytum ×uplandicum

Russian comfrey | Boraginaceae | zones 3–9 | 1–4 ft × 3 ft

Comfrey is a great plant for the permaculture garden. It produces large amounts of biomass that can be used for mulching or compost. It is also a dynamic accumulator of potassium, phosphorous, calcium, copper, iron, and magnesium. Bees love the attractive flowers as much as we do. Comfrey grows densely enough that it can be used as a weed barrier. The Russian comfrey (also known as the Bocking 14) cultivar is a sterile hybrid, so it won't spread via seeds.

KEY

Light needs:
☼ full sun
part sun
shade

Water needs:
low
medium
high

Viola spp.

violets | Violaceae | zones 3–6 | 4–16 in × indef.

Violets are common in shade gardens throughout the temperate world, but few people know that the flowers are edible. Some species in particular have a very pleasant flavor. In addition to their culinary and ornamental value, some violets have medicinal uses as anti-inflammatories and treatments for headaches. Violets are dynamic accumulators of phosphorous and serve as ground covers for the shadier part of the garden.

GROUND COVERS

Basella alba

Malabar spinach | Basellaceae | zones 10–11 | 20–60 ft × indef.

Although only perennial in the warmer parts of the subtropics, Malabar spinach can be grown as an annual during the warm part of the year in much colder places. These sprawling vines can cover the soil or climb, covering the side of buildings and retaining walls. The leaves are nutritious and edible. They are fairly high in vitamins, minerals, fiber, and protein. The pink stems and green leaves commonly seen make for an impressive, productive, low-maintenance addition to the landscape.

Crassocephalum crepidioides

Okinawa spinach | Asteraceae | zones 9–11 | 1–2 ft × indef.

Okinawa spinach is a very aesthetically pleasing ground cover for tropical gardens. The upper side of the leaves is green while the underside is an attractive purple. The greens can be eaten fresh in salad or cooked. It spreads via rhizomes and, if protected from the harsh tropical sun during establishment, can cover an area quite quickly.

Ipomoea batatas

sweet potato | Convolvulaceae | zones 9–11 | 10–20 ft × indef.

Sweet potatoes can grow as climbers or as sprawling ground covers producing delicious, edible tubers. They can cover soil incredibly fast in a forest garden setting. Wherever the plant's nodes come in contact with the soil, they can put down tubers. When you are growing them for tuber crops, it can help to curl the vines manually and move the vines so that they stay within specified boundaries. This will result in the tubers appearing in a more concentrated area. The young leaves and shoots are also edible, tasty, and high in iron. These are often incorrectly labeled and sold as yams throughout North America.

Vaccinium vitis-idaea

lingonberry | Ericaceae | zones 2–6 | 6–12 in × 2 ft

 | 

Lingonberries hail from northern Europe and Asia. They are often found as a ground cover in open areas with acidic soils. They are a very attractive plant with glossy green leaves and bright red berries. The berries have a flavor similar to cranberry with sweetness improving after frost. The leaves are used for making dye. The berries are also excellent as wildlife food.

INVISIBLE STRUCTURES

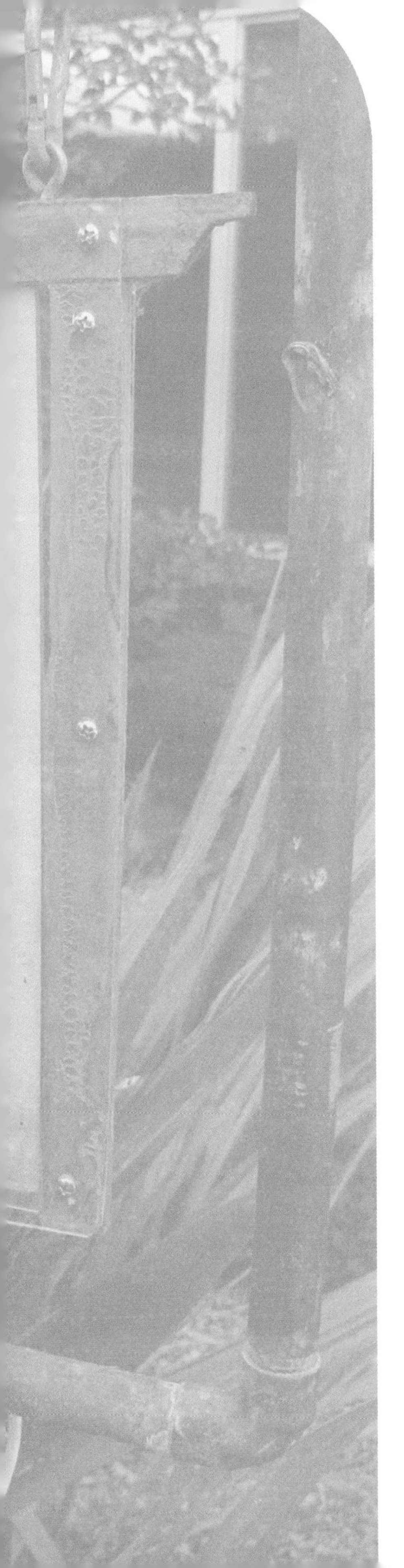

Good design doesn't stop with the landscape elements you can see. Imagine all the aspects of designing a piece of land that don't appear on the master plan. Legal issues, management and decision-making processes, social systems, and financial management are all intimately tied to a successful design. We call these aspects invisible structures. In this chapter we explore the many aspects of invisible structures in three categories: economics, social and community ties, and land access.

ECONOMICS AND FINANCIAL PERMACULTURE

The Buddhist concept of right livelihood provides a good way to think about economics from a permaculture perspective. Right livelihood means making a living doing something that doesn't harm the earth or others. We all deserve to feel good about what we do each day, not guilty because we're forced to do unethical things just to make ends meet. Therefore, we should always be thinking about how we can offer opportunities for right livelihood in our designs.

Parallel economics

We're all subject to an overarching, global economic system that does not necessarily reward us for making decisions that avoid harming others or the planet. Therefore, we must always think in terms of parallel economics, or creating systems that meet our economic needs today as well as in a more ideal future. That means the systems we design must help to move us toward our right livelihood goals while still allowing us to meet our needs today. If we miss either aspect of parallel economics, our system will likely fail.

For instance, if a business pays all its workers high wages, it may not have enough money to stay solvent. A reasonable way to design a business to avoid this issue may be to create an incentive-based wage system that rewards the hardest workers with higher wages and, ultimately, the ability to become a part owner of the company. That is how many workers' cooperatives operate. That way a business can avoid going broke in the present but still demonstrate its values. Similarly, planting a commercial farm for maximum biodiversity might ultimately result in production that

Reminders about living lightly on the land are posted at Columbia Ecovillage in Portland, Oregon.

is spread out, difficult to access and harvest, and low in volume, which might lead to labor and management becoming cost prohibitive. Perhaps creating simpler species mixes and organizing the plants into easy-to-access blocks of agroforestry crops would both increase diversity and maintain a commercially viable system.

Not all permaculture designs have income production as a major component. After all, many people creating permaculture designs for their yard aren't really looking to swap their lawn for a full-on farm. However, even if our systems aren't intended to provide income, they can (and should) certainly decrease expenses for their owners. The more we can provide for ourselves, the less money we have to spend. In turn, the less money we need, the less we need to work for someone else and the more time we have to pursue land-based endeavors. Also, the tighter we close our loops, the less we need to bring in from outside. That means we're more resilient as we're less reliant on far-off companies for our needs.

Capital and the triple bottom line

Just like building soil, a project can be designed so that it will generate more income over time as the system matures. There is certainly a lot of room to incorporate income-generating enterprises into our permaculture systems. Keep in mind that doing projects takes capital. That means if your permaculture system has no way to produce income, it may not be viable in the end. It may be fine if a particular project lacks economic opportunities, but if you design a permaculture system that can't support itself with at least some income, you may find it difficult to fund the installation of the more expensive elements of that system without external support.

Ultimately, any business or enterprise that springs from a permaculture design should be based on "triple bottom line" accounting. That means that the success of the business is measured not just by profit (the business is making money and remaining solvent) but also by planet (through its operations the business actually regenerates the earth rather than degenerates it) and people (the people working for the business, both directly and indirectly, have their needs met). Notice how the triple bottom line overlaps with permaculture ethics. It also means that our businesses can't externalize the very real costs of environmental degradation and social injustice.

Cottage industry and value-added products

Cottage industry refers to a small-scale enterprise run out of the home. This could include selling vegetables at a roadside market stand, blacksmithing old chunks of metal into hooks and latches, or operating a home nursery. For many home-scale permaculture projects, a cottage industry is the most appropriate way to create economic opportunities. When designing, part of our goal should be to create opportunities for many diverse potential cottage industries. That way users of the system will have plenty of options should they desire to earn some income on their property.

Value-added products are products that have been processed to increase their value. Selling value-added products could mean selling jams and jellies instead

A rooftop garden at the Bastille, a French restaurant in Seattle, grows a good amount of the food used below. This is not only cost effective but also creates a unique niche for the restaurant.

of fresh fruit. It could also mean following the example of Bella Farm in Vermont and making incredible pesto for sale in local markets from basil, garlic, and herbs grown on the farm. Selling the ingredients alone at wholesale prices would result in a much lower profit than selling pesto, even when you account for the extra time it takes to make the pesto. Thus small-scale, value-added products help growers to capture more income by creating products from the produce or materials they grow. Controlling the means of production as well as processing and sale means no middle man will take a portion of the profits. This also means maintaining control over much more of the value chain so you can make sure every step reflects your values and takes a triple bottom line approach.

Outside the realm of value-added food products, other materials can be produced and have value added. For instance, you can grow plants from which to produce your own fiber and craft items for both home use and sale.

PLANTS FOR FIBER AND CRAFT ITEMS

Agave sisalana
sisal
zones 9–11
excellent for fiber and cordage

Apocynum cannabinum
dogbane hemp
zones 4–9
excellent cordage plant

Bambusa textilis
weaver's bamboo
zones 9–11
used for basketry and weaving

Calamus spp.
rattan
zone 11 and warmer
used for furniture
basketry and crafts

Cavanillesia platanifolia
cuipo
zone 11 and warmer
rainforest giant with fibrous bark

Cordyline australis
cordyline
zones 9–11
provides leaf fibers for cloth, cordage, and basketry along with edible roots pith and shoots

Cyperus papyrus
papyrus
zones 9–11
used for paper and cordage and for other fiber arts
also an excellent water filtration plant

Musa basjoo
Japanese snow banana
zones 8–11
provides fiber for cloth and paper and leaves useful for tamale wrappers

Pandanus tectorius
hala
zone 13
fiber plant of the Polynesians
used for mats, hats, wall partitions, and more

Phormium tenax
New Zealand flax
zones 9–11
good for making cordage, cloth, basketry, and paper
fiber plant of the Maoris

Salix spp.
willow
zones vary
spectacular coppiceable basketry material

Thuja plicata
western red cedar
zones 6–8
bark used by indigenous people in the Pacific Northwest for buildings, basketry, and clothing

Trichostigma octandrum
hoopvine or Haitian basket vine
zones 10–12
great basketry vine with leaves that are edible when cooked

Yucca spp.
yucca
zones vary
fibers in the leaves good for making twine, baskets, and mats

Alternative currency

Alternative currency generally refers to any system of exchange that doesn't use government-issued money or go through the banking system. At its simplest, that can mean barter. Trading apples for milk, homemade baskets for nursery stock, or massage services for babysitting are all barter relationships.

Some places have actually created local exchange trading systems (LETS), both physical and digital. These currencies have value only on the local level, so they cycle and recycle value at the local level. This is one strategy for keeping value in your local community instead of sending it far away. In turn, this supports local growers, producers, and service people and creates a more vibrant local economy. For successful examples, look up BerkShares, Fourth Corner Exchange, and Ithaca Hours.

Ethical investment

Once you have managed to generate some capital, by whatever means, you can still use permaculture thinking to figure out how to put that money to use to further your goals in a way that reflects permaculture ethics.

Credit unions have been around since the nineteenth century. They offer an option for banking services (checking, savings, and credit) within an institution that is controlled by its members. Because credit unions are nonprofit cooperatives controlled by a board of directors elected by the members, they are accountable to their membership. These institutions tend to have a better record than big banks of putting people before profit since their jobs depend on successfully balancing both. A good example is the Permaculture Credit Union in Santa Fe, New Mexico—the first and only credit union in the United States that operates with permaculture values. The Permaculture Credit Union requires members to have completed a permaculture design course, to be affiliated with a permaculture institute, or to agree with permaculture ethics. Members can receive lower interest rates on loans (sustainability discounts) for purchasing things like electric cars, bicycles, home solar heating retrofits, and rainwater catchment systems.

Lending circles are groups of individuals who participate in peer-to-peer lending. The concept is frequently seen in immigrant communities where families or cultural groups pool money to help newcomers get established upon arrival in their new country. Once that money is paid back, it can be loaned out to someone else, and the former loan recipient can become part of the community who contributes money to the pool. This concept doesn't just have to exist for immigrant communities, though. Any group of people who have money to invest in their community can pool resources to support projects that reflect their values.

Sound financial management as a preparedness strategy

Managing your finances wisely ties into the concept of resilience as much as it ties to ethical investment. All the resilience concepts that apply to the rest of our permaculture design also apply to how we manage our money. For instance:

- Diversity of investments helps you be financially stable, just like biodiversity yields dynamic stability in ecosystems.

- ✻ Disaster planning in the landscape might include specific strategies for dealing with hurricanes, fires, and floods (more on this later in the chapter), but it is also important to have strategies in place in case of financial disaster (such as loss of job, major medical expenses, or economic instability). This can mean having backup plans for income generation, minimizing cash expenses, or setting up community-based insurance funds to help meet people's short-term needs in case of financial disaster.
- ✻ Just like crops such as potatoes and winter squash can be stored to provide food when it is difficult to produce it outside, one can save money for a rainy day. In this case, diversity again makes sense in terms of keeping that saved money in multiple financial instruments (such as cash, precious metals, or other valuable goods).

The fallacy of self-sufficiency

Some people refer to the goal of becoming self-sufficient, meaning they aim to provide for all their needs themselves. While perhaps a laudable goal, it isn't very realistic. Historically, very few people have actually been self-sufficient as individuals or households. Trade and interchange between individuals and family units has always taken place. This was the basis for the formation of villages and commerce thousands of years ago. Relying upon each other and working together is an inherent part of the human experience. Thus, striving for self-sufficiency at a community scale makes much more sense and is much more likely to succeed than trying to do everything yourself.

SOCIAL PERMACULTURE

A good friend of ours once said, "We know how to pump water, grow food, and build nontoxic buildings. There are hundreds of books about those topics. What we really need to do is figure out how to get along with one another." There's a lot of truth in that statement. Social systems are one of the biggest obstacles to broad-scale uptake of permaculture design thinking. Here we present a wide variety of ways that permaculture design applies to social systems.

Community living

We are all members of many communities such as families, households, neighborhoods, towns, groups of employees, and cultural groups. Within all of these groups are norms for behavior and interaction. Whether thinking about these existing groups or new groups that might be forming as part of your permaculture design, it is important to spend time thinking about the cultural components that form the basis of a well-functioning group.

Governance is one important aspect to discuss for many of these groups. Who is in charge of what? What powers and responsibilities do those people have? How much latitude do they have to act on the group's behalf without checking in first? These kinds of questions are important to discuss, especially for groups newly

Sometimes the best way to get organized as a group is to use big boards and move around pieces of paper. This is how Dave and the Bullocks plan for their three-week permaculture design course each year.

forming. The potential for social dysfunction is high if groups don't have clearly defined expectations and responsibilities.

One system of governance that deserves attention from those forming new groups (such as homeowner associations, land management groups, or ecovillages) is sociocracy or dynamic governance. Although it appears somewhat hierarchical in organization, sociocracy actually allows for people on the ground to provide feedback and information to the higher levels in the hierarchy. Action can be initiated from any point in the organizational structure. This is often lacking in traditional, autocratic hierarchies like those that might be found in a corporation. With sociocracy, leadership is key, but autocratic control is not. Sociocratic groups are organized into circles at each level in the structure. Each circle selects an individual to represent them to the circle above. Circles can also choose who they want to represent them in lower circles. In that way each circle is directly connected to those above and below, and power is distributed, not concentrated. At the top is effectively a board of directors selected by the membership.

In addition to governance, decision making is a critical point to spell out clearly from the start. Having a clear decision-making process from the beginning can help groups avoid a lot of headaches. The key thing to remember is that each decision-making structure has an appropriate time and place. Using the best technique for a given situation can result in a lot less frustration and wasted time for all involved.

We're all familiar with individual decision making, where someone higher up in a hierarchy makes a decision and everyone else follows it. The military chain of command is an example. While it may sound off-putting from a permaculture perspective, this decision-making process is actually the best for emergency situations such as fires, disasters, and medical emergencies as it results in decisions turning into action in the quickest way.

For group decision making, most of us are familiar with voting. Majority (or some other proportion) rule is a sensible and efficient way to make decisions when issues aren't contentious (for example, when picking which movie to show at an outreach event). Consensus decision making involves everyone in the group being able to come to a point where they do not object to a decision. This can be time consuming and difficult (or impossible) to achieve, yet in some cases this may be the most appropriate way to make decisions.

Sociocracy uses consent decision making, which is a bit different. In consent decision making, decisions have attached to them a term or trial period and a method of evaluating their success. The group revisits those decisions later to see if they are working. That means making an initial decision to do something is not the end of the line, so less discussion tends to be needed and a general of attitude of "Let's see how it works" prevails.

Village design

Permaculture design doesn't just apply to single family homes and yards or farms in the country. We can also use permaculture design tools to help plan developments at the village scale. All the same principles apply. The part that tends to increase quite a bit in complexity is the social systems. Here we provide you with some advice for those situations where you're designing an entire community.

When we work to design neighborhood- and village-scale developments, we find that we can generally all agree upon a few inherent goals. First, we want thriving, active, safe communities. Most of us want to live in the kind of place where everyone knows their neighbors and the kids can run around outside without us fearing for their safety. Coming together for social time with the people in our community is the icing on the cake. However, if we truly seek sustainability, we want to design places where people don't just come together to socialize but also to work, plan, and connect.

By looking for opportunities to design interdependence into our communities, we can create situations where the benefits of interacting with our neighbors outweigh the benefits of not doing so (even for that neighbor we don't really like that much). By incentivizing cooperation and collaboration with neighbors, we also create stronger, more familiar, more resilient communities. For instance, if neighbors collectively grow staple crops like corn, squash, and potatoes and everyone receives a share of the harvest based on the hours they put into caring for the crops, cooperation benefits everyone.

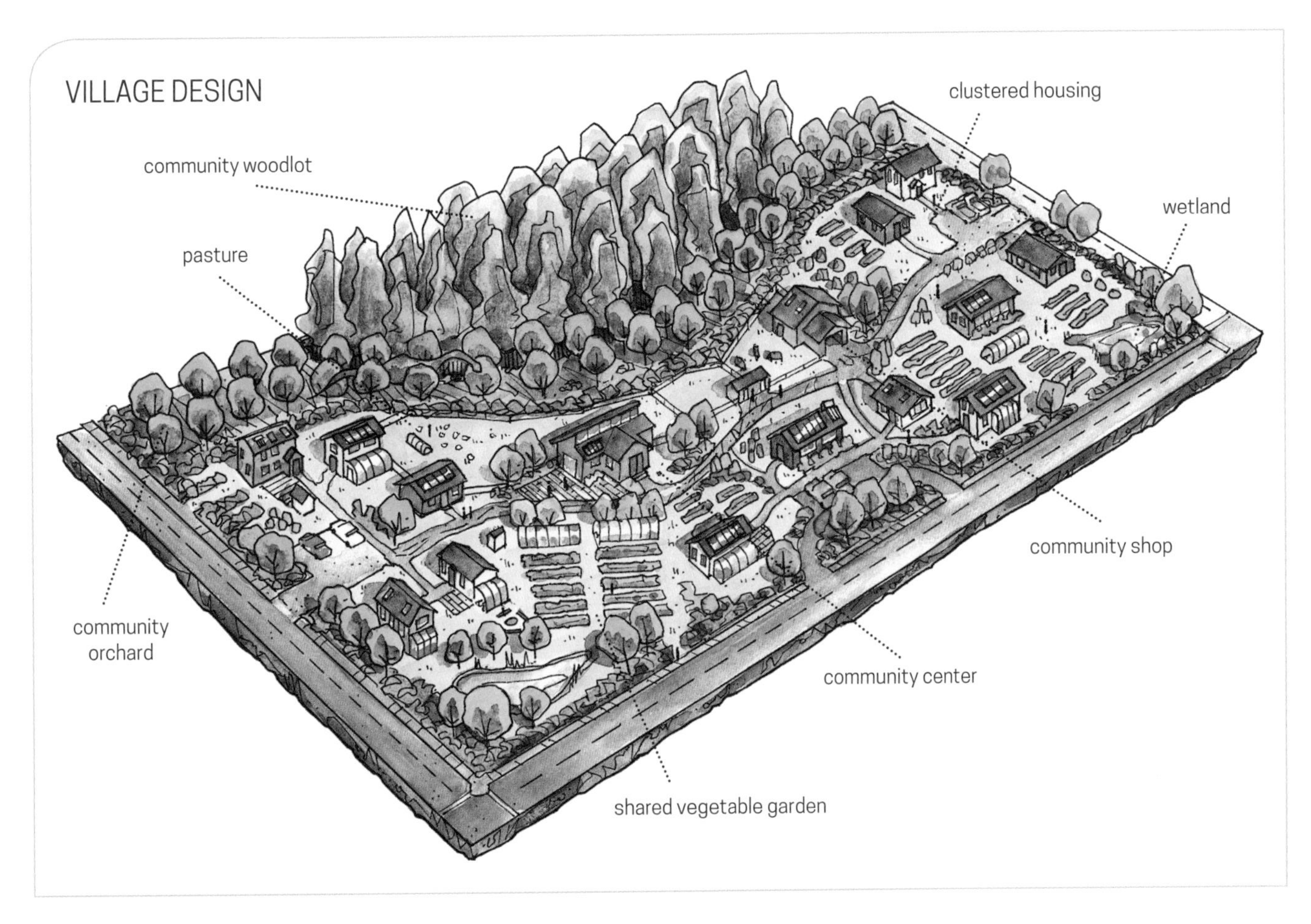

Permaculturist Andrew Millison has dedicated a portion of his front yard in Corvallis, Oregon, to community space constructed with used concrete or urbanite. This creates an inviting place to interact with neighbors and a great place for his son Rio to climb around.

Another overarching goal is the integration of everyone's individual systems. That means paying close attention to scale and taking advantage of the benefits. Imagine if everyone in a suburban neighborhood dedicated half of their backyard to intensive vegetable production. The space in several backyards could be enough square footage for an urban farmer who was also a resident of the community to grow the food for those houses and sell the surplus elsewhere.

We may also choose to approach integrating systems from the perspective of matching up needs and yields. If all the houses in your village had composting toilets and someone collected the humanure and properly composted it, that could mean a lot of high-quality organic material for ornamental and timber plantings and right livelihood for one or more people. What other wastes in your neighborhood could be conglomerated and turned into a valuable product if the scale were just right?

Designing at the neighborhood or village scale can also allow people to share the cost of expensive things that not everyone needs. For example, every four houses could share a washer and dryer. Most people don't do laundry every day. Each household could use the machines on a different day of the week. Repairs could either be accomplished by someone in one of the houses or paid for from a pooled maintenance fund. Either way, at the neighborhood scale this would mean that fewer total laundry machines were needed. We can use this type of thinking with all kinds of things such as lawn mowers and even renewable energy infrastructure. Residents of a small ecovillage in Iowa share access to a grid-tied renewable energy system (wind and solar). If any money is owed at the end of any given month, whoever used the most juice has to pay the entire bill for the neighborhood. That's a good incentive to go easy on the electricity usage.

Although the design of any village depends on the people and land involved, a few things are always important to consider. The big question is, How can we design the physical environment to set the stage for a thriving community?

A good place to start is with the topography. Imagine a bunch of houses placed around a cone-shaped hill. Everyone looks away from everyone else, and out of sight often means out of mind. Now imagine houses placed around a valley. You may be able to see your neighbors across the valley. You might even hear them occasionally. This sets the stage for people to interact much more than the first scenario.

Also, consider adding some sort of shared community infrastructure to your list of elements. Community centers, kitchens, pavilions, and parks all fit the bill. In fact, when you are planning your implementation strategy, consider constructing the community infrastructure before individual homes. That way, future residents will develop a culture of using that infrastructure and coming together as a group. Some pioneering types may even choose to live in the community structures while their individual homes are being built. Also, consider putting community infrastructure on one of the nicest spots on the property (in terms of views, access to water, and such) so that everyone can share in those things that make the property special. This is more egalitarian than letting the person with the most money have the best view.

This bike storage area at Columbia Ecovillage in Portland, Oregon, provides a great solution for people with small residences. Residents are more likely to use bikes in communities that are designed to be bike friendly.

Another feature of the design that can set the stage for community to develop or prevent it is access. Spend a good amount of time thinking about how to lay out the access in the community. Community generally doesn't happen within a vehicle, so strive to design a walkable community. Village Homes in Davis, California, is a great example where all streets run past the homes' small fenced backyards and end in cul-de-sacs. The homes' unfenced front yards face walking paths that connect every part of the village.

Designing cluster housing can also create a community feel in a neighborhood. With cluster housing, dwellings are grouped into small clusters instead of spread evenly across the property. This creates little groups of neighbors with modest yards. The extra space in between clusters can be used for community benefit, whether as parkland, vineyard, orchard, or community garden space. This allows each dwelling to have elements of both community interaction and privacy, an important consideration for most people. Through a well-designed access system, residents of each cluster could come together at some sort of community infrastructure element so that they encounter each other frequently. Ask yourself what you can do to both encourage and remove obstacles to community happening.

Never doubt that a small group of thoughtful, committed citizens can change the world. Indeed, it is the only thing that ever has.

—MARGARET MEAD

Building community

Figuring out how to organize social systems within your permaculture project is important, but permaculture design also encourages thinking about building community outside the bounds of your project. Why is this important? First, remember that having a thriving, active community around you is among the best ways to build resilience. After all, these are the people you may need to call on for help during a disaster. Also, knowing the people in your community and having positive interactions with them is one of the best ways to prevent crime and live a more positive life. You can even work with others in your community toward common goals for the betterment of all.

So how can permaculture have an impact on your neighborhood, village, town, or bioregion? The Transition movement is a great example. This movement has spread across the world and motivated people to begin interacting with their local communities in a different way. In essence, the Transition movement exists to help local communities take steps toward addressing the twin issues of peak oil and climate change. The idea that you can't address one without addressing the other is key, and creating a resilient, low-energy world is the goal. The Transition movement offers communities a set of steps they can follow to get others on board to collectively address those problems and build resilience.

In Portland, Oregon, the City Repair Project gets people from a neighborhood together to celebrate and paint an intersection.

Another example of a project that has done a lot to activate the local community is the City Repair Project in Portland, Oregon. The project seeks to reclaim common spaces that can become nodes for community interaction. The idea is that in many modern cities, urban plazas and gathering places for people are too uncommon. If the places for interaction don't exist, people can't connect and a real sense of community doesn't develop. One project of City Repair is called Intersection Repair. Local people from a neighborhood have a celebration and paint an intersection, not only activating the community but also slowing down traffic. This has become codified in Portand's ordinances and it is now spreading to other cities across North America.

City Repair also spawned the Village Building Convergence, which takes place annually. People from all over come to Portland for several days of learning and celebration. Each day people can participate in a variety of educational community projects, from sculpting cob benches to planting forest gardens to working on Intersection Repair sites. Speakers and events are scheduled in the evenings. As they say, "It's a bit like a conference, a bit like a festival, and a bit like a neighborhood improvement day," and people love it.

Legal considerations

Another part of the social sphere that permaculturists need to be aware of is the potential legal issues they may run into when implementing a permaculture design. It is important to pay close attention to the zoning, permitting, and land use laws in your area. You must also educate yourself about the rules and CC&Rs of your condominium or homeowners association. Be aware that these rules are often created based on a set of desired outcomes different from your own and in response to something unethical having been done in the first place.

By and large, laws and rules stop the least ethical people from doing bad things but also tend to prevent people from doing things better than the standard way. For instance, composting toilets, load-bearing straw bale houses, and greywater systems are not permitted in many locations even though plenty of precedents demonstrate the success of these systems when they are done right. You need to decide how to interact with these legal issues on a case-by-case basis. You can ignore them and fly under the radar, follow the rules and compromise in your design, or work to change the laws that prevent the best practice.

DESIGN FOR DISASTER

The unfortunate truth is that most of us will face a disaster at some point in our lives. How well prepared you are for that disaster can be the difference between a minor disruption to your day-to-day activities and a major game changer that reorients your priorities toward basic survival. Permaculture offers us a sensible way to approach those preparations that begins with designing for disaster.

Disaster situations are characterized by events happening that are outside your control. Through good design, you can minimize that loss of control and avoid or mitigate many potential disasters. This is one of the points where it is absolutely critical to refer back to our principle of job redundancy and resilience and think through the scenarios asking, What if . . . ?

By setting yourself up to not just survive but continue to thrive in the face of adversity (not just survival but "thrival"), you have a couple of positive effects. First, by making sure that you and your family are taken care of, you also ensure that you won't require as much emergency relief assistance, which can then be used by those less fortunate. Second, by making sure you are taken care of, you build capacity for yourself to help others in emergency situations. That relates directly to our ethical mandate to care for people. Realistically, you would always rather be on the helping side than the needing-help side.

You should focus your planning on the disasters most likely to befall you. That starts with resilient design. In your site analysis and assessment, you should have identified the most likely situations you may face. We describe a few common disaster types here to give you an idea of how to design your way out of a lot of potential trouble.

STOCKPILING

Without being obsessive hoarders, we want to make certain that we design systems that have space for storage of a limited supply of essentials in case of emergency. This includes nonperishable food, water, toilet paper, flashlight batteries, and the like. That means we need to design food systems that yield storable surpluses (for example, keeper apples, grains, beans) and include ample storage spaces, such as root cellars and pantries, in our structures. For stockpiling, choose foods with good storage qualities that you use regularly anyway so you can rotate stock in your day-to-day life but still have surplus for emergencies. Ideally, you should have enough surplus around at any given time to last for three days. As your system develops, you can slowly build up to a stockpile that can supply your needs for up to a month.

This subfloor root cellar provides an excellent spot to store extra nonperishable foods in bulk, both for everyday use and for emergencies.

Earthquakes

If you're in an earthquake-prone zone, all structures must be built to withstand extreme stresses. Generally, that means they should either be really flexible or really firmly anchored and strong. You also might want to consider lower buildings with shapes that tolerate shaking (such as curves and domes). In these situations it is especially important to avoid ponds on steep slopes since earth tremors can actually get the water sloshing and spill it over the edge or damage the dam. Elevated water tanks also need to be designed with earthquakes in mind. Water heaters should be strapped to walls.

Landslides

In some areas, the soils are prone to landslides for a number of reasons. In these cases it is crucial to make sure steep slopes stay vegetated (or get revegetated quickly). You also may want to avoid putting your home or other crucial infrastructure at the bottom of a cleared slope to avoid being buried if the land does move. If the slopes cannot be stabilized due to underlying geology, you may want to find a different site.

Floods and tsunamis

Many people around the world live in flood-prone areas. In general, if your land is flat and surrounded by low hills, you can probably assume you're in a floodplain. Even if it hasn't flooded in recent memory, you should plan for eventual flooding or tsunamis that may come with climate change. Proven strategies include building houses on stilts or on floats tied to a well-anchored pylon. If a nearby river seems prone to flooding regularly, you can help to avoid erosion and soil loss by establishing vegetation on the riverbanks. The right vegetation will knit the banks together, and trees and shrubs can prevent large debris from washing through your living room. Of course, the best strategy may be to avoid putting homes in floodplains altogether and instead use these extremely fertile areas for crop production that may benefit from occasional flooding.

Blizzards and snow emergencies

If blizzards threaten to cut you off from access to supplies for long periods, you should design your buildings with multiple backup heating options. It is also important to have ample storage space for essential goods to get you through the emergency.

This straw bale noise barrier at the Solar Living Institute in Hopland, California, rests on cages of rock called gabions. During floods, the water can run underneath the wall without ruining the straw bale structure.

Tornadoes

In areas prone to tornadoes, consider designing buildings with basements and/or root cellars where you can wait it out. Avoid building on hilltops as you may be more susceptible to the effects of the tornado than if you're located on a midslope or a valley bottom.

Fires and wildfires

In areas with danger of wildfire, pay close attention to your sectors. Where is the fire likely to come from? Cutting firebreaks, planting fire-resistant vegetation (such as cactus, sedum, many euphorbias), or locating ponds between you and the fire sector can help to protect your structure from damage. Consider less flammable construction techniques than wood, such as stone or cob. If your home does burn, it can be helpful to have outbuildings elsewhere on the property where you can live while you rebuild. Also, carefully plan your water sources to make sure you have several backups that can be used for firefighting. Note that shallow-buried PVC pipe can melt in a fire, so don't make such pipe part of your fire-suppression strategy. Really think through the details of what would happen if a fire were to come through.

Volcanic eruptions

People have lived in volcanic regions for a long time. No matter how you prepare, the possibility of a wipeout always exists, so one of the strategies is to build low-cost temporary structures. If they need to be replaced, at least you didn't invest much in the first place. You can also choose to site essential elements on hills and build low rock walls around the uphill side of the property. Either way, when the lava is coming it's probably best to go elsewhere for a few days.

Other disasters

Among the most common disasters that could befall many of us is loss of the income that we rely on to pay our mortgage or develop our permaculture sites. Following the permaculture principle of obtaining a yield can help you make sure you are always diversifying your yields. In general, producing more with your permaculture system also means you'll need to spend less money elsewhere. We've all heard Ben Franklin say, "A penny saved is a penny earned," and it is true with permaculture systems as well. Sound financial planning and saving for a rainy day take place outside the landscape but are essential pieces of making your permaculture dream a reality.

Finally, it's never too soon to get out of debt. With obvious instability in our global economic system, prioritizing getting out of debt will make you more resilient in the long run because your likelihood of losing the land you're investing in goes way down. If you're in debt (or going into debt), start by coming up with a time-bounded, achievable way of paying it off.

Civil unrest and war aren't fun to think about, but they can also be a reality. The likelihood of facing these situations increases where socioeconomic stresses are more extreme. The best ways to deal with these situations are by way of overarching strategies that can serve as essential parts of your plan for dealing with all the rest of the disasters we've covered.

Keys to resilience

In any disaster, having a strong community that comes together to help one another is a key to resilience. This means at the very least getting to know your neighbors and checking on them periodically. Community-scale disaster planning is even better. Make sure you know your neighbors and actively participate in your community. The people around you may be the ones you need to rely on in a bind.

One other way to prepare is to invest in your own skill sets so that you can provide essential functions in an emergency. Whether you take advanced emergency medical training, learn to operate a HAM radio, or figure out how to make fire by rubbing sticks together, you make yourself a more valuable member of your community both during an emergency and in everyday life.

ACCESS TO LAND

For many aspiring permaculture designers, gaining access to land is a huge hurdle. At least in the United States, we have a shrinking middle class. That means many people just can't afford to buy a piece of land to put together their permaculture dream. Luckily there are some creative ways to think about land access in our permaculture design toolbox.

Land trusts

A land trust refers to a situation where one person or entity owns land on behalf of a beneficiary. Conservation land trusts have the goal of conserving some function on a piece of land such as wildlife habitat or agriculture. In these cases,

(left) Learning about your local wild edible species is a valuable skill and a great starting point for learning more about the natural world. "Old-growth" permaculturist Sam Bullock has stayed an active learner his whole life, exploring everything from off-grid energy systems to community living arrangements to mushroom identification.

(right) This community bulletin board in the Sellwood neighborhood of Portland, Oregon, serves as a place for neighbors to get information to one another.

certain rights can be separated from land ownership so that an organization like the Nature Conservancy can purchase the rights to develop a piece of land while not purchasing the land. Community land trusts usually exist to serve a specific goal such as affordable housing or providing community garden spaces. Becoming part of a community land trust can be one way to gain access to land.

Leases

Leasing is another way to gain access to land. Conversely, if you have land, offering leases can be a great way to get help with your project and help someone out at the same time. For example, let's say we plant some nut trees at their mature spacing on 40-foot centers in a zone 4 part of our property. Instead of managing weeds in all the space in between, we might choose to lease out the space in between to a like-minded berry or vegetable grower for the next seven to ten years (until the nuts begin bearing). This means the lessee gains the benefit of land access without needing to purchase it and the lessor gains the benefit of some income and help managing the property. If the right lessee is chosen, benefits could also include net soil improvement in the area and infrastructure improvements.

Other ways to gain access to land

Aside from these fairly formal ways of obtaining access to land, you can always just ask a neighbor as well. Putting up a flyer at the co-op, speaking up at a community meeting, or just generally putting the word out can often result in offers to have a garden in someone's yard. The fact is that lots of land in urban and suburban areas goes unused. Showing initiative and gumption can be all it takes to gain access to a little plot for yourself. In these situations, the key is to make sure you go above and beyond what you might do if it were your own property in terms of tidiness and aesthetics. The goal is to delight the person who allowed you to use their land, not cause them concern. Sharing fresh strawberries goes a long way as well!

Guerilla gardening is another approach you can choose. As you may guess, this means growing food on a vacant property without asking permission first. This can also mean sneaking pumpkin seeds into the landscape outside of an office building or grafting 'Gala' and 'Fuji' scions onto the flowering crabapples along the street. In some cases, this may be totally appropriate and you'll be applauded for your contribution. In other cases you could get in trouble, so it's up to you to judge how to approach this issue. Either way, if you're going to be a guerilla gardener, we suggest cultivating a sense of nonattachment, as otherwise it can be heartbreaking to swing by the office building to harvest your pumpkins and discover that they've all been cleared out by the landscaping crew or blasted with herbicides.

Seattle's first community garden, the Picardo P-Patch, does a great job of providing space for city dwellers who need access to land for growing food.

INVISIBLE STRUCTURES STARTING POINTS

	Easiest	More involved	Most involved
Finances, capital, and investment	Join the Permaculture Credit Union.	Form a lending circle with friends.	Invest in other people's permaculture projects.
	Start saving money.	Create a business plan.	Own your own supply chain (vertical integration).
	Pay down debts.	Sell produce.	
	Barter with friends and neighbors.	Sell value-added products.	
	Diversify investments.	Use alternative currency.	Start an alternative currency in your area.
Social sphere, community	Assess the skill sets in your neighborhood.	Participate in a neighborhood organization.	Start a neighborhood organization.
	Participate in your local Transition movement.	Participate in Intersection Repair.	Launch a local Transition movement.
	Share small-scale community infrastructure (such as free boxes, cob benches).	Share medium-scale community infrastructure (such as washing machines, tool libraries).	Share large-scale community infrastructure (such as wind generators, constructed wetlands for wastewater treatment).
	Flex your design to obey laws and rules.	Push the boundaries on laws and rules.	Work to change laws and rules.
	Be prepared for a three-day emergency.	Be prepared for a seven-day emergency.	Be prepared for a thirty-day emergency.
			Start or move to an ecovillage.
Land access	Rent a community garden plot.	Lease land.	Join a land trust.

A FINAL WORD

How wonderful it is that nobody need wait a single moment before starting to improve the world.

—ANNE FRANK

People often report that after learning about permaculture and putting it into action, they feel healthier and happier because they have a new holistic perspective, feel more connected to where they live and the people around them, and feel a sense of pride and passion about how they are spending their time. Remember the transitional ethic? Anyone can take simple steps to get started with permaculture.

So where do you go from here? You now have a basic understanding of the pieces of the permaculture design system and a smattering of techniques for getting closer to sustainability, but there is so much more to learn. Moving forward, we recommend focusing your energy in two areas. First, continue to research and experiment with the aspects of the book that most piqued your interest. If forest gardening is your thing, learn more. If you were really fascinated by managing water with earthworks, go deeper there. Second, focus some of your energy on the topics that you felt were intimidating or beyond your ability. You don't have to become an expert in these areas, but by bringing up your baseline level of understanding you will become more well-rounded with a solid base of generalist knowledge. That's what makes a good permaculturist. Simultaneously improving your understanding of the whole and specializing where you feel passionate are the ticket to success. The references and resources at the end of the book provide further opportunities for you to continue your learning along these pathways.

We also recommend pondering the following questions now that you've read the book and could have a different perspective:

- What does your version of paradise look like?
- Do you feel good about yourself in the sense of both a healthy body and a healthy mind?
- How do you spend your time?
- Do you work more than you want and feel financial pressures?

Permaculture invites you to imagine a life where you are healthy and happy and have all of your needs met. With this book in hand, you're on your way to that life. We hope to see you there.

This permaculture homestead's garden demonstrates that functional spaces can also be beautiful.

Metric Conversions and Hardiness Zones

Conversion tables

inches	centimeters
¼	0.6
½	1.3
1	2.5
2	5.1
3	7.6
4	10
5	13
6	15
7	18
8	20
9	23
10	25
feet	**meters**
1	0.3
2	0.6
3	0.9
4	1.2
5	1.5
10	3
20	6
30	9
40	12
50	15
60	18
100	30

Plant hardiness zones

Average annual minimum temperature

Zone	Temperature (°F)	Temperature (°C)
1	Below-50	Below-46
2	-50 to -40	-46 to -40
3	-40 to -30	-40 to -34
4	-30 to -20	-34 to -29
5	-20 to -10	-29 to -23
6	-10 to 0	-23 to -18
7	0 to 10	-18 to -12
8	10 to 20	-12 to -7
9	20 to 30	-7 to -1
10	30 to 40	-1 to 4
11	40 to 50	4 to 10
12	50 to 60	10 to 16
13	60 to 70	16 to 21

To see the U.S. Department of Agriculture Hardiness Zone Map, visit planthardiness.ars.usda.gov/phzmweb/.

For Canada, go to planthardiness.gc.ca/ or sis.agr.gc.ca/cansis/nsdb/climate/hardiness/index.html.

Resources

Agroforestry Research Trust, agroforestry.co.uk
UK-based temperate agroforestry organization directed by Martin Crawford.

agroforestry.net
Pacific Islands resource for tropical and subtropical agroforestry. Many excellent reference books by Craig Elevitch available for download.

Alderleaf Wilderness College, alderleafwildernesscollege.com
Educational programs in the Pacific Northwest relating to permaculture design, nature awareness, and wilderness skills.

American Bamboo Society, bamboo.org
Information clearinghouse for bamboos worldwide.

Aprovecho Sustainability Education Center, aprovecho.net
Permaculture and appropriate technology education center in Cottage Grove, Oregon.

Blue Future Filters, bluefuturefilters.com
US-based company offering slow sand filters for a variety of scales and situations.

Bullock's Permaculture Homestead, permacultureportal.com
Family homestead offering permaculture education opportunities in the Pacific Northwest.

California Rare Fruit Growers, crfg.org
Excellent membership organization for amateur fruit growers in California-like climates.

Centre for Alternative Technology, cat.org.uk
Education and demonstration site for a wide range of appropriate technologies and sustainability techniques.

Chaikuni Institute, chaikuni.org
Permaculture research and education center in the P eruvian Amazon.

City Repair, cityrepair.org
Portland, Oregon–based organization facilitating placemaking activities in neighborhoods.

Cob Cottage Company, cobcottage.com
Courses in construction with cob.

Darren Doherty, heenandoherty.com
Australian permaculturist offering education and design services for broadscale landscapes. Specializes in keyline design.

ECHO, echonet.org
Tropical hunger relief organization focused on agroforestry and integrated systems.

Edible Forest Gardens, edibleforestgardens.com
Dave Jacke's website with information on forest gardening.

Fungi Perfecti, fungi.com
Company founded by Paul Stamets, offering mushroom cultivation education and resources.

Garden Organic, gardenorganic.org.uk
Organization providing research and education relating to organic gardening.

Growing Power, growingpower.org
Will Allen's urban agriculture organization offering education and resources.

***Home Power* magazine, homepower.com**
Excellent resource for home-scale power systems.

Liberation Ecology blog, liberationecology.org
An excellent blog exploring many aspects of permaculture design.

Linnaea Farm, linnaeafarm.org
Educational farm in coastal British Columbia, Canada, offering training in permaculture and production farming.

Maya Pedal, mayapedal.org
Appropriate technology organization in Guatemala offering pedal-powered solutions.

Mud Girls Natural Building Collective, mudgirls.wordpress.com
Women's natural building collective in British Columbia, Canada, offering courses, training, and building services.

Native Seeds/SEARCH, nativeseeds.org
Amazing organization dedicated to preserving heirloom seeds from the desert Southwest.

New Forest Farm, newforestfarm.net
Mark Shepard's production-focused commercial farm showcasing permaculture and keyline design techniques.

North American Fruit Explorers, nafex.org
Membership organization for amateur fruit growers in North America.

Northern Nut Growers Association, nutgrowing.org
Membership organization for amateur and commercial nut growers in North America.

NRCS Soil Survey Online, websoilsurvey.nrcs.usda.gov
Good starting point for region-specific soil research in the United States.

N.W. Bloom, nwbloom.com
Jessi Bloom's design-build company in western Washington, United States.

Oasis Design, oasisdesign.net
Art Ludwig's design company offering design, education, and information resources relating to water systems and greywater management.

Occidental Arts and Ecology Center, oaec.org
Permaculture and sustainability education center near Santa Rosa, California.

Pattern Literacy, patternliteracy.com
Toby Hemenway's website with excellent articles on a wide variety of permaculture topics.

***Permaculture Activist* magazine, permacultureactivist.net**
US-based permaculture magazine.

Permaculture Credit Union, permaculturecu.org
US-based credit union offering financial resources for people interested in permaculture ethics.

***Permaculture* magazine, permaculture.co.uk**
UK-based, internationally focused permaculture magazine.

Permaculture Research Institute, permaculturenews.org
Australian clearinghouse for permaculture information.

Permies online forums, permies.com
Online permaculture community that can help answer questions or recommend resources.

Planet Repair Institute, planetrepair.wordpress.com
Placemaking demonstration and education site in a Portland, Oregon, urban neighborhood.

Plants for a Future database, pfaf.org
Excellent resource for permaculture plant information. Includes edibility and medicinal ratings.

Polyface Farms, polyfacefarms.com
Joel Salatin's site predominantly dedicated to innovative sustainable livestock production.

Project Bona Fide, projectbonafide.com
Tropical permaculture education and demonstration site on Isla Ometepe in Nicaragua.

Rainwater Harvesting, harvestingrainwater.com
Brad Lancaster's website with lots of good information on rainwater management.

Rancho Mastatal Environmental Learning Center and Lodge, ranchomastatal.com
Excellent permaculture education site in northwestern Costa Rica with a focus on natural building.

Regenerative Design Institute, regenerativedesign.org
Penny Livingston-Stark's permaculture education and design organization in Bolinas, California.

Savory Institute, savoryinstitute.com
Excellent online resource for information about Allan Savory's holistic management system and rangeland management.

Seed Savers Exchange, seedsavers.org
US-based organization dedicated to preserving heirloom annuals and perennials.

Sociocracy.info
Comprehensive website and archive for information on sociocracy.

Solar Energy International, solarenergy.org
Educational resource for renewable energy projects. Excellent courses and workshops.

Terra Phoenix Design, terraphoenixdesign.com
Whole systems design firm run by Dave Boehnlein, Paul Kearsley, and Doug Bullock in western Washington.

The Farm intentional community, thefarm.org
Intentional community and sustainability education site in Tennessee.

Transition Network, transitionnetwork.org
Information clearinghouse for the Transition movement worldwide.

Tropical Forage Interactive Selection Tool, tropicalforages.info
Forage species profiles appropriate for livestock systems in tropical locations.

Verge Permaculture, vergepermauclture.ca
Calgary, Canada–based permaculture design and education organization.

Whole Systems Design, wholesystemsdesign.com
Permaculturist Ben Falk's demonstration site and design company based in Vermont.

Yestermorrow Design Build School, yestermorrow.org
Excellent design-build school in Vermont focused on natural building.

Suggestions for Further Reading

Permaculture and Design

Alexander, Christopher, et al. 1977. *A Pattern Language*. New York: Oxford University Press.

Alexander, Christopher. 1979. *The Timeless Way of Building*. New York: Oxford University Press.

Aranya. 2012. *Permaculture Design*. East Meon, Hampshire, UK: Permanent Publications.

Bane, Peter. 2012. *The Permaculture Handbook*. Gabriola Island, British Columbia, Canada: New Society Publishers.

Benyus, Janine M. 1997. *Biomimicry: Innovations Inspired by Nature*. New York: HarperCollins.

Brown, Azby. 2009. *Just Enough: Lessons in Living Green from Traditional Japan*. Tokyo: Kodansha International.

Falk, Ben. 2013. *The Resilient Farm and Homestead*. White River Junction, Vermont: Chelsea Green.

Hemenway, Toby. 2009. *Gaia's Garden: A Guide to Home-Scale Permaculture*, 2nd ed. White River Junction, Vermont: Chelsea Green Publishing.

Holmgren, David. 2002. *Permaculture: Principles and Pathways Beyond Sustainability*. Hepburn, Victoria: Holmgren Design Services.

Holzer, Sepp. 2004. *Sepp Holzer's Permaculture*. White River Junction, Vermont: Chelsea Green.

Jacke, Dave, and Eric Toensmeier. *Edible Forest Gardens*. 2 vols. White River Junction, Vermont: Chelsea Green Publishing, 2005.

Madigan, Carleen, ed. 2009. *The Backyard Homestead*. North Adams, Massachusetts: Storey.

McHarg, Ian L. 1969. *Design with Nature*. New York: Natural History Press.

Meadows, Donella H. 2008. *Thinking in Systems*. White River Junction, Vermont: Chelsea Green.

Meitzner, Laura S., and Martin L. Pierce. 1996. *Amaranth to Zai Holes: Ideas for Growing Food Under Difficult Conditions*. North Fort Myers, Florida: ECHO.

Mollison, Bill. 1988. *Permaculture: A Designer's Manual*. Tyalgum, Australia: Tagari.

———. 1999. *Permaculture Two: Practical Design for Town and Country in Permanent Agriculture*. Tyalgum, Australia: Tagari.

Mollison, Bill, and David Holmgren. 1990. *Permaculture One: Perennial Agriculture for Human Settlements*. Tyalgum, Australia: Tagari.

Mollison, Bill, and Reny Mia Slay. 2009. *Introduction to Permaculture*, 2nd ed. Tyalgum, Australia: Tagari.

Morrow, Rosemary. 2006. *Earth User's Guide to Permaculture*, 2nd ed. Sydney, Australia: Kangaroo Press.

Reid, Grant W. *Landscape Graphics: Plan, Section, and Perspective Drawing of Landscape Spaces*, revised ed. New York: Watson-Guptil, 2002.

Shein, Christopher, with Julie Thompson. 2013. *The Vegetable Gardener's Guide to Permaculture*. Portland, Oregon: Timber Press.

Watkins, David. 1993. *Urban Permaculture*. East Meon, Hampshire, UK: Permanent Publications.

Woodrow, Linda. 2007. *The Permaculture Home Garden*. New York: Penguin.

Water

Evenari, Michael, et al. 1971. *The Negev: The Challenge of a Desert*, 1st ed. Cambridge, Massachusetts: Harvard University Press.

Lancaster, Brad. 2006. *Rainwater Harvesting for Drylands and Beyond*, vol. 1. Tucson, Arizona: Rainsource Press.

———. 2008. *Rainwater Harvesting for Drylands and Beyond*, vol. 2. Tucson, Arizona: Rainsource Press.

Ludwig, Art. 2005. *Water Storage: Tanks, Cisterns, Aquifers, and Ponds*. Santa Barbara, California: Oasis Design.

Matson, Tim. 2002. *Earth Ponds A to Z*. Woodstock, Vermont: The Countryman Press.

———. 2006. *Landscaping Earth Ponds: The Complete Guide*. White River Junction, Vermont: Chelsea Green.

Yeomans, P. A. 1993. *Water for Every Farm: Yeomans Keyline Plan*. Edited by Ken B. Yeomans. Queensland, Australia: Griffin Press.

Waste

Jenkins, Joseph. 2005. *The Humanure Handbook*, 3rd ed. Grove City, Pennsylvania: Joseph Jenkins.

Logsdon, Gene. 2010. *Holy Shit: Managing Manure to Save Mankind*. White River Junction, Vermont: Chelsea Green.

Ludwig, Art. 2009. *Create an Oasis with Greywater*, 5th ed. Santa Barbara, California: Oasis Design.

Energy

Pahl, Greg. 2003. *Natural Home Heating: The Complete Guide to Renewable Energy Options*. White River Junction, Vermont: Chelsea Green.

Solar Energy International. 2004. *Photovoltaics: Design and Installation Manual*. Gabriola Island, British Columbia, Canada: New Society Publishers.

Shelter

Chiras, Daniel D. 2000. *The Natural House: A Complete Guide to Healthy, Energy-Efficient, Environmental Homes*. White River Junction, Vermont: Chelsea Green.

Evans, Ianto, Michael G. Smith, and Linda Smiley. 2002. *The Hand-Sculpted House: A Practical and Philosophical Guide to Building a Cob Cottage*. White River Junction, Vermont: Chelsea Green.

Hren, Stephen, and Rebekah Hren. 2008. *The Carbon-Free Home*. White River Junction, Vermont: Chelsea Green.

Kahn, Lloyd. 2004. *Homework: Handbuilt Shelter*. Bolinas, California: Shelter Publications.

Racusin, Jacob Deva, and Ace McArleton. 2012. *The Natural Building Companion: A Comprehensive Guide to Integrative Design and Construction*. White River Junction, Vermont: Chelsea Green.

Wolverton, B. C. 1996. *How to Grow Fresh Air: 50 Houseplants That Purify Your Home or Office*. New York: Penguin Books.

Food Production

American Horticultural Society. 1999. *Plant Propagation*. New York: DK Publishing.

Bubel, Mike, and Nancy Bubel. 1991. *Root Cellaring: Natural Cold Storage of Fruits and Vegetables*, 2nd ed. North Adams, Massachusetts: Storey.

Coleman, Eliot. 2009. *The Winter Harvest Handbook*. White River Junction, Vermont: Chelsea Green.

Crawford, Martin. 2010. *Creating a Forest Garden*. Devon, UK: Green Books.

———. 2012. *How to Grow Perennial Vegetables*. Devon, UK: Green Books.

Dirr, Michael A., and Charles W. Heuser, Jr. 1987. *The Reference Manual of Woody Plant Propagation*. Athens, Georgia: Varsity Press.

Elevitch, Craig R., ed. 2004. *The Overstory Book: Cultivating Connections with Trees*, 2nd ed. Holualoa, Hawaii: Permanent Agriculture Resources.

———, ed. 2011. *Specialty Crops for Pacific Islands*. Holualoa, Hawaii: Permanent Agriculture Resources.

Elevitch, Craig R., and Kim M. Wilkinson, eds. 2000. *Agroforestry Guides for Pacific Islands*. Hawaii: Permanent Agriculture Resources.

Fern, Ken. 1997. *Plants for a Future: Edible and Useful Plants for a Healthier World*. East Meon, Hampshire, UK: Permanent Publications.

Fukuoka, Masanobu. 1978. *The One-Straw Revolution*. New York: New York Review Books.

Gershuny, Grace. 1993. *Start with the Soil*. Emmaus, Pennsylvania: Rodale Press.

Gilman, Steve. 2002. *Organic Soil Fertility Management*. White River Junction, Vermont: Chelsea Green.

Gladstar, Rosemary. 2012. *Medicinal Herbs: A Beginner's Guide*. North Adams, Massachusetts: Storey.

Gough, Robert, and Cheryl Moore-Gough. 2011. *The Complete Guide to Saving Seeds*. North Adams, Massachusetts: Storey.

Harker, Donald, et al. 1993. *Landscape Restoration Handbook*. Boca Raton, Florida: Lewis Publishers.

Heistinger, Andrea. 2013. *The Manual of Seed Saving: Harvesting, Storing, and Sowing Techniques for Vegetables, Herbs, and Fruits*. Portland, Oregon: Timber Press.

King, F. H. 2004. *Farmers of Forty Centuries: Organic Farming in China, Korea, and Japan*. New York: Dover.

Kourik, Robert. 2005. *Designing and Maintaining your Edible Landscape Naturally*. White River Junction, Vermont: Chelsea Green.

Lewis, Daphne, and Carol Miles. 2007. *Farming Bamboo*. Raleigh, North Carolina: Lulu Press.

Lowenfels, Jeff, and Wayne Lewis. 2006. *Teaming with Microbes: A Gardener's Guide to the Soil Food Web*. Portland, Oregon: Timber Press.

Lyle, Susanna. 2006. *Fruit and Nuts*. Portland, Oregon: Timber Press.

Martin, Franklin W., and Ruth M. Ruberté. 1979. *Edible Leaves of the Tropics*, 2nd ed. Mayagüez, Puerto Rico: U.S. Department of Agriculture.

Phillips, Michael. 2011. *The Holistic Orchard: Tree Fruits and Berries the Biological Way*. White River Junction, Vermont: Chelsea Green.

Randall, Bob. 1999. *Year Round Vegetables, Fruits and Flowers for Metro Houston*, 12th ed. Houston, Texas: Year Round Gardening Press.

Rodale's Illustrated Encyclopedia of Organic Gardening. 2002. New York: DK Publishing.

Schafer, Peg. *The Chinese Medicinal Herb Farm*. 2011. White River Junction, Vermont: Chelsea Green.

Shepard, Mark. 2013. *Restoration Agriculture*. Austin, Texas: Acres U.S.A.

Smith, J. Russell. 1950. *Tree Crops: A Permanent Agriculture*. Washington, D.C.: Island Press.

Solomon, Steve, with Erica Reinheimer. 2013. *The Intelligent Gardener*. Gabriola Island, British Columbia, Canada: New Society Publishers.

Stamets, Paul. 2005. *Mycelium Running*. New York: Ten Speed Press.

Sunset editors. 2010. *Western Garden Book of Edibles*. Menlo Park, California: Sunset.

Toensmeier, Eric. 2007. *Perennial Vegetables*. White River Junction, Vermont: Chelsea Green.

Torgrimson, John, ed. 2009. *Fruit, Berry and Nut Inventory*, 4th ed. Decorah, Iowa: Seed Savers Exchange.

van Wyk, Ben-Erik. 2005. *Food Plants of the World*. Portland, Oregon: Timber Press.

———. 2004. *Medicinal Plants of the World*. Portland, Oregon: Timber Press.

Animals

Bernstein, Sylvia. 2011. *Aquaponic Gardening: A Step-by-Step Guide to Raising Vegetables and Fish Together*. Gabriola Island, British Columbia, Canada: New Society Publishers.

Bloom, Jessi. 2012. *Free-Range Chicken Gardens*. Portland, Oregon: Timber Press.

Conrad, Ross. 2007. *Natural Beekeeping: Organic Approaches to Modern Apiculture*. White River Junction, Vermont: Chelsea Green.

Cranshaw, Whitney. 2004. *Garden Insects of North America*. Princeton, New Jersey: Princeton University Press.

Hutchinson, Laurence. 2005. *Ecological Aquaculture: A Sustainable Solution*. Meon, Hampshire, UK: Permanent Publications.

Savory, Allan, and Jody Butterfield. 1999. *Holistic Management*, 2nd ed. Washington, D.C.: Island Press.

Walliser, Jessica. 2013. *Attracting Beneficial Bugs to Your Garden: A Natural Approach to Pest Control*. Portland, Oregon: Timber Press.

Xerces Society. 2011. *Attracting Native Pollinators*. North Adams, Massachusetts: Storey.

Invisible Structures

Abrams, John. 2005. *Companies We Keep*. White River Junction, Vermont: Chelsea Gree.

Deppe, Carol. 2010. *The Resilient Gardener*. White River Junction, Vermont: Chelsea Green.

Francis, Mark. 2003. *Village Homes: A Community By Design*. Washington, D.C.: Island Press.

Kellogg, Scott, and Stacy Pettigrew. 2008. *Toolbox for Sustainable City Living*. New York: South End Press.

Krishnamurti, Jidda. 1992. *On Right Livelihood*. San Francisco: HarperCollins.

MacNamara, Looby. 2012. *People and Permaculture*. Meon, Hampshire, England: Permanent Publications.

McKibben, Bill. 2007. *Deep Economy: Economics as if the World Mattered*. London, England: Oneworld Publications.

Schumacher, E. F. 1973. *Small Is Beautiful: Economics as if People Mattered*. New York: Harper & Row.

Acknowledgments

We owe our deepest appreciation to our incredibly knowledgeable mentors, friends, and heroes. Thank you for being the pioneer species in this movement where your hard work, dedication, and generosity have helped prepare the path we now walk: the Bullock family, Toby Hemenway, Dave Jacke, Penny Livingston, Marisha Auerbach, Andrew Millison, Jude Hobbs, Brad Lancaster, Tom Ward, Allan Savory, Buckminster Fuller, Mark Lakeman, Will Allen, Joel Salatin, and Vandana Shiva.

Thanks to our friends and reviewers Terry Phelan, Robin Haglund, John Valenzuela, Simon Walter-Hansen, Zsofia Pastor, Tad Hussey, Lacia Lynn-Bailey, Matt Curtis, and James Most. Special thanks to Michael Street—much appreciation for providing us with quality tunes, caffeine, calories, positive perspective, and laughter during the final push on this project. And we especially thank the folks who were generous enough to share their time and spaces for us to learn from and photograph to share with readers.

Jessi thanks her boys, Noah and Micah Kenney, for their patience while Mom has been on a mission to help heal the earth; and she thanks all of those who helped make that happen. Special thanks to Susan McKinley, Michael Stark, Zac Kopra, Fernando Lopez, and Tenney Kerr for their continual support.

Dave thanks supporters Yuko Miki, Michael Becker, Michael Judd, Scott Godfredson, Bruce Hostetter, and Christopher Shanks.

Paul thanks friends and supporters Seth June, Arunas Oslapas, Melissa Oscarson, Dawn and Nicole Kimberling, and the Kearsley crew: Patti, Doug, Kelly, Dan, Laura, Monica, and Justin.

Thanks to Lorraine Anderson and to Juree Sondker and the Timber Press crew who put up with us. Last, thanks to the plants and animals that kept us fed during this endeavor and the pets that kept us entertained: Spanky the genius pup and Rocky the wonder cat.

Greens grown in this small container are as beautiful as they are functional.

Index

Photo and illustration credits

Ravenel Bisbee, pages 90, 112 left

Dave Boehnlein, pages 4–5, 17, 21, 27 left, 32, 39, 43, 74 left, 94, 109, 136, 140, 142 left, 147, 150, 157 upper right, 160, 166 bottom, 175 left and right, 179 right, 181, 185, 193 upper left and lower right, 198, 200, 211, 217 upper right, 231 right, 238, 239 left, 241 left, 293

William Brown, pages 144, 213 bottom left

Doug Bullock, page 161

Scott Godfredson (Exos Design), pages 73, 164, 217 upper left, 261, 305

Paul Kearsley, pages 19, 33, 35, 39 bottom, 41, 46, 49, 52, 53, 54, 55, 56, 60, 61, 62, 66, 73 bottom, 76, 78, 82, 85, 86, 101, 106, 110, 117, 125, 126, 127, 134, 150, 151, 153, 158, 164, 167, 168, 169, 170, 178, 186, 190, 196, 201, 204, 208, 217, 230, 232, 233, 235, 247, 256, 258, 264, 299

Jason Knight (Alderleaf Wilderness College), page 268

Mark Lakeman (Communitecture), page 98

Brendon McKeon (Costa Rica Center for Natural Living), page 154

Andrew Millison, pages 88, 148, 163 left, 299

Kyle Thomas Ryan, page 207

Chris Shanks (Project Bona Fide), pages 213 bottom right, 214 bottom

Carmin Thomas, page 138

All other photos and illustrations are by Jessi Bloom.

Michael Street

Alaine Sommergren

JESSI BLOOM is an award-winning ecological landscape designer, professional horticulturalist, and certified arborist. She is lead designer and owner of N.W. Bloom—EcoLogical Landscapes, known for innovation in sustainable landscape design, construction, and maintenance. Jessi is committed to educating others and now spends much of her time teaching, consulting, and speaking nationwide in addition to designing landscapes. Her best-selling book *Free-Range Chicken Gardens* (Timber Press, 2012) has been praised for being informative and inspiring, changing the way people integrate animals into their landscapes. Recognition for her work includes awards from the Washington State Department of Ecology, the American Horticultural Society, *Pacific Horticulture* magazine, *Sunset* magazine, the Washington State Nursery and Landscape Association, and the Washington Association of Landscape Professionals. She lives north of Seattle with her sons on her permaculture homestead.

DAVE BOEHNLEIN serves as the education director at Bullock's Permaculture Homestead on Orcas Island, Washington. He is also the principal and a founder of Terra Phoenix Design, where he helps clients around the globe achieve their sustainability goals through integrated master planning. His freelance teaching services are highly sought after by universities, nonprofits, and other organizations. He seeks to mainstream permaculture design with integrity. In addition, Dave is passionate about plants, especially weird but useful ones. Ultimately, Dave just wants to make the world a better place and eat really good fruit while doing it.

Dave Boehnlein

PAUL KEARSLEY, a skilled designer with a passion for art and nature, collaborates with like-minded professionals to create inspiring models of sustainability. He has been an integral part of both institutional- and residential-scale permaculture design companies. As art director for Terra Phoenix Design, he has worked with world-class designers from California to Peru. On a local scale, Paul helped create Homestead Habitats LLC, a landscape contractor that specializes in designing and installing edible landscapes for urban and suburban homes. He currently teaches design courses at Western Washington University's College of Engineering.